Neutrino Mass and Gauge Structure of Weak Interactions
(Telemark, 1982)

AIP Conference Proceedings
Series Editor: Hugh C. Wolfe
Number 99
Particles and Fields Subseries No. 30

Neutrino Mass and Gauge Structure of Weak Interactions

(Telemark, 1982)

Edited by
Vernon Barger and David Cline
University of Wisconsin

American Institute of Physics
New York 1983

L.C. Catalog Card No. 83-71072
ISBN 0-88318-198-3
DOE CONF- 820971

TABLE OF CONTENTS

Neutrino Mass Experiments

Majorana Neutrinos/Relic Neutrinos

Cosmology/Astrophysics

e/μ Asymmetry

Structure of Weak Interactions/Neutrinos and Symmetries

A LIMIT ON THE MASS OF THE ELECTRON NEUTRINO FROM STUDIES OF INNER BREMSSTRAHLUNG

H.L. Ravn[1)], J.V. Andersen[2)], G.J. Beyer[1*)], G. Charpak[1)], A. De Rújula[1)], B. Elbek[3)], H.Å. Gustafsson[4)], P.G. Hansen[2)], B. Jonson[1)], P. Knudsen[3)], E. Laegsgaard[2)] and J. Pedersen[3)].

EUROPEAN ORGANIZATION FOR NUCLEAR RESEARCH

1) CERN, Geneva, Switzerland.

2) Institute of Physics, University of Aarhus, Aarhus, Denmark.

3) Tandem Accelerator, Niels Bohr Institute, Risø, Denmark.

4) Institute of Physics, Technical University of Lund, Sweden.

ABSTRACT

Sources of ^{193}Pt and ^{163}Ho decontaminated from other radioactivities by a factor of > 10^{13} have been produced at the ISOLDE facility on-line to the CERN 600 MeV Synchrocyclotron.

The ^{193}Pt was used to verify the theory for inner bremsstrahlung electron capture decay (IBEC). The Q_{EC} value is 56,3 ± 0,3 keV, and the data make it possible to give an upper limit on the electron neutrino mass of 500 eV (90% C.L.). For ^{163}Ho the partial M electron capture half life was determined to be (4.0 ± 1,2) · 10^4 y, from which we deduce a half life of (7 ± 2) · 10^4 y. Together with the independently measured ^{163}Ho-^{163}Dy mass difference of 2.3 ± 1.0 keV they imply an upper limit of m_{ν_e} of 1.3 keV.

*) Visitor from Zentralinstitut für Kernforschung, Rossendorf bei Dresden, DDR.

1. INTRODUCTION

The best limits on the electron anti-neutrino mass, result from the analysis of the shape of the tritium beta spectrum. The most recent experiments have given $m_{\bar{\nu}_e} < 55$ eV/c^2 [1], $14 < m_{\bar{\nu}_e} < 46$ eV/c^2 [2] and $m_{\bar{\nu}_e} < 65$ eV/c^2 [3]. Improvements of this technique are in progress.

The present upper limit of 4,1 keV on the mass of the electron neutrino is set by Beck and Daniel [4] who analyze the positron spectrum of ^{22}Na. To make more restrictive experiments by means of positron decay seems to be difficult, since the dominance of the electron capture process (EC) at low decay energies excludes that a β^+ emitter comparable to the tritium case can be found.

Our neutrino mass measurement programme is based on a new technique proposed by De Rújula [5] who pointed out that the neutrino mass information present in the high energy part of the tritium beta spectrum is available in a similar way near the end point of the continuous electromagnetic radiation (inner bremsstrahlung (IB)), accompanying EC beta decay. From the theory of electron capture developed by Glauber and Martin [6] it can be shown that internal bremsstrahlung at low energies arises mainly from p-capture. The resonant nature of the process leads to important enhancements of the photon intensities corresponding to p-s x-ray transitions.

Provided that a particular nucleus with an IB end point close to a resonance and with a reasonable half life can be found, greatly improved limits on the electron neutrino mass can be obtained. It is furthermore probable, but not certain, that the IBEC technique could be developed to become superior to the tritium experiments. From the chart of the nuclides we have chosen the two radioactive nuclei ^{193}Pt and ^{163}Ho as the best cases for an experimental approach.

In the following, we summarize the results obtained during the 18 months that have elapsed since the decision was made to embark on this programme. The isotope ^{193}Pt with a Q_{EC} value of 56.6 ± 0.3 keV (our value) is used for the first detailed test of the De Rújula-Glauber-Martin theory and it provides a new and improved limit : $m_{\nu_e} < 500$ eV. A second experiment has determined Q_{EC} and $T_{1/2}$ for ^{163}Ho, which right now seems to be the most promising candidate for a serious neutrino-mass experiment. Finally, the prospects for further experiments are discussed.

2. SOURCE PREPARATION

The long half lives of ^{193}Pt and ^{163}Ho and the low intensity of the IB spectrum require sources with a level of radioactive contamination well below 10^{-13} in order to avoid disturbing photon impurities in the spectra. Since the studied radiation is of low energy, it was necessary to produce the samples in a carrier free form so that very thin (monoatomic) layer sources could be prepared. The best method to achieve these requirements is to combine a mass separation with a radiochemical separation.

2.1 The ^{193}Pt Source

Sources of ^{193}Pt can be produced directly at the ISOLDE on-line mass separator as decay product of ^{193}Hg obtained by bombarding a Pb target with the 2.4 μA beam of 600 MeV protons from the CERN synchrocyclotron. Unfortunately this material contains a 4.3 day isomer of ^{193}Pt and 10,2 day ^{188}Pt. To avoid the losses in a Pt mass separation or one year of cooling, this source was prepared by means of a rather complex carrier free radiochemical separation from spent ISOLDE Pb targets, cooled for more than three years. In such a target, one can estimate that the only radioactive Pt that remains is a total of 10^{16} ^{193}Pt atoms produced directly or via decay of gold during a typical ISOLDE running period. In addition, it contains an equal amount of a variety of other long-lived activities.

The targets consist of a steel container from which the bulk amount of Pb was removed through vacuum sublimation. About 1 kg of Pb was treated in this way. The material remaining in which the platinum is enriched, was then treated in the 40 step radiochemical separation, procedure shown in Fig. 1. A sample of $5 \cdot 10^{14}$ ^{193}Pt atoms was isolated with a decontamination factor of 10^{13} or better.

2.2 The ^{163}Ho Sources

The ^{163}Ho was produced at ISOLDE by spallation of Ta in a specially developed version [7] of the high temperature Ta target ion source assembly, earlier used for Yb production [8]. By exchanging the original Ta powder material against rolls of 10 μm Ta foils, all rare earths were allowed to diffuse to the tungsten surface ionization source of the mass separator. In this way, an increase in yield by a factor of 3 was obtained since the sample collected at mass 163 contained all the isobars which decay to ^{163}Ho. In order to achieve maximum source purity, the resolution of the ISOLDE mass separator was raised to $M/\Delta M$ (FWHM) = 1500 by closing down the source outlet opening to 0,2 mm. A typical rare earth mass spectrum as observed at the focal plane of the separator is shown in Fig. 2. The mass 163 was

RADIOCHEMICAL SEPARATION OF PLATINUM

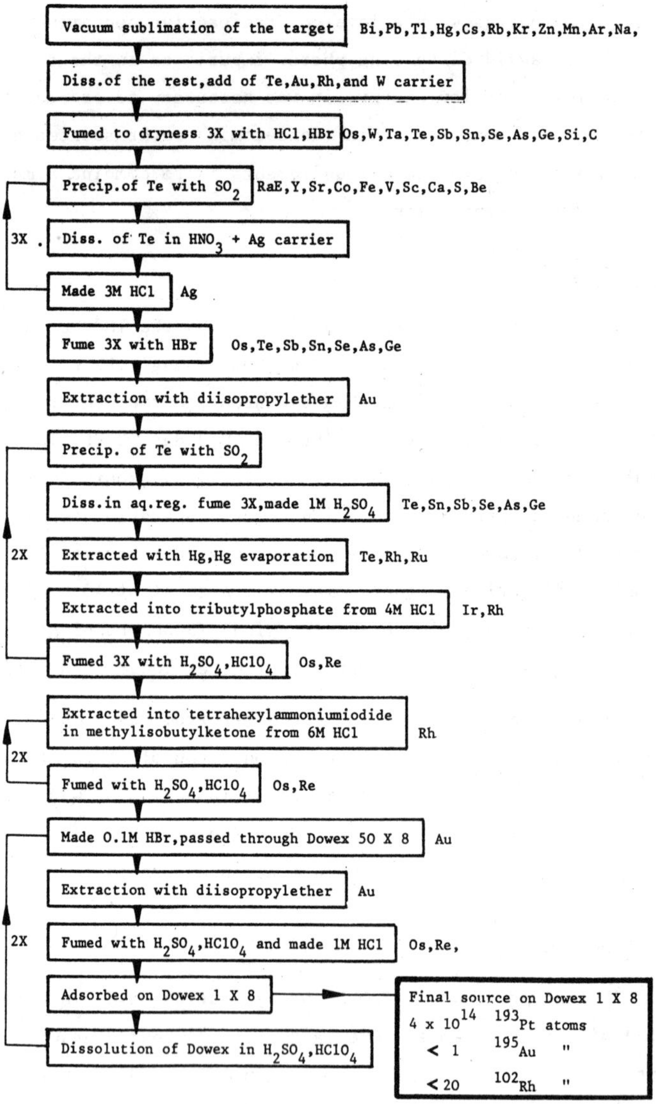

Fig. 1 : Radiochemical separation of ^{193}Pt from 1 kg of lead. Note the many steps used for separating the 10^{15} ^{102}Rh and ^{195}Au atoms from the platinum. In fact, no efficient single separation step which gave a decontamination factor of more than 70 was found.

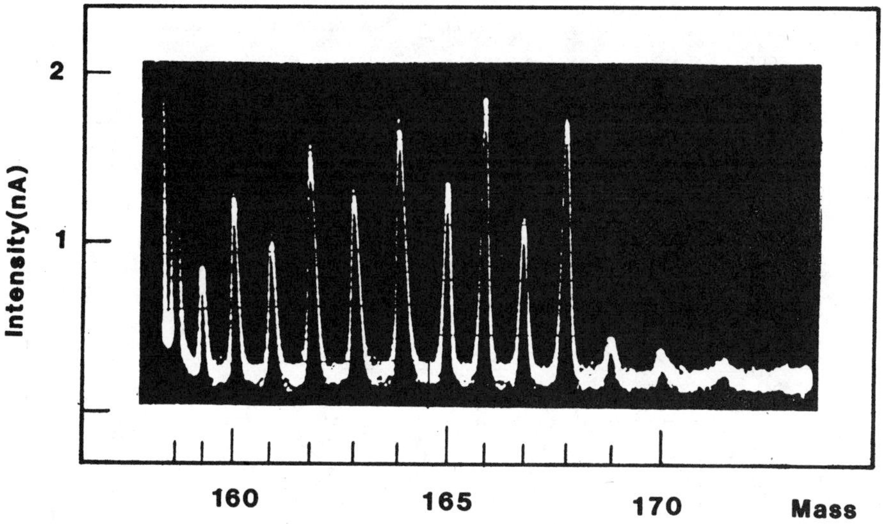

Fig. 2 : Rare earth mass spectrum obtained at ISOLDE by irradiation of a 122 g/ cm^2 Ta target with a 2,2 μA 600 MeV proton beam.

was directed out through an external beam line and collected in a Faraday cup with secondary electron suppression. From the integrated current, the absolute number of atoms in the 10 μm Ta collector foil was calculated to be $1{,}36 \cdot 10^{15}$. After some weeks of cooling, the main contaminant (determined by gamma spectroscopy) was $1{,}4 \cdot 10^{12}$ atoms of ^{147}Eu collected as the molecular side band $^{163}(^{147}Gd^{16}O)^+$. Another $4 \cdot 10^9$ atoms of both ^{167}Tm and ^{169}Yb were collected from the impurity level of 10^{-5} caused by the finite resolution of the mass separator. This number allows us to estimate the upper limit of the 1200 y ^{166m}Ho contamination in the source to be 10^8 atoms.

A further purification was obtained by a procedure based on wet chemistry. The collector foil was dissolved in the presence of 3 mg of lanthanum carrier and microgram amounts of other rare earths, except holmium. For tracing purposes a small amount of short-lived ^{160}Ho was added. The separation was carried out by precipitation steps followed by ion-exchange chromatography. The holmium fraction was cleaned twice, again by the same procedure. From the shape of the elution curves (see Fig. 3), it was estimated that each step gave a decontamination of a factor 10^3 from the heavier rare earths (such as Tm, Yb) and considerably more from the lighter ones. Thus the total decontamination from other elements offered by the combination of mass separation and chemistry was of the order of 10^{14}-10^{15}. The yield in the chemical procedure was 44%.

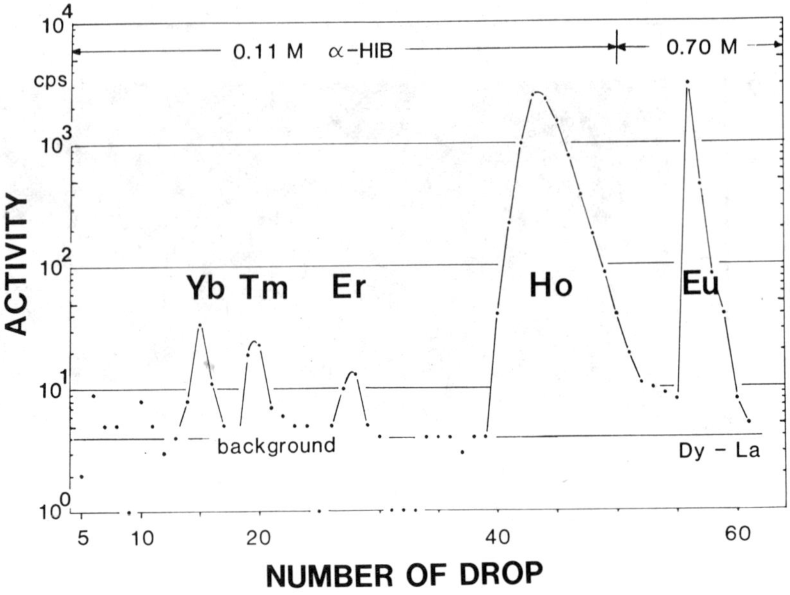

Fig. 3 : Ion exchange chromatogram of the first Ho separation on an 10 × 0,5 cm Aminex A5 column. The elution scale and the resolution as calibrated by means of the activity signal obtained from the added ^{160}Ho tracer and the impurities discussed in the text. Only the three drops at the peak of the Ho curve were used in the subsequent separation.

Two thin sources were prepared from this stock of radioactivity by vacuum evaporation from a rhenium ribbon at 1800° C. The first (source A) was made before the ^{160}Ho tracer had decayed, so that the yield could be determined : the result of 49% corresponds to $1.9 \cdot 10^{13}$ atoms in the central part of a 64 cm^2 tantalum backing.

A second thin source (source B) was prepared by evaporating a sample of 14.6% of the original stock onto a $0.1 \cdot 20 \cdot 20$ mm^3 beryllium foil. In an experiment with a Sm tracer and the same geometry, the evaporation yield was measured to be 35%, which would correspond to $3.1 \cdot 10^{13}$ atoms.

3. DETAILED TEST OF THE THEORY FOR RADIATIVE P CAPTURE BY MEANS OF ^{193}Pt.

Previous measurements to test the Glauber-Martin theory were confined to the energy region above the 1s pole which consequently showed relative small effects due to p-capture (see e.g. ref. 9). With the extension of the Glauber-Martin theory by De Rújula, further tests of the theory were needed, especially data for energies below the 1s pole. An isotope with ideal parameters for this purpose is ^{193}Pt, discovered by Hopke and Naumann [10]. They determined the decay to be a pure ground-state transition with a Q-value of 61 ± 3 keV, which falls short of the pole corresponding to the K binding energy of 76,1 keV, so that this pole will dominate the inner bremsstrahlung spectrum.

3.1 Spectrometric measurements on ^{193}Pt

The ^{193}Pt source was mounted on a 5 μm mylar foil and placed in a 3 mm diameter hole at the center of a 1 mm thick Cd plate, which served as a barrier to cross-talk between the two detectors of the spectrometer. The bremsstrahlung was measured with an intrinsic Ge detector with 200 mm^2 surface and a thickness of 6 mm. The counting efficiency of this detector was 13% over the energy region of interest. A Si detector was employed for the detection of the L x-rays from the Pt decay, and the source was sandwiched between the two detectors. The IBEC spectrum was recorded both in singles and in coincidence with L x-rays. The whole set-up was built into a graded shielding consisting of lead, cadmium and copper.

3.2 Discussion on the ^{193}Pt x-ray and IB Spectra

The non radiative EC is dominated by capture from the s-states. The Iridium singles L x-ray spectrum from ^{193}Pt shown in Fig. 4 becomes quite complex due to

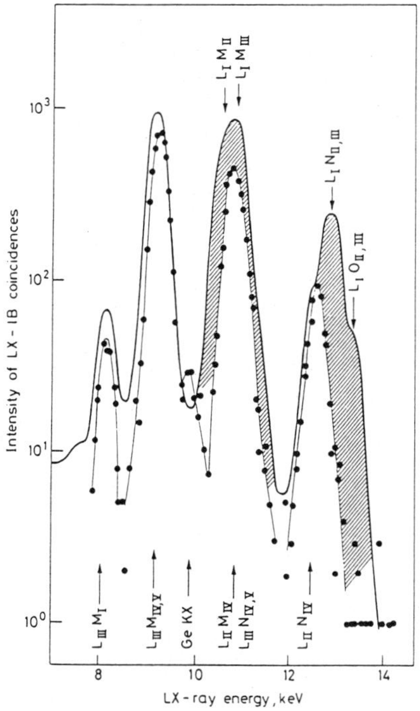

Fig. 4 : Iridium L x-rays from ^{193}Pt observed with a Si(Li) counter in singles (heavy line) and in coincidence (black points) with internal bremsstrahlung. The line components originating directly in the L_I sub-shell are identified in the upper part of the figure; below the dominant lines from L_{II}, L_{III} vacancies are identified. Note that the L_I lines are suppressed by about two orders of magnitude in the coincident spectrum; this is direct proof that the bremsstrahlung leads to population of the 2p states only. (The weak coincident line at 9.9 keV is a germanium x-ray due to cross talk from the detector used for observing the bremsstrahlung.)

the Coster-Kronig transitions and only the high energy part is of reasonably pure 2s origin. In the IB L x-ray coincidence spectrum also shown in Fig. 4, this part is seen to disappear proving that low energy IB originates entirely in P-wave capture. The limit of 1.5% for the 2s contribution obtained from the absence of the L_I lines is the 12-14 keV group corresponds approximately to the theoretically expected value.

The singles bremsstrahlung spectrum could be normalized to the L x-rays and turned out to agree exactly with the absolute values from calculations [5] using the "improved" expressions, in which the poles have been shifted to reproduce the experimental x-ray energies. Similar agreement is obtained for the $2p_{3/2}$ bremsstrahlung spectrum measured in coincidence with the 9.14 keV line which is pure $L_{III} \rightarrow M_{IV}, M_V$. Leaving aside a small correction for feeding of this line through Coster-Kronig transitions the coincident spectrum represents pure $2p_{3/2}$ radiative capture. The intensity can be put to an absolute scale on the basis of the number of L x-ray gates; again the agreement is quantitative (Fig. 5). From the extrapolated end-point we derive $Q_{EC} = 56.6 \pm 0.3$ keV.

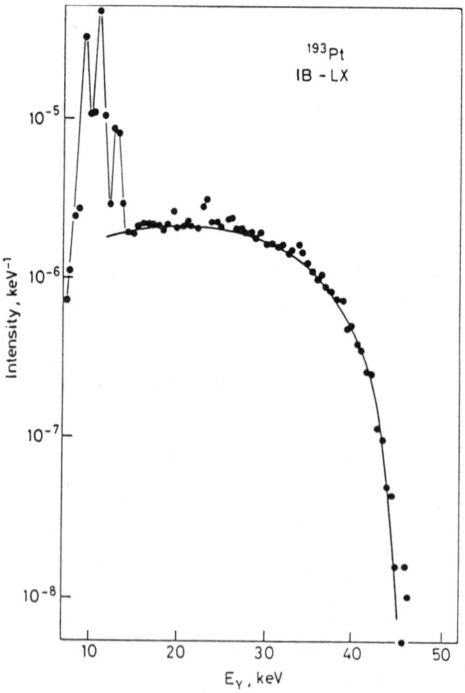

Fig. 5 : Internal bremsstrahlung from ^{193}Pt observed in coincidence with (all) L x-rays. The spectrum has been corrected for randoms but, owing to errors in the subtraction, weak L x-rays remain. The lines at 23 and 26 keV arise from fluorescent excitation of the Cd source holder. The spectrum has been brought to an absolute scale on the basis of the number of L x-rays gates and it has been corrected for the tail of the line but not for finite resolution. The theoretical curve is the sum of the $2p_{1/2}$ and $2p_{3/2}$ components, also given on an absolute scale.

Finally we examine the shape in more detail by dividing out the phase-space factor. The result (Fig. 6) coincides exactly with the calculation for 2p capture; the 2s capture would correspond to a constant shape-factor plot.

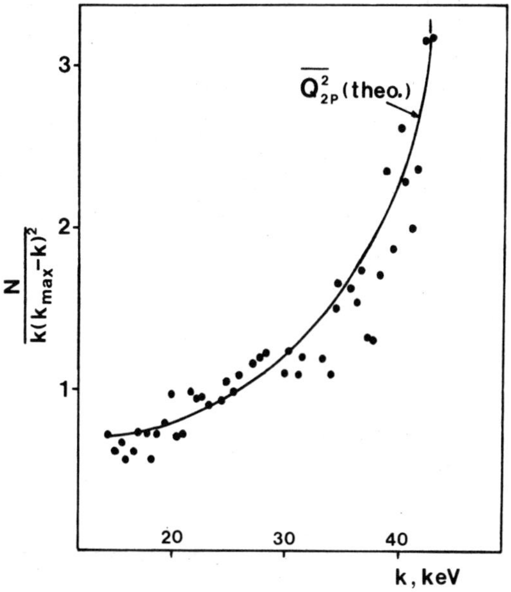

Fig. 6 : Shape-factor plot of the coincident 2p inner-bremsstrahlung (IB) spectrum from Fig. 5 and the corresponding theoretical curve. (The 2s bremsstrahlung would in this representation be constant ≅ 0.01.)

In conclusion, the comparison of the ^{193}Pt IBEC spectra with theory furnishes excellent agreement which, for the absolute intensities, is even quantitative.

3.3 A New Limit on m_{ν_e} Derived from the ^{193}Pt Data

From our present data we may already improve the previous limit of 4100 eV (67% confidence level) set on the electron-neutrino mass by Beck and Daniel [4]. Fig. 7 shows our data for the 2p spectrum transformed in a manner similar to the usual Kurie plot. The theoretical curves correspond to different assumed values of m_{ν_e}. As this experiment is essentially statistics limited one finds from a simple analysis that it sets an upper limit of 500 eV for a 90% confidence level. The case of ^{193}Pt with an end-point for the 2p IBEC spectrum of 45,3 keV is not an ideal case for a precise determination of m_{ν_e}. Although it should be relatively easy to improve this experiment further, it would hardly be worth the effort as the other radioactive isotope, ^{163}Ho, offers much more promise for a precise determination of the electron-neutrino mass.

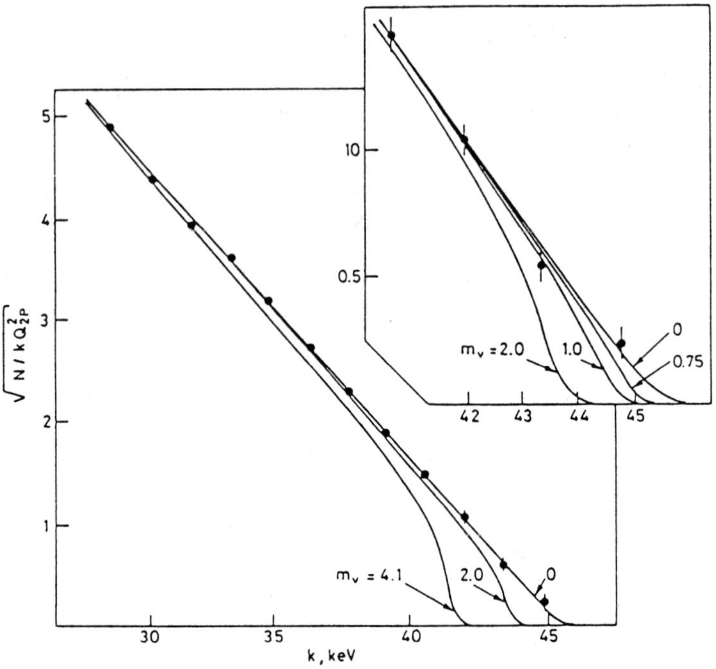

Fig. 7 : The high-energy part of the IB spectrum in coincidence with the 9.1 keV L x-ray (Fig. 4). The spectrum which thus represents essentially the $2p_{3/2}$ component, has been transformed similarly to a "Kurie plot". Theoretical curves are shown for different assumed values of the neutrino mass; the curves have been corrected for the detector resolution of 0.50 keV.

4. DETERMINATION OF THE ^{163}Ho DECAY PARAMETERS

The long lived isotope ^{163}Ho was discovered by Naumann et al. [11] who detected [12] radiations of approximately 1.3 keV which were interpreted as Auger electrons and x-rays following M capture. This assignment was recently confirmed through a measurement [13] with a high-resolution Si(Li) photon spectrometer. The result and the limit [12] set on the L capture-branch brackets the Q-value between the limits 2.05 < Q < 9.1 keV. These limits are, however, clearly inconsistent with the half-life of 33 ± 23 y [12] since the upper limit of the Q-value can be shown to correspond to a lower half-life limit of 150 y. In the following we shall report on a half-life determination of ^{163}Ho by counting M-shell radiations from carefully prepared sources containing a known number of atoms. We have also determined the Q-value by a novel technique based on single-nucleon transfer reactions.

4.1 Surface Analysis of the ^{163}Ho Source

By counting the two sources A and B (prepared as described in section 2.2) in a multiwire proportional counter, the relative count rates allows us to estimate the number of ^{163}Ho atoms on source B to be $4.5 \cdot 10^{13}$. An entirely independent check on the contents of source B was obtained by means of a highly sensitive surface-analysis technique : elastic Coulomb scattering of alpha particles in back angles at relatively low energy offers a sensitive probe for surface layers and a good mass resolving power arising from the kinematic energy loss in the collision. The experiment, carried out with a beam of 3.5 MeV alpha particles from the Aarhus single-stage Van de Graaff accelerator, shows (Fig. 8) a well-resolved peak at the holmium mass position with an intensity (integrated over the source) of $5.0 \cdot 10^{13}$ atoms. There is thus excellent agreement with the Faraday-cup and chemical-yield data. On the basis of these results, we stress that a consistent scale for the number of ^{163}Ho atoms on the sources has been obtained by two entirely independent techniques.

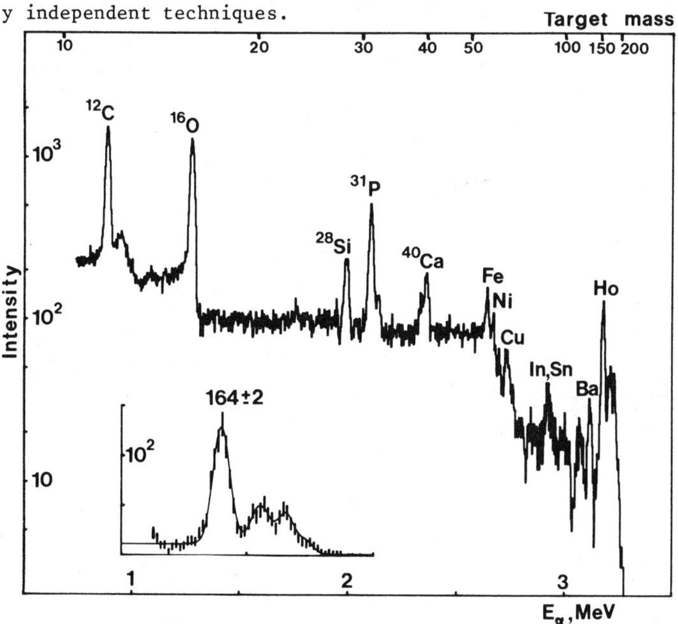

Fig. 8 : Energy spectrum of 3.5 MeV alpha particles scattered from the centre of foil B through 170°. Surface impurities such as holmium appear as clearly resolved peaks, whilst bulk impurities give rise to a step function which reflects the energy loss of the alpha particle inside the target. The energy scale was calibrated by scattering the beam from a thin target of gold. The inset shows that the high-energy region is dominated by a peak at mass 164 ± 2 with weaker peaks around 183 and 203, presumably Ho, Ta/Re, and Au/Pb, respectively. The intensity of the holmium peak, integrated over the area of the foil, gives an independent check on the number of atoms in the source.

4.2 The ^{163}Ho Half-Life and Q-Value

By means of single wire and multiwire proportional counters, both the M-Auger and M x-ray decay rates were determined for the two precisely "weighed" sources. Typical spectra are shown in Fig. 9.

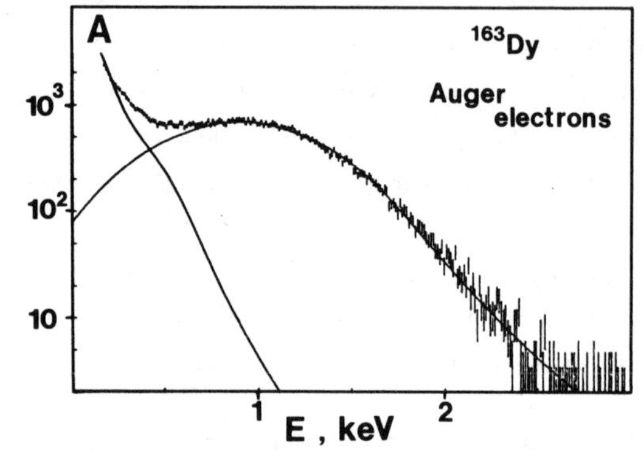

Fig. 9 : Energy spectrum of electrons from source B measured with a cylindrical central wire proportional counter filled with a helium-methane mixture. The lines indicate a fit to the background measured with the source removed and a fit to the M-Auger peak.

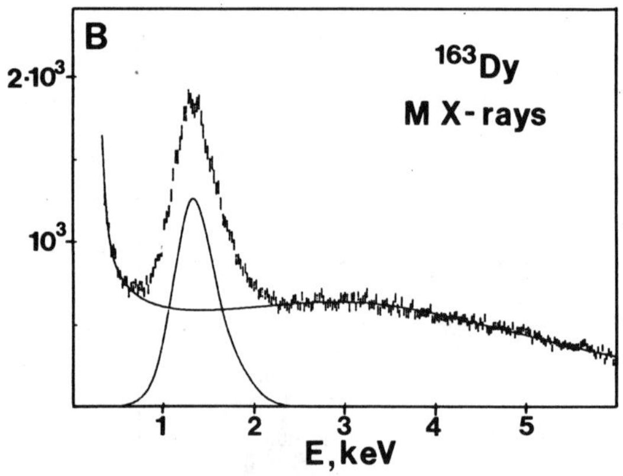

Fig. 10 : Spectrum of M x-rays from ^{163}Ho measured through a 2 μm mylar foil to absorb Auger electrons. After correction for background, only a broad peak near 1.3 keV remains. The fit shown uses the experimental energy resolution determined with Al x-rays and a fine structure corresponding in energy to the one expected theoretically.

The results from the two sources are consistent, and the M x-ray and electron measurements agree, proving thus that the sources were thin enough to permit an efficient counting of 1 keV Auger electrons.

Finally, from the various measurements given in Table 1, a partial M capture half-life for each source can be calculated. Since the measured values serve as a safeguard against systematic errors, it would have little meaning to quote a best value, and we choose instead to represent the data by a single number*)

*) The recent paper by Yasumi et al. [15] determined the number of atoms in a ^{163}Ho source by proton-induced x-ray emission (PIXE) and the disintegration rate by M x-ray counting. From their data one derives a partial M-capture half-life of 51.000 y in good agreement with our value.

Table 1 : Determinations of the partial M-capture half-life of ^{163}Ho

Source	No of atoms $(\cdot 10^{-13})$	M disintegration rate (s^{-1})	$t^M_{1/2}$ (y)
A	1.9	8.2 *c)*	$5.1 \cdot 10^4$
B	4.5 *a)*	32.4 *d)*	
	5.0 *b)*	27.4 *e)*	$3.0 \cdot 10^4$ *g)*
		21.6 *f)*	

a) From the ratio of counting rates B/A = 2.34 measured with a MWPC.

b) From surface analysis by Rutherford scattering (see Fig. 8).

c) M Auger electrons counted with a MWPC.

d) M Auger electrons counted with a He-filled single-wire proportional counter.

e) M x-rays counted with an argon-filled single-wire proportional counter; 2 μm mylar absorber with calculated transmission for M x-rays of 48%. Assumed M fluorescence yield 0.98% (Fig. 9).

f) As (*e*) but 4 μm mylar absorber, calculated transmission 29%.

g) Based on (*a*) and (*d*).

$t^M_{1/2} = (4.0 \pm 1.2) \cdot 10^4$ y. It is possible to convert the measured M-capture transition probability into an estimate of the phase-space factor for the ^{163}Ho decay by capture for the M_I shell

$$\phi(M_I) = [Q-E(M_I)] \left\{[Q-E(M_I)]^2 - m_\nu^2 c^4\right\}^{1/2} = (0.53 \pm 0.10 \text{ keV})^2,$$

where $E(M_I)$ is the M_I binding energy and m_ν is the electron-neutrino rest mass. The details of this calculation have been given elsewhere [14]. Essentially, it relies on the fact that the only other unknown quantity is the nuclear matrix element which, owing to a fortunate coincidence, can be fixed precisely by reference **) to the corresponsing transition in the nucleus ^{161}Ho. Assuming $m_\nu = 0$ and taking the known value $E(M_I) = 2.05$ keV, we thus find Q = 2.58 ± 0.10 keV.

**) The essential difference originates in the so-called pairing correction factor $\mu_p^2\mu_n^2$, which was calculated theoretically for the two cases. We take the opportunity to point out that the two values were cited in the wrong order in ref. 14; thus, the value for ^{163}Ho is 0.354.

4.3 Limit on m_{ν_e} from the Phase Space Factor and the Nuclear Reaction Q-Value

It is possible to obtain the quantity Q (here defined as the mass difference ^{163}Ho-^{163}Dy) from nuclear reactions. Such an experiment is being performed at the Niels Bohr Institute Tandem Accelerator and its associated multigap magnetic spectrometer. The measurement was based on the two reactions (here d and t denote deuterons and tritons)

$$^{163}\text{Dy}(d,t)^{162}\text{Dy}$$

and

$$^{162}\text{Dy}(^{3}\text{He},d)^{163}\text{Ho}.$$

Reference lines were obtained by elastic scattering of the projectiles from the same targets. The novel feature of the experiment was that the outgoing ^{3}He were detected as singly charged ions so that they followed a path through the spectrograph close to that of the tritons. The Q value emerges as the sum of the energy differences of the two line pairs d,d and t,$^{3}He^{+}$ after a correction for differences in recoil energy. The fact that several excited states are populated in the reactions aids in the analysis. The preliminary value [14] for Q is 2.3 ± 1.0 keV, but further improvements in the precision are to be expected. Combination of this result with the value of $\phi(M_I)$ discussed in subsection 4.2 gives a limit [14] of 1.3 keV in m_{ν_e}. This limit has already been superseded by the bremsstrahlung experiment reported in subsection 3.3 of the present paper.

5. CONCLUDING REMARKS

The two first phases of our experimental programme have now been completed. Using ^{193}Pt we have shown that experiment and theory agree extremely well for low-energy bremsstrahlung. We find the factor 100 enhancement demanded for this case by the 1s resonance effect and we show for the first time that low-energy IB is of electric dipole nature since it gives rise to p capture. The ^{163}Ho experiment has determined the partial half-life to be 40.000 y, to which corresponds a calculated total half-life of (7 ± 2) · 10^{3} y. With a Q value only 530 eV away from the M_I resonance point, the case of ^{163}Ho remains at present the best candidate for an electron-neutrino-mass experiment.

For the next phase of the experiments we are attempting to develop Auger-electron bremsstrahlung coincidences with high-resolution gas counters and, in parallel, calorimetric (or total-absorption) spectrometry based on solid-state detectors. The second technique is, of course, especially interesting since it will exploit the large enhancement of the electron-ejection processes [16]. With luck the first results could be available during the next months, and if all continues to go well. resolution will be the main concern from then on.

With our much too imcomplete knowledge of nuclear data, it is obviously too early to be sure that ^{163}Ho will remain the best candidate, and further studies of nuclear parameters are clearly needed. During the last few weeks we have been investigating the weak EC branch in the decay of ^{157}Tb. This branch to a 54.5 keV excited state seems to have a Q value just below the L_I resonance point, but we do not yet know whether it is near enough to make the case of practical value for our purpose.

REFERENCES

[1] K.E. Bergkvist, Nucl. Phys. B39 (1972) 317; Phy. Scripta 4 (1971) 23.

[2] V.A. Lyubimov, E.G. Novikov, V.Z. Nozik, E.F. Tretyakov and V.S. Kosik, Phys. Lett. 94B (1980) 266.

[3] J.J. Simpson, Phys. Rev. D23 (1981) 649.

[4] E. Beck and H. Daniel, Z. Phys. 216 (1968) 229.

[5] A. De Rújula, Nucl. Phys. A374 (1982) 619.

[6] R.J. Glauber and P.C. Martin, Phys. Rev. 104 (1956) 158; P.C. Martin and R.F. Glauber, Phys. Rev. 109 (1958) 1307.

[7] P. Hoff et al., High Temperature Tantalum Foil Targets, to be published.

[8] H.L. Ravn, Phys. Rep. 54 (1979) 201.

[9] M.H. Biavati, S.J. Nassif and C.S. Wu, Phys. Rev. 125 (1962) 1364.

[10] P.H. Hopke and R.A. Naumann, Phys. Rev. 185 (1969) 1565.

[11] R.A. Naumann, M.C. Michel and J.C. Power, J. Inorg. Nucl. Chem. 15 (1960) 195.

[12] P.K. Hopke, J.S. Evans and R.A. Naumann, Phys. Rev. 171 (1968) 1290.

[13] C.L. Bennett, A.L. Hallin, R.A. Naumann, P.T. Springer, M.S. Witherell, R.E. Chrien, P.A. Baisden and D.H. Sisson, Phys. Lett. 107B (1981) 19.

[14] J.U. Andersen, G.J. Beyer, G. Charpak, A. De Rújula, B. Elbek, H.Å. Gustafsson, P.G. Hansen, B. Jonson, P. Knudsen, E. Laegsgaard, J. Pedersen and H.L. Ravn, Phys. Lett. 113B (1982) 72.

[15] S. Yasumi, G. Rajasekran, M. Ando, F. Ochiai, H. Ikeda, T. Ohta, P. Stefan, M. Maruyama, N. Hashimoto, M. Fujioka, K. Ishii, T. Shinozuka, K. Sera, T. Omori, G. Iwazawa, M. Yagi, K. Masumoto and K. Shima, Measurement of the Mass of the Electron Neutrino using the Electron Capture Decay Process of the Nucleus, paper presented at the Lake Balaton Conference, June 1982, to be published.

[16] M. Lusignoli, Neutrino masses. Proc. Eighth Int. Workshop on Weak Interactions and Neutrinos, Javea Alicante, Spain, 5-11 September 1982, to be published.

AN EXPERIMENT TO STUDY THE β-DECAY OF FREE ATOMIC AND MOLECULAR TRITIUM

T. J. Bowles, R. G. H. Robertson, M. Maley, J. C. Browne, T. Burritt,
J. Toevs, M. Stelts, J. F. Wilkerson
J. Helfrick,[a] D. Knapp,[b] and A. G. Ledebuhr[c]

Los Alamos National Laboratory, Los Alamos, NM 87545, U. S. A.

An apparatus is described which will allow the measurement of the β-decay of free tritium atoms and molecules. It consists of an RF dissociator, a long cylindrical decay region open at both ends, a guide field, and a magnetic spectrometer.

There is some interest in determining whether the electron neutrino has mass. Recently, Lyubimov et al.[1] reported a measurement of the β-spectrum of ^3H which shows conclusive evidence for an antineutrino mass between 14 and 46 eV at the 99% confidence level. Despite careful study, no substantial flaw has been detected in their procedure. Nevertheless, many scientists would like to see a confirming experiment. Much of the concern revolves around the use of a solid source (tritiated valine, an amino acid) for which one may not know the

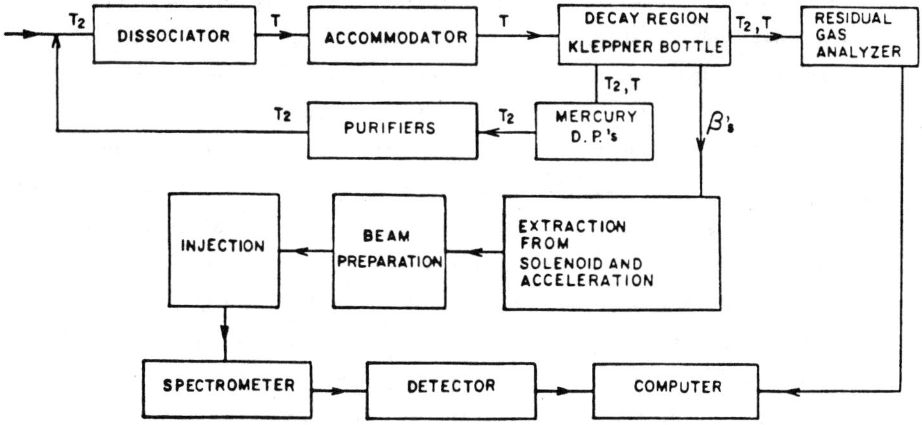

Fig. 1 Functional diagram of experiment.

atomic and molecular final states of the ^{3}He daughter atom, the scattering and energy loss of β's in the source, and the shape of the background near the end-point as well as one would like.

The ideal source would be free tritium nuclei, but this turns out to be impractical owing to space charge limitations. The next best thing, free tritium atoms, may form the basis of a practical source for which detailed and accurate calculations of the atomic final states and electron energy losses can be performed. Recent advances in the production of dense gases of spin-polarized hydrogen encourage us to believe that a free-atom tritium source of adequate strength can be constructed. A functional plan of the experiment is shown in Fig. 1.

Molecular tritium at 300-mT pressure enters a Pyrex discharge tube cooled to 77° K (LN_2). The molecules are dissociated in an RF discharge and emerge through a small orifice into a cylindrical decay region ("Kleppner bottle")[3] whose walls are coated with a thin layer of Pyrex glass, which inhibits recombination. The maximum length of this decay region is set by the recombination rate, which has not been measured for tritium at present, and by the molecular fraction which can be tolerated. Recombination rates have been measured for atomic hydrogen on pyrex[6] and shown to be approximately 2.5×10^{-5} (probability of recombination) per wall collision at 120 K. The two competing processes for recombination are due to atomic hydrogen in the source recombining on chemisorbed and physisorbed atoms on the wall. These processes are independent of the isotopic mass and depend only on binding energies, which are identical to first order for hydrogen and tritium. An equilibrium density of atomic T is built up as established by the influx and the conductance of the tube. The "Kleppner bottle" we are designing is 200-cm long with a 5-cm i.d. This provides an equivalent source thickness integrated along the axis of 2×10^{15} atoms/cm^2 with a molecular fraction of about 15% assuming 95% dissociation fraction for the source.

The Kleppner bottle is placed in a solenoidal magnetic field of about 1 kG with a small axial gradient. Betas ($B\rho < 463$ Gauss-cm) spiral about the field lines. At one end of the solenoid a pinch coil with a peak field of about 3 kG reflects most of the βs that start with a velocity component directed towards that end; as a result about 90% of the βs reach the weak-field end of the solenoid. There they are extracted and accelerated through a potential of 20 kV.

An important feature of our experiment is that this energy gain is never compensated by deceleration later in the apparatus: the entire decay region floats at -20 kV. There are two advantages in this: first, 18-keV electrons from the decay region are raised to 38 keV and are well above the energy of any βs from tritium that may find its way into the beam transport or spectrometer; and, second, the spatial component of phase space is reduced. Thus, not only is the background in the region of the shifted end point far lower than it would be at 18 keV, but the emittance of the beam is improved. We might also remark that this idea is equally applicable to solid sources. The price paid is, of course, that higher resolving power is required in the spectrometer.

One of the most interesting aspects of the problem is extraction of the βs from the solenoidal field into a field-free region at the object of the spectrometer. Given that in order to enter the spectrometer, all βs must pass through a collimator of radius r_c at an angle less than ψ_s, we find two general theorems: 1. There is a maximum radius in the solenoid R_4 beyond which no β can originate and still enter the spectrometer; and 2. The maximum fraction of rays originating within R_4 transmitted to the spectrometer is

$$\frac{6x\rho + 2\rho^2}{3R_4^2}$$

where $x = \frac{r_c p'}{p} \sin\psi_s$ and ρ is the (maximum) radius of orbits in the field. The electron momenta before and after acceleration are p and p', respectively. The first result follows from conservation of the canonical angular momentum and the second from application of the Poincaré invariant.

The significance of the first theorem is, of course, that any electrons originating from tritium adsorbed on the walls of the Kleppner bottle can be rejected absolutely, independent of aberrations or imperfections in the optical system. This is most important, because a monolayer of adsorbed tritium represents 4 orders of magnitude more activity than the gaseous source. However, from measurements on hydrogen[6] we estimate that the equilibrium density of tritium on the pyrex walls at 120 K will be less than 10^{-4} of that in the source region. A schematic diagram of the entire source is shown in Fig. 2.

Acceleration and extraction is accomplished by grids and coils. Spectrum distortion can result from tritium decaying in the acceleration region, and the

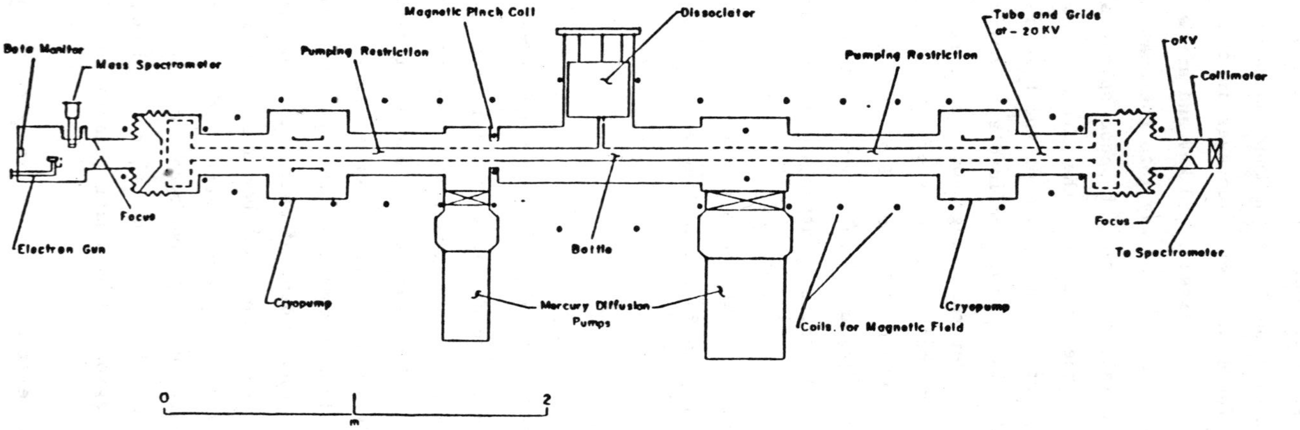

Fig. 2 Source and extraction system showing location of coils.

volume is minimized by defining it with grids. In the present design, the fraction of detected decays occurring in the acceleration gap is of order 10^{-6}. Plane, parallel grids form a lens with infinite focal length, which does not introduce the relativistic aberration. Scanning the spectrum can be performed in several ways, but the preferred method is to vary the potential of the Kleppner bottle and maintain all other potentials and fields fixed. While this causes a (calculable) variation in extraction efficiency, there is significant advantage to presenting the extraction lens, spectrometer and focal plane detector with fixed energy particles.

The spectrometer is modeled on the toroidal design of Tretyakov.[4]

Table I

Spectrometer Design Parameters

Source - Image Distance	5.00 m
Orbit apogee	0.75 m
Inner cylinder radius	0.29 m
Entrance angle	$24 \pm 4.5°$
Exit angle	$90 \pm 15°$
Resolving power (base)	5×10^{-4}
Luminosity	0.014 cm^2
Object radius	0.50 cm
Image length (base)	0.27 cm

We have modified the design in two respects: The size of the spectrometer is approximately twice that of Tretyakov's, and particles enter the field at angles between 20° and 30° rather than between 85° and 95° to the axis (Fig. 3). Matching the atomic source to the spectrometer is then greatly simplified. Detailed calculations on the spectrometer have been carried out, and Table I summarizes the important parameters. We note that the resolution limit of the spectrometer is determined by fringing fields and is calculated to be 13 eV for 38.6-keV betas at full acceptance.

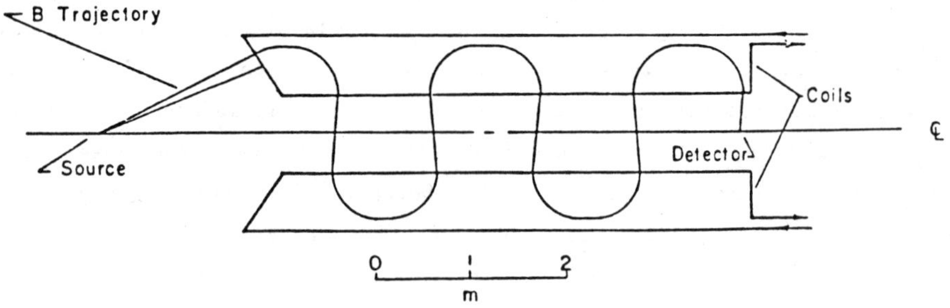

Fig. 3 Cross section of spectrometer showing coil profile and trajectory of β entering at maximum allowed angle.

The ion-optical characteristics of the system can be studied in several ways. A 40-kV electron gun is mounted at the opposite end of the source from the spectrometer and can be used to measure resolution, transmission, and aberrations. A very convenient check on the general behavior of the Kleppner bottle, extraction system, and spectrometer is to use 1.8-hr ^{83m}Kr, a daughter of 83-day ^{83}Rb. This isotope is gaseous, short-lived, and emits a 17.8-keV conversion line. By appropriate variation of the temperature of the Kleppner bottle the ^{83m}Kr source can be made to roughly mimic the distribution of the tritium. At very low temperatures, the rejection of electrons originating at the wall can be investigated.

In estimating the overall performance of the system we are hampered by lack of direct information about dissociation efficiencies and recombination rates for tritium. Our original calculations of counting rates were based on the work of Silvera and Walraven[5] on H, which gave rather modest counting rates. However, in a test setup we have produced intense beams ($\sim 10^{18}$ atoms/sec) of H and D with high dissociation fractions (>90%). Using this flux input with our "Kleppner bottle" design we estimate focal plane count rates of approximately 0.2/sec in the last 100 eV, which corresponds to a source thickness of 2×10^{15} atoms/cm^2. Higher source strengths may be available with further optimization of the system. Measured background rates in a prototype focal plane detector are of order 2×10^{-3}/sec, and Fig. 4 shows

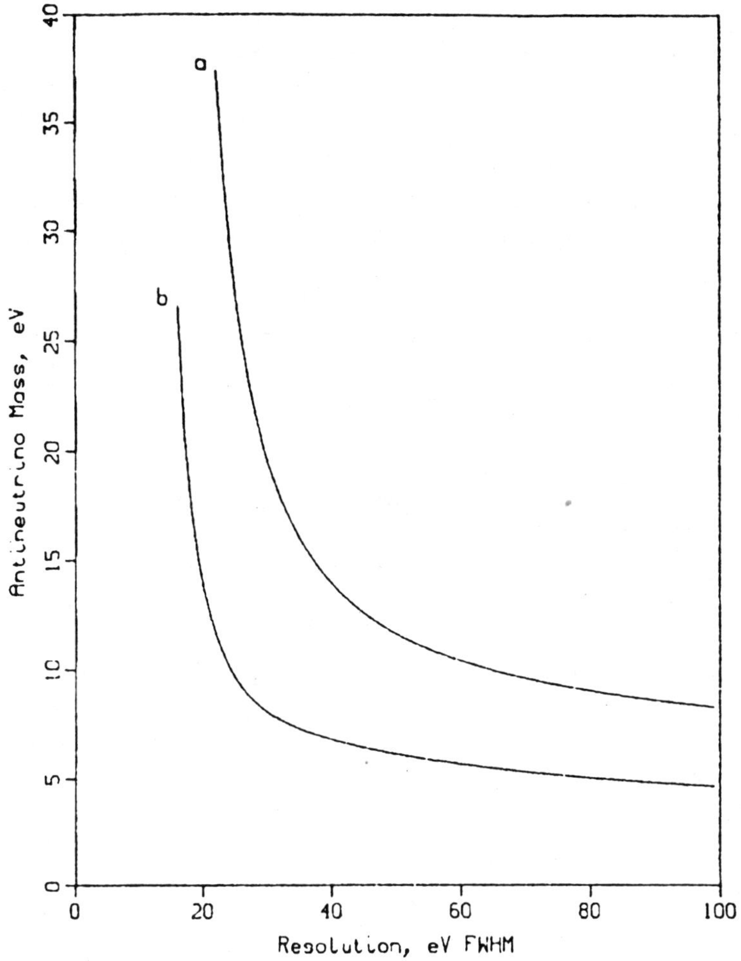

Fig. 4 Limits (2σ) on mass obtainable in 10^6 seconds running for a background rate of 2×10^{-3} sec. a) source strength 1×10^{15} cm^{-2}. b) source strength 1×10^{16} cm^{-2}.

the limits that can be placed on $m_{\bar{\nu}_e}$ in about two weeks running. The counter-intuitive behavior of these curves (namely the increase in sensitivity with decrease in resolution) reflects the improved statistical accuracy obtainable at poor resolution. However, in practice, count rate will be sacrificed for the sake of reduced likelihood of systematic error, and data will probably be taken at 30- to 40-eV resolution. Should higher source intensities be available, the spectrometer is capable of substantially better resolution.

a. University of California at San Diego
b. Princeton University
c. Michigan State University

1. V. A. Lyubimov, E. G. Novikov, V. Z. Nozik, E. F. Tret'yakov, V. S. Kozik, and N. F. Myasoedov, Zh. Eksp. Teor. Fiz. 81, 1158 (1981)

2. I. F. Silvera and J. T. M. Walraven, Phys. Lett. 74A, 193 (1979).

3. This approach was suggested to us by D. Kleppner.

4. E. F. Tretyakov, Izv. Akad. Nauk. SSR. Ser. Fiz. 39, 583 (1975).

5. J. T. M. Walraven and I. F. Silvera (to be published).

6. B. J. Wood and H. Wise, J. Chem. Phys. 66, 1049 (1962).

FEASIBILITY OF A NEW DIRECT MEASUREMENT OF THE MUON-NEUTRINO MASS*

B. Robinson
University of Pennsylvania
Philadelphia, Pa. 19104

ABSTRACT

Two possible experiments are examined for their feasibility to produce an improved measurement of the muon-neutrino mass. It is concluded that either one might result in a mass limit in the range of 50 to 100 KeV/c^2.

INTRODUCTION

The question of whether or not neutrinos have a small mass below current limits has important implications for the Grand Unified Theories (GUTS), other areas of particle physics, and cosmology. Muon neutrinos are particularly elusive because they are produced only in decays where a relatively large amount of energy is released. This is not the case for electron neutrinos, which are produced in beta decay with low momentum, leading to a much lower mass limit.

Neutrino oscillation experiments may lead to much improved limits on the neutrino mass difference, but do not measure the absolute masses and more importantly require that the mixing angle of the neutrinos is large enough to measure an effect. That is, oscillation experiments measure a combination of mixing angle and mass difference. As long as no effect is seen, no firm statement can be made about the neutrino masses.

It is, thus, highly desirable to obtain improved direct measurements on the neutrino masses in addition to oscillation experiments. Several experiments were discussed at this conference to improve the limit on the electron neutrino mass. However, there was little effort in evidence toward improving the muon neutrino mass limit.

PRESENT LIMIT

Figure 1 shows recent μ-neutrino mass limits. Little or no improvement has taken place in the last ten years. The present limit of about 500 KeV is shared by three experiments,

*This work was supported by the U.S. Department of Energy under contract EY-76-C-02-3071

each using a different experimental method. It appears that the method of Daum, et al.[1], π decay at rest, is least subject to improvement since it would require a significantly improved knowledge of the pion and muon masses. We are left with the $K_{\mu 3}$ decay of the K_L (Clark, et al.[2]) and pion decay in flight (Anderhub, et al.[3]). Improvements in these experiments seem possible with new technology currently available. Lacking any other method to obtain an improved measurement, these are the only ones that will be considered.

The ORCID

Before describing the proposed experiments in detail, I will briefly describe a new form of Cherenkov detector called the ORCID[4] (Optical Readout Cherenkov Imaging Detector) which was developed recently for use in Fermilab experiment E609 (a hadron jet experiment, see Fig. 2). This will be done because a detailed description of its performance has not yet been published and because the inclusion of this type of detector is the principle improvement proposed to the experiments to be considered.

The essential features of the ORCID have been described previously[4]. Briefly, Cherenkov light is focused by a large spherical mirror to sharp rings, nearly circular in shape. This image field is then reduced and refocused onto an image detector consisting of two image intensifiers in series and a CCD (Charged-Coupled Device) sensitive to light. Cherenkov light is reproduced, intensified and reduced in scale, at the CCD essentially as it appeared at the primary focus, in well defined rings. The positions of the rings on the image field correspond (nearly linearly) to the particles' angles through the detector. The size of a ring corresponds to the angle of the Cherenkov light cone. Particles' directions and velocities may thus be measured simultaneously. The accuracy to which they may be measured depends on the measurement of the ring position and radius. The photoelectron dispersion σ_{pe} is determined by the optics, multiple Coulomb scattering of the particle in the radiator, energy loss, and pe measurement error. But the errors in determining the ring position and radius are reduced, barring systematic measurement errors, by the square root of the number of photoelectrons on the ring, since each constitutes an essentially independent measurement.

More than one particle may be measured at the same time if its ring size or its angle are significantly (about 2 σ_{pe}) different from one another. For instance, a pion decaying with the muon going forward gives a muon with nearly the same energy and angle, but they may both be detected because the muon will give a significantly larger ring if the energy is low enough and the index of refraction of the radiator is matched to that energy (See Fig. 3).

The only major changes between the proposed detector[4] and the one actually used in E609 were the replacement of a refocusing lens with a second spherical mirror and the improvement of the gain of each image intensifier by the addition of a microchannel plate.

In beam tests it was found that an average of about 8 photoelectrons were detected for $\beta = 1$ particles (with limiting Cherenkov angle of 46 mr). The dispersion of photoelectrons around a ring was found to be 2.5 mr. The systematic error of determining the ring position and size was found to be less than 1 mr. Work is in progress to further reduce this figure in the anaylsis with improved ring shape parameters and other methods. It will be assumed in what follows that the photoelectron dispersion will be 1 mr and that systematics will be less than 0.1 mr. These are not particularly optimistic estimates considering the smaller ranges of particle angle and better optics which would be used in these experiments and refinements in the system which would be made for them.

A first order optics design has been worked out for the experiments considered here to determine its feasibility and estimate the errors. From this it was learned that a 20 m long device could be built with two 150 cm diameter mirrors with focal lengths of 3.2 m giving F/2.1 compared to the F/1.2 for the smaller mirror in the E609 design. The range of particle angles accepted, however, would be only about 4 mr compared to 150 mr in the E609 design. For a $n-1 = 4.6 \times 10^{-4}$, an 8 GeV pion would have a Cherenkov angle of 25 mr and give about 45 pe's.

π DECAY IN FLIGHT

As previously mentioned, pion decay at rest depends heavily on a very accurate knowledge of the pion and muon masses. This effect is reduced when the decays are measured in flight. In addition, if decays are measured where the muon goes nearly forward, the momentum of the neutrino in the laboratory is reduced. Since the momentum and energy are related to the mass through a quadratic sum, it is very advantageous to lower their values as much as possible. Contrary to these effects is the relative difficulty of measuring the higher energy pion and muon in the decay from which the neutrino mass is to be calculated. In particular, for very relativistic pions ($\gamma_\pi \gg 1$) undergoing nearly forward decays, the energy of the neutrino is given approximately the following expression:

$$E_\nu \simeq \frac{P^*}{2\gamma_\pi} + \frac{\gamma_\pi}{2P^*}\left(m_\nu^2 + P_t^2\right) \qquad (1)$$

where E_ν is the laboratory neutrino energy, P* is the center of mass momentum, γ_π is the relativistic gamma of the pion (E_π/m_π), m_ν is the neutrino mass, and P_t is the transverse momentum of the muon with respect to the original pion direction. The pion and muon masses enter only through P*. The effect of imperfect knowledge of P* is reduced at larger γ_π, and can be corrected by comparing data at more than one pion energy. It can also be seen easily in Eq. 1 that the P_t is a crucial measurement. Since P_t enters the equation in the same way as m_ν, it cannot be eliminated by comparing data at different energies as the first term can. If there is a systematic effect in the average P_t, it will be interpreted as a non-zero neutrino mass. This effect will be discussed later. E_ν is the other critical experimentally determined number.

An experiment to accomplish these goals is represented in Fig. 4. It consists of two magnetic spectrometers with a long ORCID in between. The ORCID has been divided into two sections for optical reasons, but the sections share the same radiator. The first spectrometer, the pion spectrometer, would measure the original pion momentum in connection with the ORCID. The first leg of the spectrometer would measure the initial direction with PWC planes. The angle (position on the image field) in the ORCID would give the final angle and thus the momentum of the pion.

In addition to the primary detectors, a threshold Cherenkov and appropriate absorber with trigger counters will identify the pion before the decay and the muon after the decay, respectively, in order to improve the triggering efficiency.

If there is a 1 mm accuracy in the PWC's over a 10 m lever arm, there will be a 0.1 mr error in the initial angle. A bend of 90 degrees could be achieved with a series of magnets with an average field of 27 KG with a 10 m radius for 8 GeV particles, the highest energy contemplated for the experiment. The final angle, in the first part of the ORCID, will be known to 1 mr/$\sqrt{N_{pe}}$. In order to keep multiple scattering low but the measurement accuracy good (i.e. the number of photoelectrons high) a moderately high index of refraction is contemplated. The ORCID should be kept near atmospheric pressure to allow thin windows to be used. For 20 m of radiator, a mixture of Freon 12 and Nitrogen for instance, with an $n-1 = 6.2 \times 10^{-4}$, there will be approximately 50 pe's detected and an average of 0.6 mr multiple scattering. If, as assumed before, the pe dispersion is 1 mr, it will dominate the error, giving 0.17 mr for the final error in pion angle (assuming 0.1 mr systematic error). Assuming that the particle path through the spectrometer will be in vacuum except for windows which can be kept thin, it will not significantly contribute to the multiple Coulomb scattering. In this case, the change in pion angle will be known to approximately 0.24 mr for a 90 degree bend, or $\Delta\theta/\theta = 0.8 \times 10^{-4}$. For an 8 GeV particle, this represents an error of 0.6 MeV/c in the momentum.

The neutrino energy is the difference of the pion and muon energies. Mirror image spectrometer systems for the pion and muon would give a total error for a single event of about 1 MeV for the neutrino energy. This is, of course, only an approximate number and will depend on the details of the final design and an optimization of the parameters, but is not likely to be more than a factor of two off. A much larger pe dispersion, which now dominates the error, is not expected since a figure close to that assumed has already been achieved. On the other hand, a much better figure is also not likely since there are several factors that would enter in just below the contemplated error.

At this point it might seem that the ORCID adds little to the experiment, since it is the reason for the introduction of the dominate measurement error in the neutrino energy. However, there are several reasons for its inclusion. First of all, it provides an accurate measure of the actual decay angle of the muon relative to the pion and, therefore, of P_t. More wire chamber planes might give the same accuracy for the angular measurement, but at roughly the same cost in material in the beam. But the ORCID also provides other information that chambers cannot. The ring sizes give a second, if less accurate, measure of the particle energies, useful in calibration and in detecting systematic effects. The dispersion of the ring provides a measure of the Coulomb scattering of the particle, since it will be broadened in the direction of scattering thus allowing at least statistical separation of Coulomb scattering from decay angle in the final sample. Also, the number of pe's in each ring gives an approximate position of decay.

Energy loss effects will be kept to a minimum by performing a consistency check between the pion and muon momenta measured and the decay angle observed in the ORCID. Landau fluctuations and large scatters from PWC wires can be eliminated in this way.

We come now to the question of the ultimate systematic error in the experiment, which is the crucial question, since at least in principle, sufficient statistics can be taken to reduce the experimental error to this level. The principle method of calibration and reduction of systematic errors would be to use pions and muons which went through the whole system, including multiplexed triggers during actual data taking. This would allow: 1) Relative calibration and monitoring of the spectrometer and ORCID calibrations, 2) Correction for small systematic effects within a device, such as any non-uniformity of ring size with position in the ORCID or in momentum with position in the spectrometers, 3) Determination of normal Coulomb scattering effects (i.e. the effect of cuts on average energy), and 4) Recalibration of the system over time. A secondary calibration and monitoring system would be the accurate measurement of the magnetic fields in the spectrometers and the index of refraction of the ORCID radiator.

The most difficult systematic effect to deal with will be in determining the average P_t of the sample. Rewriting Eq. 1 by solving for m_ν^2, we have:

$$m_\nu^2 = \frac{2P^*}{\gamma_\pi}(E_\nu - \frac{P^*}{2\gamma_\pi}) - P_t^2 \qquad (2)$$

The contribution to the error in m_ν from P_t is the same as the error in P_t for small m_ν and P_t. The P_t is essentially the pion momentum times the difference in pion and muon angles in decay. The dominant error will be in this angle. The neutrino energy comes from subtracting the pion and muon energies. E_ν vs. P_t is quadratic in P_t with a multiplying constant of $\gamma_\pi/2P^*$, independent of the neutrino mass. One is constrained in the calibration to this value. Therefore, data at different P_t's may be related, and together with a Monte Carlo of known experimental effects, be used to obtain fits to the data. One can then correct the neutrino energy distribution observed with the P_t measured.

The systematic effects in the data at fixed γ_π will show themselves, in general, in this fit and thus be amenable to correction. For instance, a systematically wrong determination of zero P_t or wrong optical correction, will show as a dependence of the P_t on the azmuthal angle of the decay. An improper energy loss correction will show as a wrong multiplying constant. The accuracy of this fit is dependent mostly on the single event measurement error and the statistics used. Note that this can only be done because of the large amount of redundancy in the experimental data and the simplicity of the kinematics. Using this technique, a 10% systematic uncertainty in the RMS P_t would give approximately a 100 KeV error in the mass from this source for a sample centered on zero P_t. This seems to be a pessimistic estimate of the systematic error for this fit due to the uncertainty in the average P_t caused by the consistency cuts, because it may be checked with straight-through particles at nearly the same energy and with no decay P_t.

The above fit determines the neutrino energy at $P_t = 0$. This is just $P^*/2\gamma_\pi + (\gamma_\pi/2P^*)\, m_\nu^2$. The first term is well measured and has an inverse dependence on γ_π allowing its separation when using data at different pion energies. The "$E_\nu = 0$" level can be found from straight-through particles to high accuracy.

The major open question in the above discussion, then remains the expected systematic uncertainty in the experiment. Until a detailed design and Monte Carlo is performed, an accurate estimate is impossible. However, the above discussion would suggest that it is likely to be below 100 KeV, which if achieved would yield a five times improvement to the current mass limit.

In order to reduce the statistical error to 100 KeV in m_ν, it would be necessary to collect approximately 10^4 useful events. A similar number of calibration events would be necessary, but could be multiplexed with the normal data and would not then add to the time of the experiment. Considering the small phase space and the probability of decay in the fiducial region, about 10^{-5} of the beam pions will generate useful events with P_t's less than 1 MeV. This means an integrated beam flux of about 10^9 pions must be taken. The ORCID currently has a live time of 10 μs and a readout time of 5 ms. This means that an instantaneous rate of 10^5 pions/sec with a trigger rate of 0.002 could be taken. That is, the data for one energy could be taken in approximately 3 hours of beam time, not including any machine duty factor. An error of 50 KeV would require 48 hours of beam time, however, since the final error in m_ν depends on the fourth root of the statistics. So without improving the single event measurement error, a mass limit below about 50 KeV would seem impractical using this method.

THE K-LONG DECAY EXPERIMENT

The method of Clark, et al.[2] to determine the μ-neutrino mass from the $K_L \to \pi\mu\nu$ decay is similar in principle to that of β decay. But, unfortunately, the energy scale of the decay is much larger. The shape of the phase space distribution is measured near the limit of zero neutrino momentum $E_\pi + E_\mu \cong E_K$. The largest effective mass of the pion and muon is decreased if the neutrino has a mass. The phase space for this mass must be zero. Also, the phase space will be suppressed for effective masses just below that limit as long as the neutrino momentum is comparable to its mass. This is shown in Fig. 5. The effective mass is calculated from the pion and muon momenta which are measured in a two-arm spectrometer. The mass scale can be derived from K_L to two (charged) pion decays, which have an effective mass equal to the K_L mass. Also, these decays can be used to accurately determine the experimental resolution and acceptance functions so that experimental errors can be folded into the phase space distribution.

The present limit of Clark, et al. was determined by momentum resolution, statistics, muon identification, and separation of $K_{\pi\pi}$ decays from $K_{\mu 3}$ decays. The principle improvement (See Fig. 6) proposed is the addition of a pair of ORCID detectors, one to each spectrometer arm as vertex detectors. The ORCID detectors would be similar to the ones previously described but would be beside one another and would be tilted with respect to the beam at an angle to accept nearly sideways (90 degree CM) decays.

All the above effects contributing to the Clark experiment can be improved upon with the addition of the ORCID detectors. The velocity and pattern recognition features of the ORCID are the most important. By their addition, $K_{\pi\pi}$ decays may be separated very well from $K_{\mu 3}$ decays. This is important because, although the branching fraction for $K_{\mu 3}$ is much larger (27%) compared to $K_{\pi\pi}$ (0.2%), the phase space near the Kaon mass for the $K_{\mu 3}$ decays is very small compared to the $K_{\pi\pi}$ decays. So missidentified $K_{\pi\pi}$ decays can contribute a large amount to the rate of $K_{\mu 3}$ decays in the area of interest. The ORCID would also be sensitive to pion decays, improving the determination of the resolution function from $K_{\pi\pi}$ decays and also the experimental resolution of the $K_{\mu 3}$ decays by eliminating events where a pion decayed. The particle identification can also be used to separate K_{e3} decays and use them in calibration since the kinematics are similar and the branching fraction (37%) is comparable.

In order to estimate the experimental errors, let us take the form of the pion-muon effective mass:

$$m_{\mu\pi}^2 = m_\mu^2 + m_\pi^2 + 2E_\mu E_\pi - 2\vec{P}_\mu \cdot \vec{P}_\pi \tag{3}$$

where all variables are in the laboratory system. The particle masses appear in the above equation, but will not enter the final mass determination if we experimentally determine the end point energy from $K_{\pi\pi}$ decays. The partial derivatives of $m_{\mu\pi}$ with respect to the experimental quantities are:

$$\frac{\partial m_{\mu\pi}}{\partial P_{\mu\pi}} = \frac{P_\mu P_\pi \sin\theta_{\mu\pi}}{m_\pi} \tag{4a}$$

$$\frac{\partial m_{\mu\pi}}{\partial P_\mu} = \frac{P_\pi}{m_{\mu\pi}} \left(\frac{\beta_\mu}{\beta_\pi} - \cos\theta_{\mu\pi}\right) \tag{4b}$$

$$\frac{\partial m_{\mu\pi}}{\partial P_\pi} = \frac{P_\mu}{m_{\mu\pi}} \left(\frac{\beta_\pi}{\beta_\mu} - \cos\theta_{\mu\pi}\right) \tag{4c}$$

Notice that for relativistic particles, $\theta_{\mu\pi}$ will be small. This means that the $\cos\theta_{\mu\pi}$ and β terms in (4b) and (4c) will cancel in first order, reducing the dependence of $m_{\mu\pi}$ on the momenta. Also notice that for nearly equal P_μ and P_π (decays near 90 deg. CM), the partial in $\theta_{\mu\pi}$, which can be rewritten approximately as P_t^2/m_π, will be roughly independent of longitudinal momentum. For example, take E_K = 8 GeV and $m_{\mu\pi} = m_K$ for a 90 degree decay. $\Delta\theta_{\mu\pi}$ is about 0.24 mr for the ORCID alone or 0.14 mr if the spectrometers are also used to determine $\theta_{\mu\pi}$. This results in a contribution to $\delta m_{\mu\pi}^2$ of 0.23 MeV^2 (0.08 MeV^2). If the spectrometers are assumed have $\Delta P_\mu = \Delta P_\pi$ = 1 MeV/c, their contribution will be 8×10^{-4} MeV^2. The single event error is then dominated by the angle measurement and $\delta m_{\mu\pi}$ = 0.5 MeV (0.3 MeV). The larger value will be assumed below.

Defining δm as $m_K - m_{\mu\pi}$, the center of mass momentum of the neutrino P* will be:

$$P^{*2} = \delta m^2 - m_\nu^2 \tag{5}$$

The phase space at a given δm will then be:

$$\Delta P = A\sqrt{\delta m^2 - m_\nu^2} \tag{6}$$

where A is defined as the solid angle acceptance divided by $4m_K$. For 20% acceptance, $A = 2 \times 10^{-4}$ MeV^{-1}. The phase space vanishes for $\delta m < m_\nu$ and is linear with δm in the limit $\delta m >> m_\nu$. Folding the experimental resolution with this phase space distribution gives the experimental distribution of events if the acceptance is known. The distribution can be normalized in the above limit and $\delta m = 0$ determined from the average m_K in $K_{\pi\pi}$ decays.

The final limit on the neutrino mass will be determined from a fit of the experimental phase space with the resolution, as determined from $K_{\pi\pi}$ decays. In order to estimate the statistics necessary to obtain a given level of sensitivity, I determined the phase space expected for a zero mass neutrino to be 1.080×10^{-4} under previously stated conditions for δm < 50 KeV and the rate for m_ν = 50 KeV as 1.057×10^{-4} under the same conditions. Assuming that a 90% confidence limit is needed, 6400 events in the data with δm < 50 KeV would be required to establish this difference in rates. This translates to 6×10^7 $K_{\mu3}$ decays and 2.2×10^8 total decays in the fiducial region or roughly 7×10^8 total K_L's. The ORCID could accept a rate of about 10^5 decays/sec or 1.5×10^6 K_L's/sec. If such a beam could be provided, only 500 seconds of beam would be necessary to be sensitive to a mass of 50 KeV. However, a more realistic beam rate is 10^3 K_L's/sec.

At this rate it would take 200 hours to obtain the necessary data, neglecting the accelerator duty factor.

The systematic errors are, as usual, difficult to estimate with any reliability. Competing decay modes have been examined for their possible effect on the measurement and have been found to have negligible effect either because of their low rate, their kinematics, or because they can be identified with very high probability as being distinct. Other systematics are essentially eliminated by the calibration data. It is likely, then, that the experiment would be statistics limited at about 50 KeV but a much better measurement would require an unreasonable amount of beam time.

CONCLUSION

Two possible experiments have been presented each of which is likely to yield a muon-neutrino mass limit in the range of 50 to 100 KeV with a reasonable amount of beam time. The questions as to which is better or if there is a better experiment than either of them are still open.

I wish to thank Peter Nemethy and Paul Seiler for helpful discussions.

REFERENCES

1. Daum, et al., Phys. Let. 74B (1978) 126.
2. Clark, et al., Phys. Rev. D9 (1974) 533.
3. Anderhub, et al., Phys. Let. 114B (1982) 76.
4. A New Type of Cherenkov Imaging Detector, B. Robinson, Physica Scripta 23 (1981) 716.

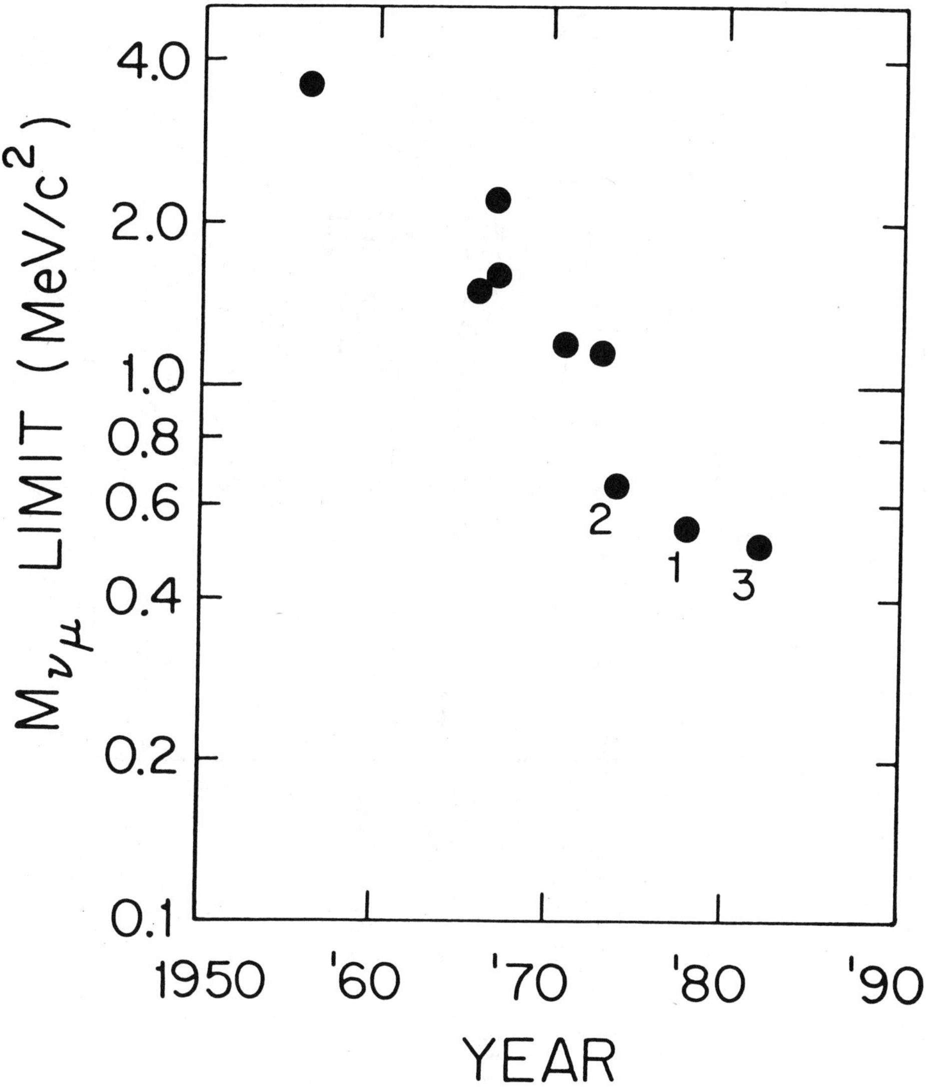

Fig. 1) Experimental limits on mu-neutrino mass (90% CL). Reference numbers for recent experiments are listed next to data points.

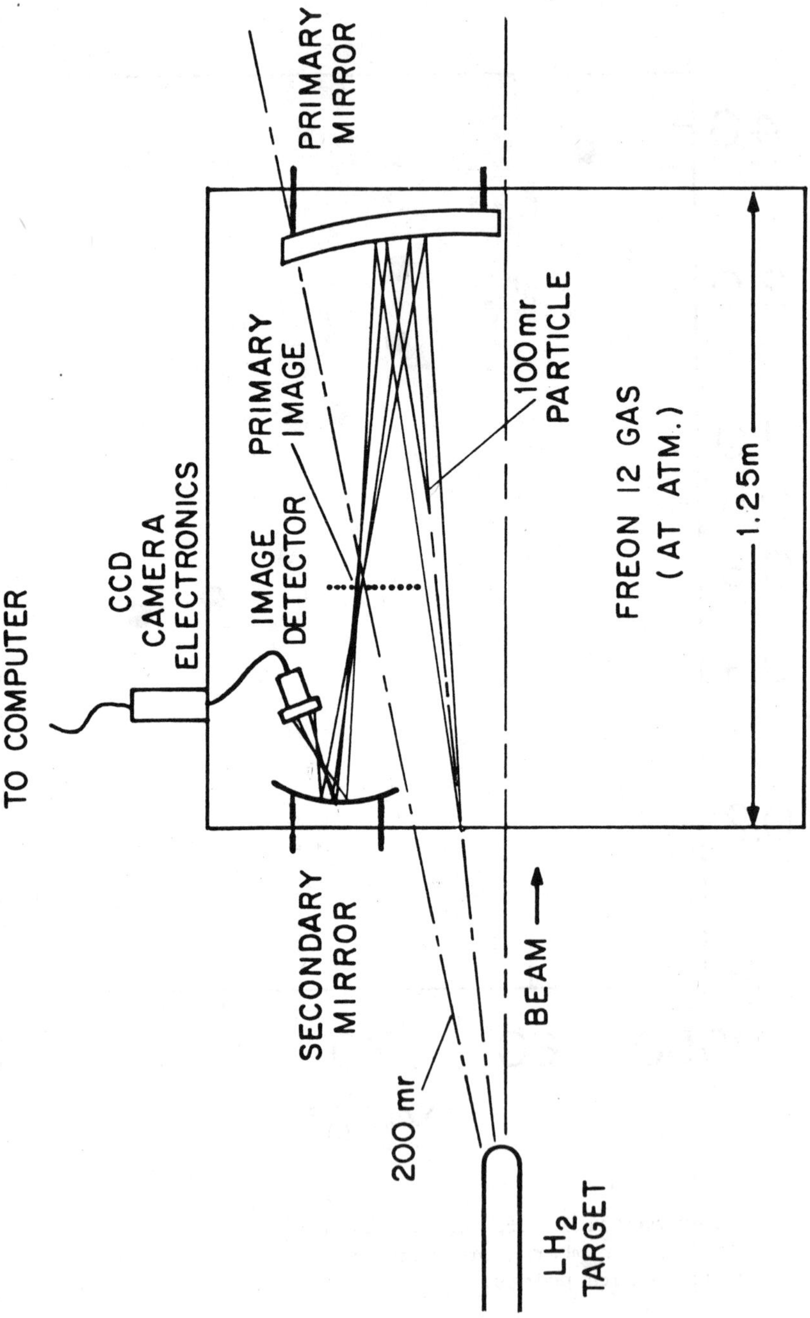

Fig. 2) ORCID used in Fermilab experiment E609. Some sample Cherenkov light paths are shown for a 100 mr particle.

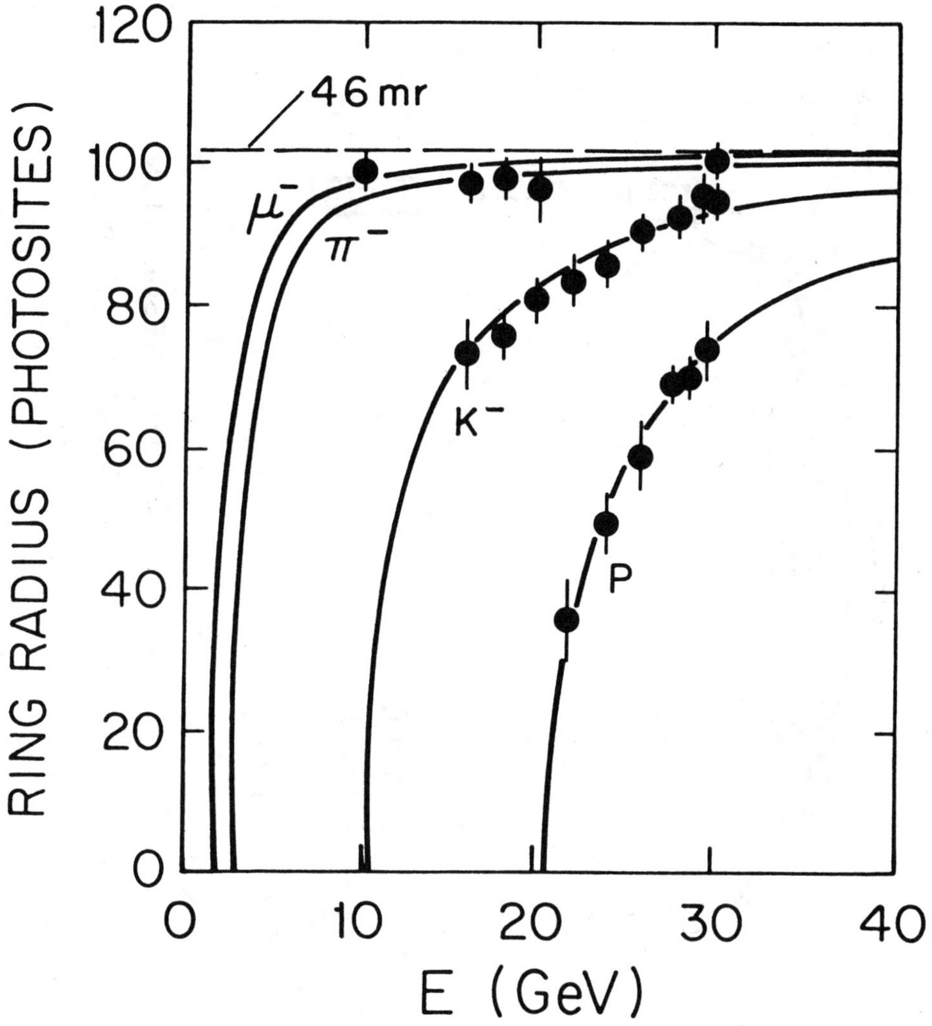

Fig. 3) Test beam measurements with ORCID. Errors shown are for single events.

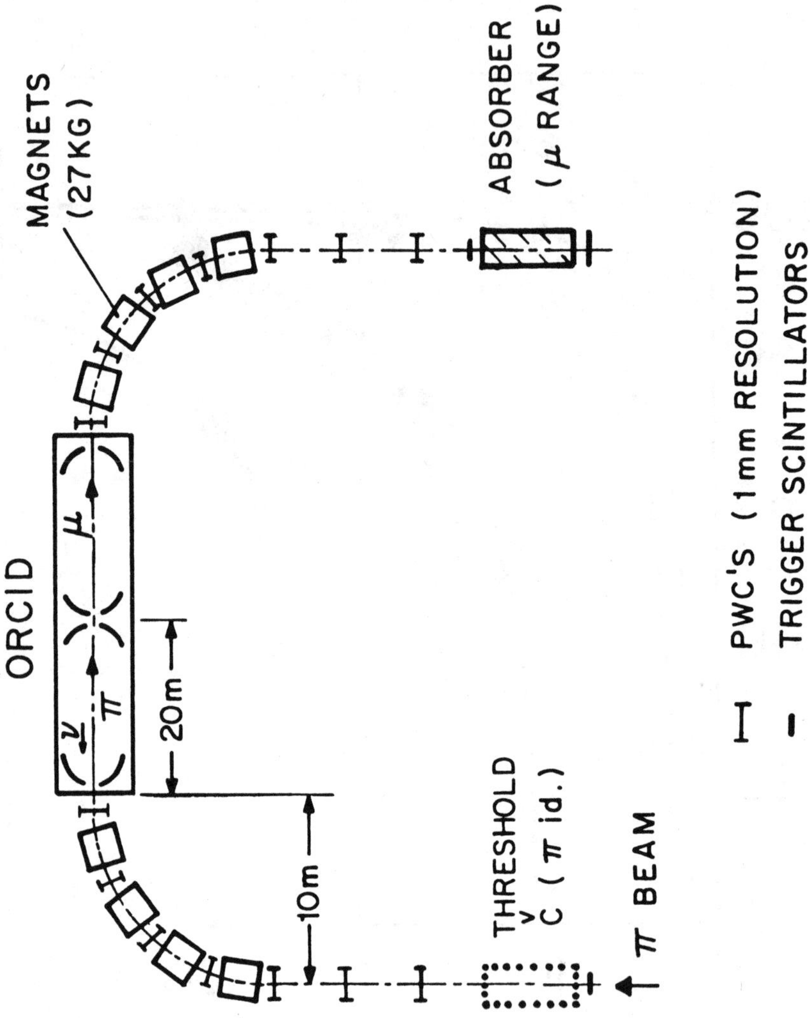

Fig. 4) Schematic view of pion decay experiment

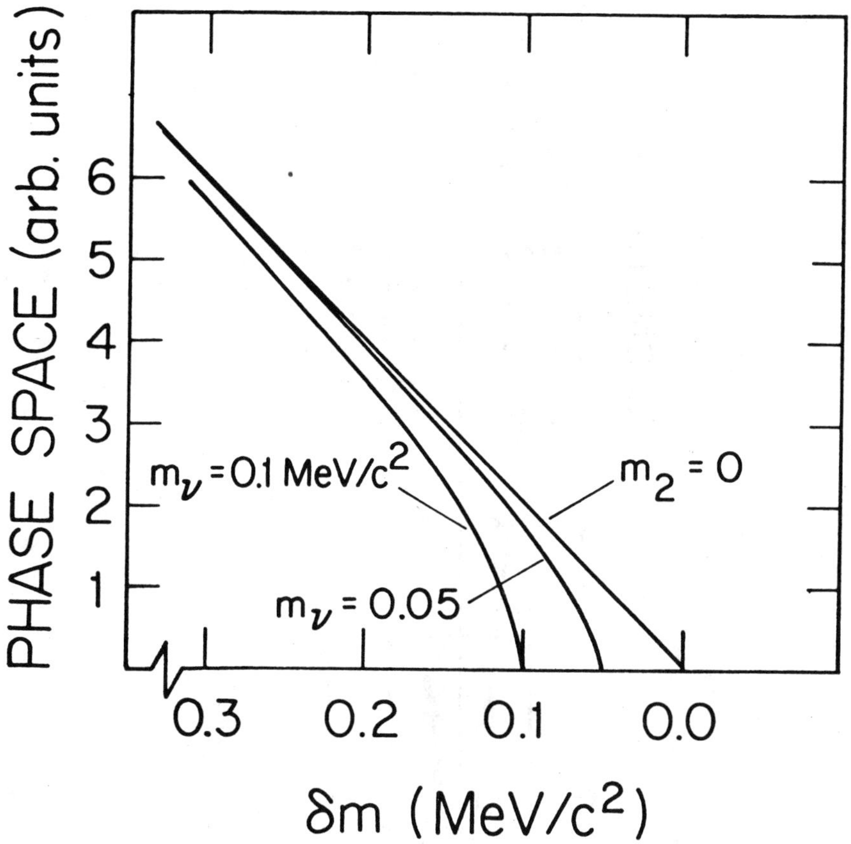

Fig. 5) Phase space distribution for K_L decay at fixed δm for massless and massive neutrinos.

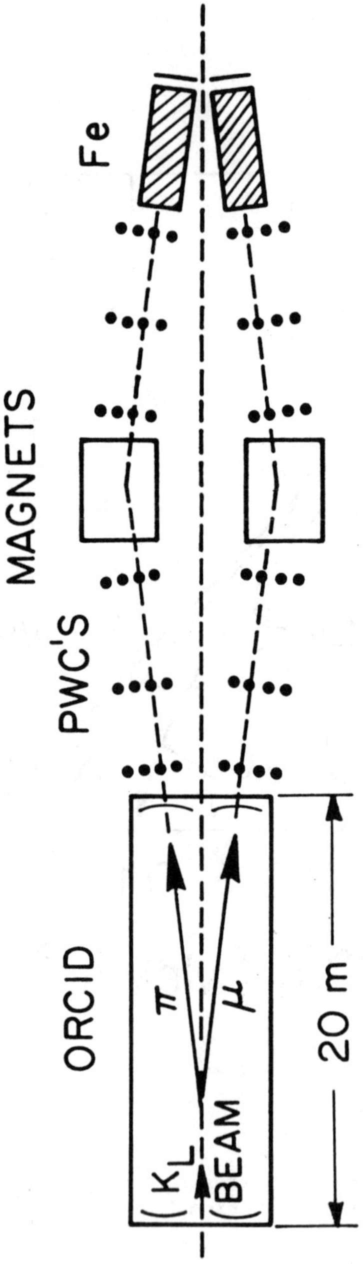

Fig. 6) Schematic view of K_L decay experiment.

KINEMATICAL ANALYSIS OF PION DECAY IN FLIGHT: A NEW METHOD FOR MEASURING THE MUON NEUTRINO MASS

Paul G. Seiler*
Labor fuer Hochenergiephysik der ETH Zurich
c/o SIN, CH 5234-Villigen, Switzerland

ABSTRACT

Forward inflight decays of pions reveal information on the muon neutrino mass. In contrast to pion decay at rest in a stop target, the result is very slightly dependent on the uncertainty of the pion mass and on the theory of the momentum loss of muons in matter. In addition, relevant calibrations can be performed.

A first test of the new method done at SIN is described.

INTRODUCTION

More than a quarter of a century has passed since the first measurements of the muon neutrino mass. But still, due to our experimental knowledge, this particle†--which is widely believed to have a small to vanishing mass--could be about as massive as the electron. Fig. 1 shows the measured upper limit of the ν_μ mass as a function of time. Six of the data points come from pion two-body decay at rest $\pi^+ \to \mu^+ + \nu_\mu$, one from muon decay at rest $\mu^+ \to e^+ + \nu_e + \bar{\nu}_\mu$, one from the three-body decay $K^o_L \to \pi\mu\nu$, and the last from pion decay in flight. In all measurements energy and momentum conservation are used and the masses of the particles other than the neutrino have to be known. Their experimental errors contribute to the final result. This contribution, however, is negligibly small in the last case ($\pi \to \mu\nu$ in flight).

Processes with three bodies in the final state allow measurements at very low neutrino momentum ($p_\nu \to 0$) where the accuracy of the neutrino mass or its upper limit is linearly dependent on the accuracy of the quantities measured. But the relative counting rates are principally low and the result is model dependent. Therefore one has concentrated on two body pion decay. Here the square of the neutrino mass is measured, which makes these experiments difficult to improve. This explains the asymptotic behavior of the curve in Fig. 1. Clearly, new ideas are needed to make a step forward.

The ETH experiment described below showed in a short period of data taking (6 weeks) that our new method--analysis of forward pion decay in flight--works.

Pion two-body decay <u>in vacuo</u> results in monochromatic muons. m_ν could be derived from the measurement of the muon energy or momentum without theoretical corrections. But previous to our experiment,[9] pions were

---

*Present address: Laboratory for Nuclear Science, Massachusetts Insitute of Technology, 51 Vassar Street, Cambridge, Mass. 02139, U.S.A.

† The muon neutrino has a not necessarily defined mass. The measured limits are valid for ν_2, the primary mass eigenstate in ν_μ, and for other ν_j if they show up in the decays investigated.

stopped to decay at rest in a target and the momentum of the muons was influenced by their passage through the target matter. Muons which leave the target show a broad momentum spectrum instead of a monochromatic line (see ref. 7, the most recent experiment of this kind.) The information on the muon momentum at decay is contained in the absolute position of the edge of this spectrum. These experiments demand a complete understanding of the slowing-down process of muons in matter and an absolute calibration of the spectrometer.

The process $\pi^{\pm} \rightarrow \mu^{\pm} + \overset{(-)}{\nu}_{\mu}$ can be analyzed in vacuo if one measures pion decay in flight. In addition, the pion momentum to work at can be chosen.

PION DECAY IN FLIGHT

As compared to the two-body decay at rest, two additional observables have to be measured: the pion momentum p_{π} and the $\pi\mu$ decay angle θ. Fig. 2 shows the relative effect of the squared neutrino mass on the neutrino momentum as a function of the pion momentum, both measured in the laboratory frame of reference. Fig. 3 shows the influence of the currently known accuracy of pion and muon mass on the neutrino momentum, which is computed from pion momentum assuming $m_{\nu} = 0$. Also shown is the contribution from uncertainty in the θ measurement for certain experimental conditions. Both figures indicate that it is promising to do the experiment at a few 100 MeV/c. In this range of the pion momentum the relative effect is increased, the contribution from Δm_{π} (and Δm_{μ}) is decreased, and the contribution from $\Delta\theta$, which enters about

$$p_{\pi}{}^{3} \cdot \overline{\theta} \cdot \Delta\overline{\theta}, \tag{1}$$

can be controlled. In addition to the enhancement of the relative effect, pion decay in flight really results in a monochromatic neutrino momentum line. In the following we describe a calibration method which allows us to understand shape and position of that line.

DERIVING m_{ν} FROM PION DECAY IN FLIGHT

One possible method of evaluating the neutrino mass from pion decay in flight is the following.

In the process $\pi^{\pm} \rightarrow \mu^{\pm} + \overset{(-)}{\nu}_{\mu}$ at small θ (forward decay), the pion momentum p_{π}, the muon momentum p_{μ}, and the decay angle θ are measured event by event.

For $\theta = 0$,

$$p_{\nu} \equiv p_{\mu} - p_{\pi} \tag{2}$$

is the measured neutrino momentum and

$$p_{\nu}^{0} \equiv p_{\mu}^{0} - p_{\pi} \tag{3}$$

the neutrino (and muon) momentum calculated from the measured p_{π} for $m_{\nu} = 0$. The difference

$$\delta p_\nu \equiv p_\nu - p_\nu^0 \tag{4}$$

is (for an ideal apparatus) the effect of a finite neutrino mass:

$$p_\nu - p_\nu^0 \simeq \left(dp_\nu/d(m_\nu^2)\right) \cdot \langle m_\nu^2 \rangle \tag{5}$$

(expansion of p_ν in m_ν^2).

This equation still holds for small angles θ: The dependence on θ of p_ν and p_ν^0 cancels out in the difference δp_ν. From δp_ν and its uncertainty

$$\Delta(\delta p_\nu) \tag{6}$$

we derive an upper limit on m_ν.

Fig. 4 shows schematically the behavior of p_ν and δp_ν. The peak at the right side (solid line) shows the measured $p_\nu = p_\mu - p_\pi$ distribution. The dashed line is the same distribution, but shifted by δp_ν due to a finite neutrino mass. The asymmetry of the line comes from the contribution of the accepted finite decay angles. The peak on the lefthand side, the distribution $p_2 - p_1$, comes from undecayed particles, which are treated as if they were decay events (i.e., measuring the momentum of each particle twice). The width of the distribution shows the resolution of the spectrometer. The offset is due to momentum loss in the detectors between first and second measurements.

Let's summarize the possibilities mentioned above:

- Undecayed particles can be used to measure the momentum loss in the detectors and the resolution function of the spectrometer.
- Undecayed particles ($\theta \equiv 0$) can be used to derive the resolution function of the apparatus, which measures the decay angles.
- The distribution of the measured decay angles can be unfolded since resolution function and production spectrum (two-body kinematics) are known. The effect of the true decay angles of the accepted events can then be folded into the calibration line for $\theta = 0$ events to reproduce the line shape of the measured neutrino momentum of decay events.

We conclude that the neutrino momentum line in pion decay in flight can be understood by means of the calibration techniques described. An improvement of the experiment can be achieved by improving resolution and calibration.

Note that it is most important to keep the systematic error of momentum differences small. This can be accomplished by measuring pion and muon momentum with the same spectrometer.

RESUME OF THE ETH EXPERIMENT

We have performed an experiment[9] to determine an upper limit on the muonic neutrino mass from pion decay in flight in the forward direction at the Swiss pi meson factory SIN. The apparatus is shown in Fig. 5. The pion momentum is $p_\pi \simeq 350$ MeV/c ($p_\mu \simeq 355.7$ MeV/c). The pion and muon momentum are measured in the same magnetic spectrometer. The spectrometer consists of a 210° Fe-magnet, four identical spark chambers (SC), and their magnetorestrictive readouts. SC2 and SC3 measure the diameter of the nearly circular particle trajectories at 180°. This arrangement minimizes the influence of multiple scattering by 180° focusing. SC1 and SC4 are added to determine the injection angle. The decay angle is determined by the four proportional chambers DKA...DKD and the decay point by time of flight (TOF) measurement (plastic scintillation counters Z1, Z2); a TOF variation of 110 ps corresponds to a variation of 1 m in the position of the decay point. The single event angular resolution of the decay angle is σ_θ = .83 mrad. The time resolution, including a small variation in the length of the beam path, is σ_{TOF} = 140 ps.

The figure-eight beam arrangement allows several calibrations using pions and muons which traverse the apparatus without decaying ("throughgoing" particles). In addition to the one mentioned in chapter 3, the TOF spectra of these pions and muons are used to monitor the stability of the TOF measurement. The interval of events with their decay point within the decay region can be defined relative to the calibration peaks.

Fig. 6 shows the momentum difference line for a certain class of calibration events and the neutrino momentum line of the decay events. We simulate the shape of the calibration spectra by calculating the probability for a particle to hit one or more chamber wires and/or window supporting grids, and folding the resulting momentum loss spectrum with the very narrow momentum loss distribution from straggling and the resolution of the spectrometer. (The particles in the satellite peak hit at least one supporting grid.) The resolution, the heights of the main and of the satellite peak, and the position of the spectrum are the four free parameters of the simulation. For the decay events the solid line is the simulated and fitted spectrum. Its shape is calculated as follows: From the distributions of the pion momenta and the physical decay angles θ, one gets the distribution of p_ν, which we convolute with a simulated calibration spectrum.

RESULT

Exploiting the calibration methods described and taking into account the measured field inhomogeneities in the spectrometer as well as the latest values of the mass errors of pion and muon, the ETHZ experiment led to the following result:

$$\langle m^2_{\nu_\mu} \rangle = -0.14 \pm 0.20 \; (\mathrm{MeV/c^2}) \quad (68\% \text{ C.L.}) \tag{7}$$

which yields the upper limit

$$m_{\nu_\mu} \leq 0.50 \text{ MeV/C} \quad (90\% \text{ C.L.}) \tag{8}$$

The main contributions to $\Delta\langle m^2_\nu\rangle$ come from the nonlinearity of the chambers in the spectrometer, the limited statistics, and the uncertainty of the average decay angle. These points can be improved, especially since our experiment was the first test of the method. Note that for the first time the uncertainty of the pion mass Δm_π has negligible influence on the final result.

POSSIBLE IMPROVEMENTS

To improve the pion decay in flight experiment, the uncertainties mentioned above should be reduced. This is most easily done with the statistical error. The ETH result is based on 10^3 decay events and 2×10^4 calibration events. Many more events should be taken. This would, of course, reduce the statistical error and, by dividing the data into subsamples, give a better understanding of the systematic error.

The two other contributions can be reduced by applying the laser calibration of drift chambers[10] which we developed for this problem but unfortunately could not apply anymore.

In addition, a magnetic spectrometer with bigger radius, chambers of a different type, work at slightly higher pion momentum, a more stringent cut of the decay angle, and a better understanding of acceptance effects would help.

Altogether, an improvement by a factor 10 in sensitivity to m_ν^2 is possible.

The author wishes to thank the whole ETH m_{ν_μ} group,[9] especially Drs. H. B. Anderhub and H. Hofer.

REFERENCES

1. W. H. Barkas et al., Phys. Rev. 101, 778 (1956).
2. G. Bardon et al., Phys. Rev. Lett. 14, 449 (1965).
3. R. E. Shafer et al., Phys. Rev. Lett. 14, 923 (1965).
4. Booth et al., Phys. Lett. 26B, 39 (1967).
5. Bakenstoss et al., Phys. Lett. 36B, 403 (1971).
6. A. R. Clark et al., Phys. Rep. D9, 533 (1974).
7. M. Daum, G. H. Eaton, R. Frosch, H. Hirschmann, J. McCulloch, R. C. Minehart, E. Steiner, Phys. Rev. D20, 2692 (1979).
8. Lu et al., Phys. Rev. Lett. 45, 1066 (1980).
9. H. B. Anderhub, J. Boecklin, H. Hofer, F. Kottmann, P. LeCoultre, D. Makowiecki, H. W. Reist, B. Sapp, P. G. Seiler, , Phys. Lett. 114B, 76 (1982).
   H. B. Anderhub, Messung einer oberen Grenze der Masse des Myon-Neutrinos aus dem Pion-Zerfall im Fluge, LEH/ETH Zurich thesis 6985, unpublished.
10. H. B. Anderhub, M. Devereux, P. G. Seiler, NIM 166, 581 (1979).
11. Particle Data Group, Review of particle properties, LBL-100 Revised, UC-34d, April 1982.

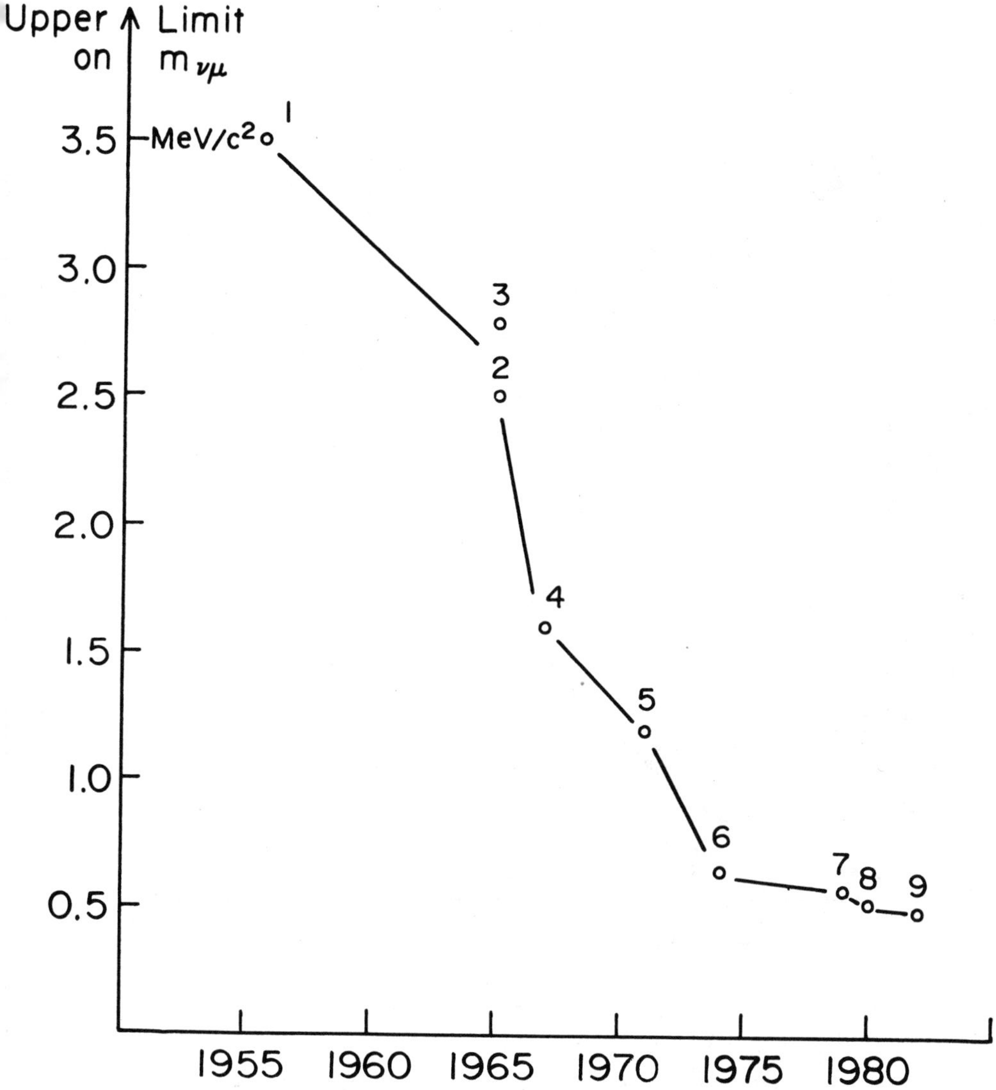

Fig. 1 Measured upper limits on m_ν (or the appropriate mixture of mass eigenstates) as a function of time. Point 1 corresponds to ref. 1, point 2 to ref. 2, and so on.

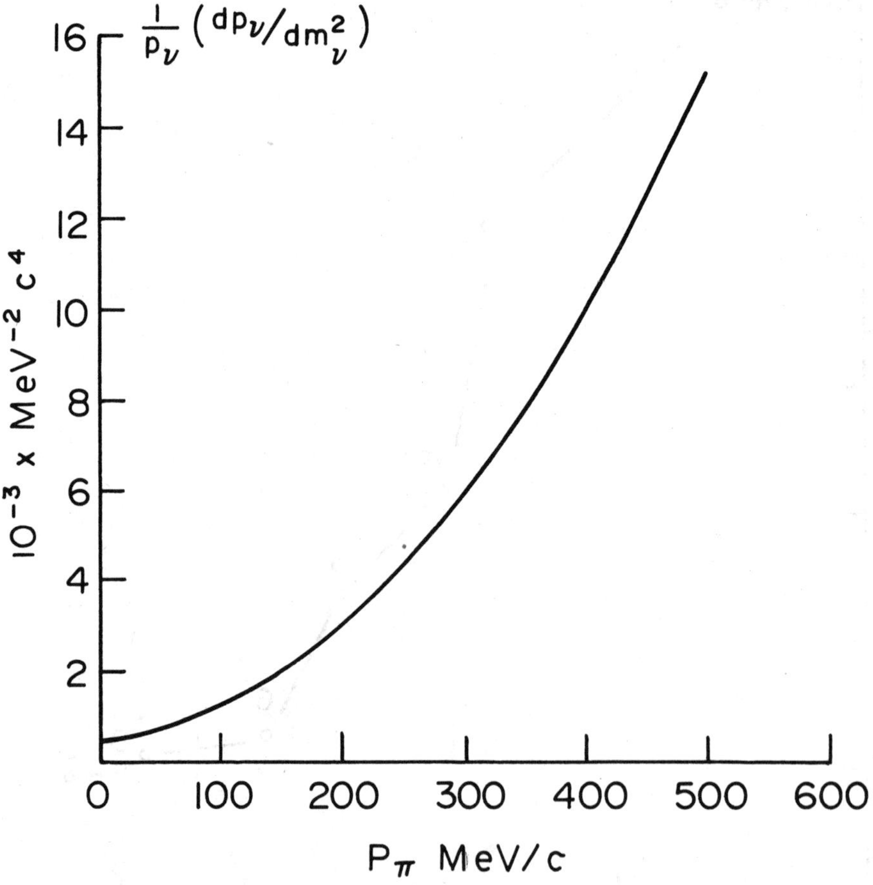

Fig. 2 **Relative effect of m_ν^2 on the neutrino momentum as a function of the pion momentum.**

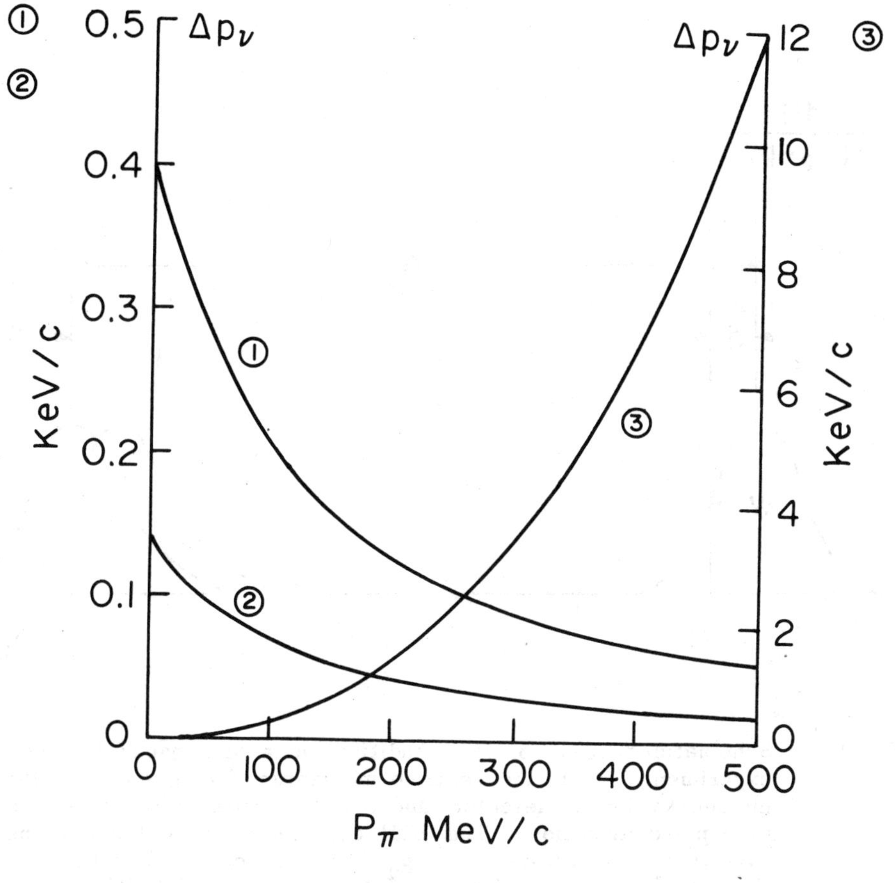

Fig. 3 Influence of the mass errors[11] $\Delta m_\pi = 0.7\ \mathrm{KeV/c^2}$ (1) and $\Delta m_\mu = 0.18\ \mathrm{KeV/c^2}$ (2) as well as the influence of $\Delta\theta$ (3) on the neutrino momentum p_ν. For 3 we assume: $\bar{\theta} = 1$ mrad;

$$\Delta\bar{\theta}^2 = 4 \cdot 10^{-5} + \frac{40}{(p_\pi \beta_\pi)^2}\ \mathrm{mrad}^2.$$

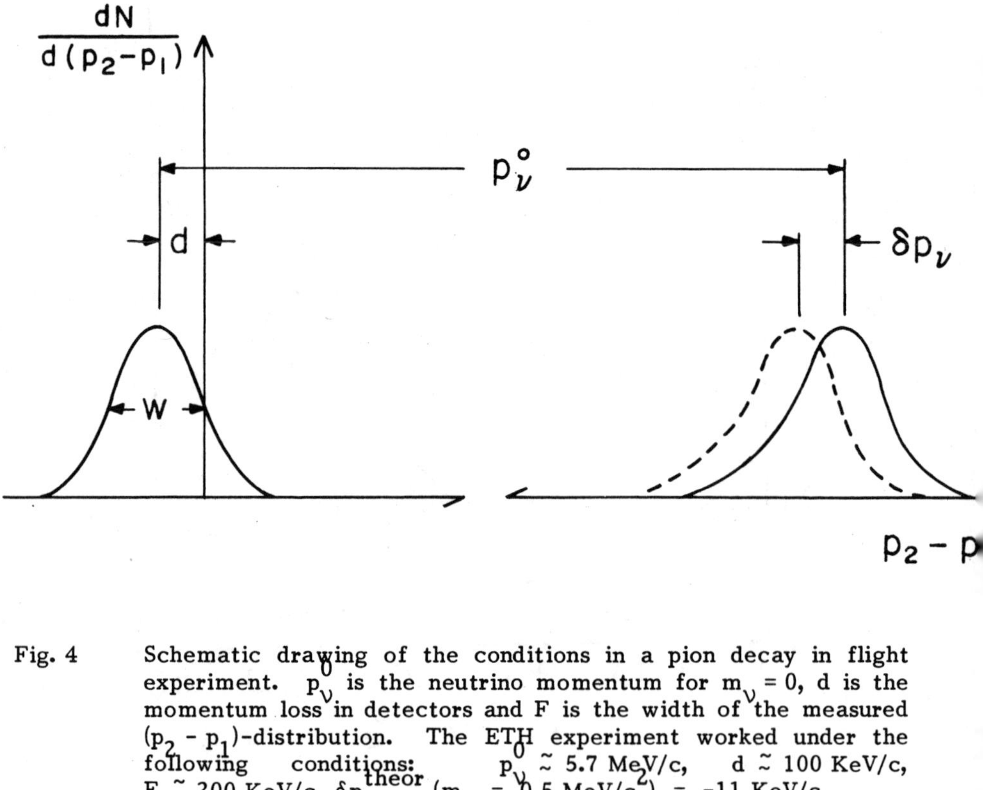

Fig. 4 Schematic drawing of the conditions in a pion decay in flight experiment. p_ν^0 is the neutrino momentum for $m_\nu = 0$, d is the momentum loss in detectors and F is the width of the measured $(p_2 - p_1)$-distribution. The ETH experiment worked under the following conditions: $p_\nu^0 \simeq 5.7$ MeV/c, $d \simeq 100$ KeV/c, $F \simeq 200$ KeV/c, δp_ν^{theor} $(m_\nu = 0.5\ \mathrm{MeV}/c^2) = -11$ KeV/c.

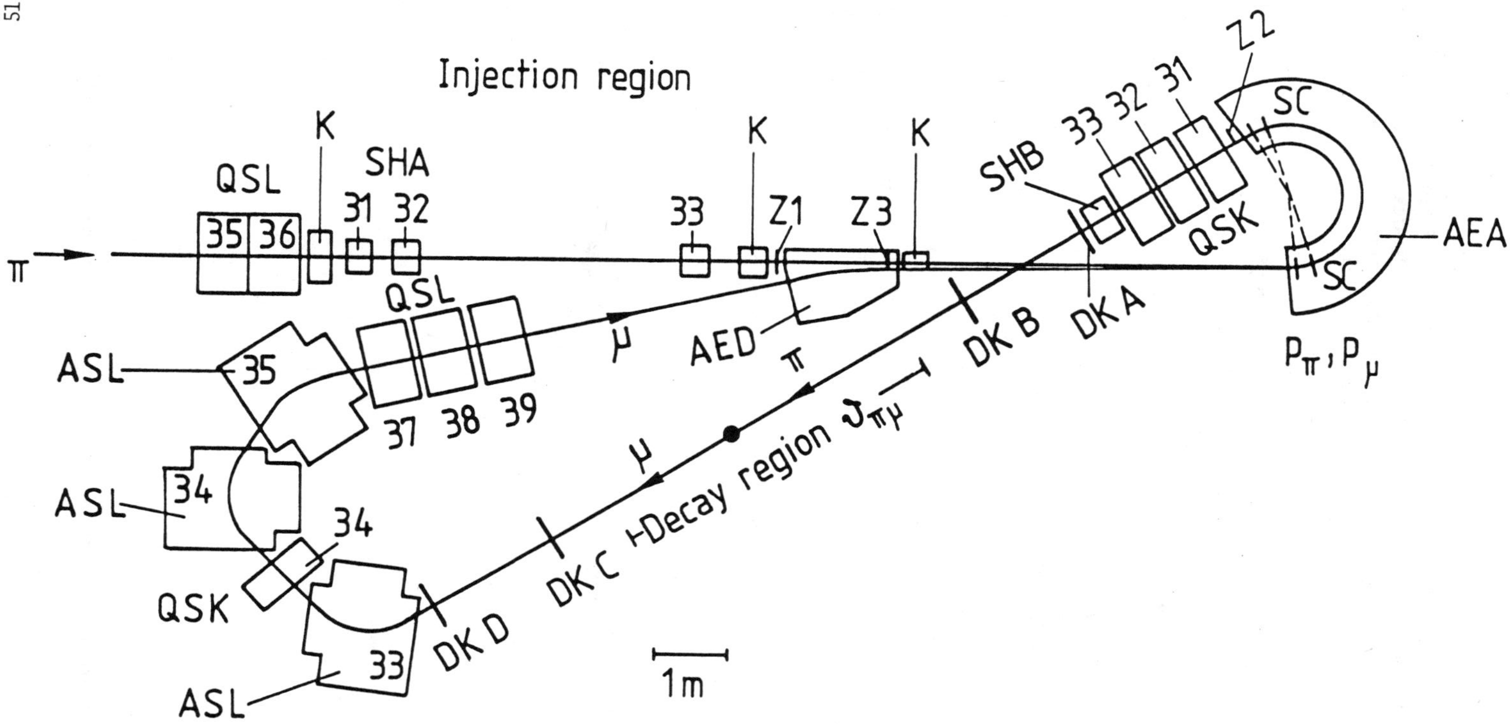

Fig. 5 Experimental arrangement of the ETHZ experiment. AEA: magnetic spectrometer with chambers SC. Z1, Z2: TOF counter. DKA,...DKD: MWPC to measure the decay angle. QS, AS, SH: conventional beam elements. AED: septum magnet.

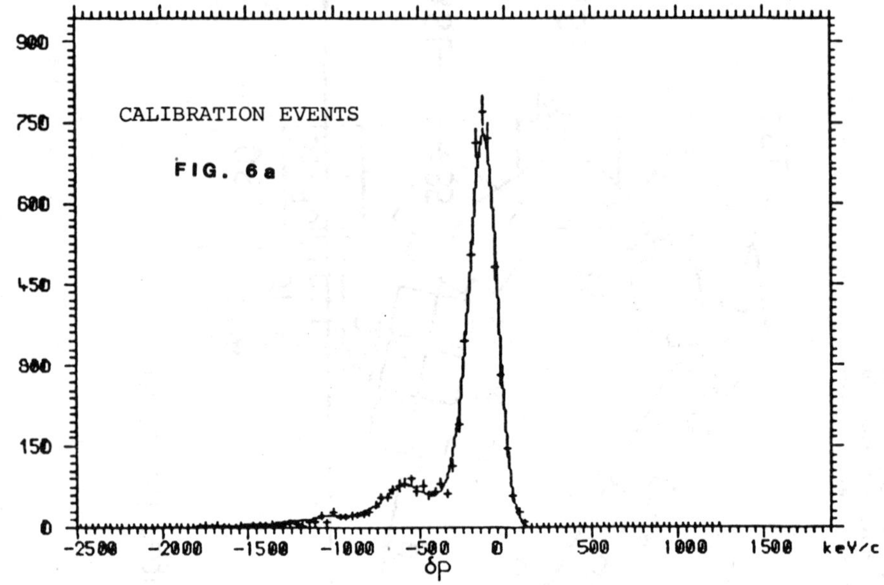

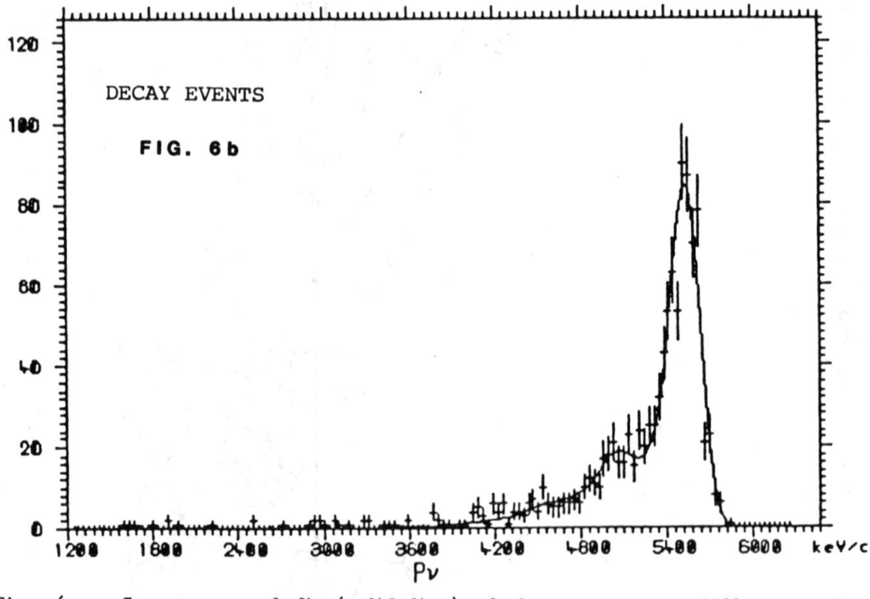

Fig. 6a Spectrum and fit (solid line) of the momentum difference for a certain class of calibration events.

6b Spectrum and fit of the measured neutrino momentum of true decay events.

EXPERIMENTAL CONSTRAINTS ON THE MASSES AND MIXINGS OF MAJORANA NEUTRINOS

Lincoln Wolfenstein
Carnegie-Mellon University, Pittsburgh, PA 15213

ABSTRACT

Data on neutrino oscillations, the end-point spectrum of beta-decay, and the rate of neutrinoless double beta-decay can be combined to give constraints on the masses and mixings of Majorana neutrinos. An illustrative example is given and its consistency with present theoretical models is discussed.

There is no fundamental theoretical reason for the neutrino to have a mass. However, if the neutrinos are massive there do exist strong theoretical reasons for believing they are Majorana particles. In this talk I will first discuss the phenomenology associated with the existence of several flavors of massive Majorana neutrinos. I will then turn to a review of some of the theoretical ideas about the Majorana mass matrix.

If there are three generations of leptons, the neutrino mass is described by a 3 x 3 matrix connecting $\nu_{\alpha L}(\alpha=e,u,\tau)$ to $\nu^c{}_{\alpha R}$, where $\nu_{\alpha L}$ is the left-handed neutrino and $\nu^c{}_{\alpha R}$ is the right-handed antineutrino. We shall assume CP invariance since no present experiments bear on the question of CP violation.[1] The matrix is then a real symmetric matrix defined by 6 real numbers. These determine the three mass eigenvalues m_1, m_2, m_3 and three mixing angles. In addition to these it is also necessary to specify two discrete parameters with values of ±1 that determine the relative CP eigenvalues of the Majorana fields.[2] For example, these may be chosen as η_{12} and η_{13}, where $\eta_{12} = -1$ signifies that ν_1 and ν_2 have opposite CP eigenvalues. It is only recently that we realized the importance of η_{12} and η_{13}. The way you come across them is that when you use an orthogonal transformation to diagonalize the mass matrix some of the eigenvalues may turn out negative. The η_{ij} are just the relative signs of these mass eigenvalues. In order to get positive masses one must define the Majorana field to be either $(\nu+\nu^c)$ or $(\nu-\nu^c)$ depending on the sign of the mass eigenvalue; thus the negative mass eigenvalues end up giving opposite CP eigenvalues.

There are four types of experiments of current interest that help to constrain the mass matrix:

(1) neutrino oscillation
(2) direct measurements of endpoint spectra
(3) neutrinoless double beta-decay
(4) decay anomalies; that is, the observation of or limit on a rare decay mode involving a heavy neutrino.

I will concentrate on the first three of these, which are being considered in detail in sessions of this conference. The fourth has been discussed extensively by Shrock.[3] The main point I want to emphasize is that the different experiments are sensitive to different features so that by combining them it may be possible to place severe constraints on the mass matrix.

Let me first discuss double beta-decay since it is here that one must be particularly careful. What do we really measure when we constrain "m(ν_e)" in double beta-decay experiments? The answer is[2]

$$"m(\nu_e)" = \sum_n (0_{ne})^2 \eta_{1n} m_n \qquad (1)$$
$$\equiv M_{ee}$$

Here 0_{ne} are the elements of the orthogonal mixing matrix that connect ν_e to the eigenstates ν_n; that is $\nu_{nL} = \sum_{\alpha=e,\mu,\tau} 0_{n\alpha} \nu_{\alpha L}$. The η_{1i} are ±1 discussed before. The important point is that in the case of a minus sign there is a cancellation between different terms in the sum. What the last equality tells us is that "m(ν_e)" is not an eigenvalue of the mass matrix but rather a diagonal element M_{ee} in the original form of the matrix with rows and columns defined by lepton flavor. This result can be obtained directly by carrying out the calculation without ever diagonalizing the mass matrix as illustrated in Fig. 1. To lowest order in neutrino mass this is an adequate method; the x in the figure represents the mass insertion which must connect ν_e to $\bar{\nu}_e$. In the extreme case that M_{ee} equals zero the mass matrix connects ν_e only to $\bar{\nu}_\mu$ and/or $\bar{\nu}_\tau$ so that double beta-decay must vanish.

To illustrate how the different experiments are to be combined in an analysis I will consider the case in which ν_e is a mixture of two eigenstates:

$$|\nu_e\rangle = \cos\theta\ |\nu_1\rangle + \sin\theta\ |\nu_2\rangle$$

There are then two masses (m_1, m_2), one mixing angle θ, and one relative sign η_{12}. As sample experimental data I use:

(1) The recent data from Gosgen[4] limiting neutrino oscillations that cause the disappearance of $\bar{\nu}_e$.

(2) The Russian H^3 data[5] giving $m(\nu_e) \approx 30$ ev.

(3) Double beta-decay analysis[6] yielding "m(ν_e)" < 15 ev.

The neutrino oscillation data provides correlated constraints between $(m_2^2 - m_1^2)$ and $\sin^2 2\theta$ as shown in Fig. 2. Combining these with the H^3 data, it is evident that there are two sets of solutions for the parameters:

A. $\sin^2 2\theta \gtrsim .17$

$m_1 = m_2 = 30$ ev

$|m_1 - m_2| \lesssim 3.10^{-4}$ ev

B. $\sin^2 2\theta \lesssim .17$ ($\theta \lesssim 12°$)

$m_1 = 30$ ev

m_2 undetermined

Note that the neutrino oscillation constraints make clear what we are measuring in the H^3 decay experiment. If there is very little mixing we measure m_1; if there is more than a little mixing both masses are essentially equal on the scale of interest. From our earlier discussion it is clear that the double beta-decay result is not inconsistent with the H^3 data provided $\eta_{12} = -1$, in which case we require from Eq. (1)

$$|m_1 \cos^2\theta - m_2 \sin^2\theta| < 15 \text{ ev} \tag{2}$$

Combining Eq. (2) with the previous results we are left with two very limited possibilities shown by the shaded regions in Fig. 2. The values of m_1, m_2, and θ for these regions are:

A. $\sin^2 2\theta > \frac{3}{4}$

$m_1 \approx m_2 = 30$ ev; $|m_1 - m_2| \lesssim 3 \cdot 10^{-4}$ ev

B. $\sin^2 2\theta \lesssim .17$

$m_1 \approx 30$ ev

$45 \text{ ev} > (m_1 + m_2)\sin^2\theta > 15$ ev

Let me emphasize again that this is only an illustrative example of how the data is to be combined.

A few comments and qualifications about these two regions are in order:

A. In the limit that $m_1 = m_2$ and $\theta = 45°$, the two Majorana particles merge into a single Dirac particle, thus completely eliminating neutrino oscillations and double beta-decay. This is what I have labeled[7] an abnormal Dirac neutrino because the right-handed component that goes with ν_{eL} turns out to be some combination of $\bar{\nu}_{\mu R}$ and $\bar{\nu}_{\tau R}$. We may note in passing that if $m_1 = m_2$ but $\theta \neq 45°$, then the two Majorana particles merge into what I have called a pseudo-Dirac particle and for this case neutrinoless double-beta decay does not vanish.[8]

B. Eqs. (1) and (2) must be modified[9] for values of $m_2 >$ 10 MeV because it is no longer possible to ignore m_2^2 in the virtual neutrino propagator $(m_2^2+q^2)^{-1}$. This increases somewhat both upper and lower limits on $\sin^2\theta$ for a given value of m_2 but does not change the qualitative results. Cosmological and astrophysical considerations[10] require that ν_2 be unstable for $m_2 > 100$ ev. In a three generation model the most likely decay made is $\nu_2 \to \nu_1 + e^+ + e^-$. Combining the astrophysical constraints on the lifetime with the theoretical limit resulting from $m_2^2 \sin^2 2\theta < 45$ ev in Eq. (6b), one finds a lower limit on m_2 of about 20 MeV.[11] Independent of cosmology, additional constraints[3] between m_2 and $\sin^2\theta$ may be obtained from nuclear beta-decay spectra and the study of $\pi \to e\nu$ and $K \to e\nu$. Present data is not sufficient to further constrain the allowed region with two possible exceptions: (a) a recent analysis[12] by Simpson of the H^3 spectrum may rule out the allowed band in the neighborhood of m_2 = 1 kev; (b) a recent experiment[13] by Bryman et al on the $\pi \to e\nu$ spectrum almost rules out the allowed region for values of m_2 between 40 and 70 MeV. Future experiments may be able to rule out most or all of region B.

I now turn to theoretical ideas about neutrino mass. Most of these are inspired by grand unified theories (GUT) and were already discussed in my talk here two years ago[14]; so I will be brief. The basic idea is that we can explain the small value of the neutrino mass if it acquires its mass by the Majorana mechanism, unlike all other elementary fermions which have Dirac masses. The Majorana mass term is characterized by $|\Delta I_3| = 1$ (since it connects the ν_L with $I_3 = \frac{1}{2}$ to $\nu^c{}_R$ which has $I_3 = -\frac{1}{2}$) and $|\Delta L| = 2$, where I_3 is the weak isospin and L is lepton number. If we assume that the weak isospin is violated only by the vacuum expectation value v_2 of Higgs doublets then

$$m_\nu \propto (fv_2)^2 \sim m_D^2$$

where f is a typical Higgs doublet coupling and m_D is a "normal" Dirac mass (like that of a quark or a charged lepton) given by fv_2. We then assume that $\Delta L = 2$ is associated with some new high mass scale M and by dimensional arguments obtain the result

$$m_\nu \sim m_D^2/M \qquad (3)$$

The smallness of m_ν is then explained by the large value of M.

Within the SU(5) GUT there are no right-handed partners for ν_L so that the only possible mass terms are of the Majorana type linking the $\nu_{L\alpha}$ to their antiparticles $\nu^c{}_{R\beta}$. However, within the simplest SU(5) the neutrinos remain massless; it is necessary to enlarge the Higgs sector to produce a Majorana mass. The situation with the SO(10) GUT is different. This theory is left-right symmetric so that each $\nu_{L\alpha}$ has a right-handed partner $N_{R\alpha}$ and we must have a Dirac mass term. In order to explain why the neutrino does not have a normal mass, it was proposed[15] that N obtains a very large Majorana mass M. The Dirac mass term then causes a very little mixing of this heavy N into the light neutrino with the consequence that m_ν is given by Eq. (3). Thus within the simplest SO(10) the neutrinos must be Majorana particles with non-zero mass. However it is possible to obtain massless neutrinos in SO(10) if a seventeenth neutral fermion is added to each generation.[16]

A model inspired by SU(5) was given by Zee.[17] It has the unusual feature of two almost degenerate neutrinos as in Solution A. In fact the mass matrix has zero diagonal elements so that M_{ee} is zero and neutrinoless double beta-decay vanishes. (In general the theory does not have a conserved lepton number so that the vanishing of double beta-decay is not exact. It can be shown that one must go to third-order in the neutrino mass to get a non-vanishing double beta-decay rate, which obviously will be negligible.) However the model has also an almost massless neutrino; to fit the neutrino oscillation data it is necessary to assume arbitrarily that the admixture of this neutrino into ν_e is very small.

The phenomenology of the neutrino mass matrix in SO(10) is the subject of many papers. In general it is found that there is a mass hierarchy with ν_e consisting primarily of the lightest neutrino and ν_τ of the heaviest neutrino. The mixing of ν_e with ν_μ is small and with ν_τ very small. Thus the best chance for finding neutrino oscillations is expected to be between ν_μ and ν_τ. Recently Chang and Pal[18] have found that under reasonable assumptions η_{12} or η_{13} or both are negative. Thus there is some cancellation in the calculation of the "m(ν_e)" for double beta-decay in Eq. (1). However, if one assumes that all neutrino masses are below the cosmological limit of 100 ev, then one expects from the mass hierarchy discussed above that "m(ν_e)" is bound to be extremely small in any case. Chang and Pal notice that by fine tuning they can also find a solution in their SO(10) model which resembles Solution A in having two almost degenerate neutrinos with almost complete cancellation in the calculation of double beta-decay.

In conclusion we have emphasized how it is possible to combine the data from different types of experiments to constrain the parameters of the Majorana mass matrix. In particular, if "m(ν_e)" derived from limits on neutrinoless double beta-decay is smaller than m(ν_e) determined from the end-point spectrum of beta decay, very tight constraints as illustrated in Fig. 2 may be found. While such solutions may seem extremely unlikely they are not completely inconsistent with present theoretical models.

This research was supported in part by the U.S. Department of Energy.

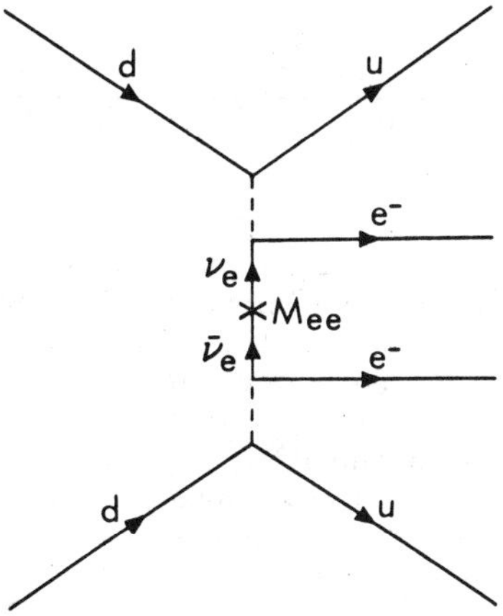

Fig. 1. Double beta-decay diagram using the flavour representation for the neutrinos.

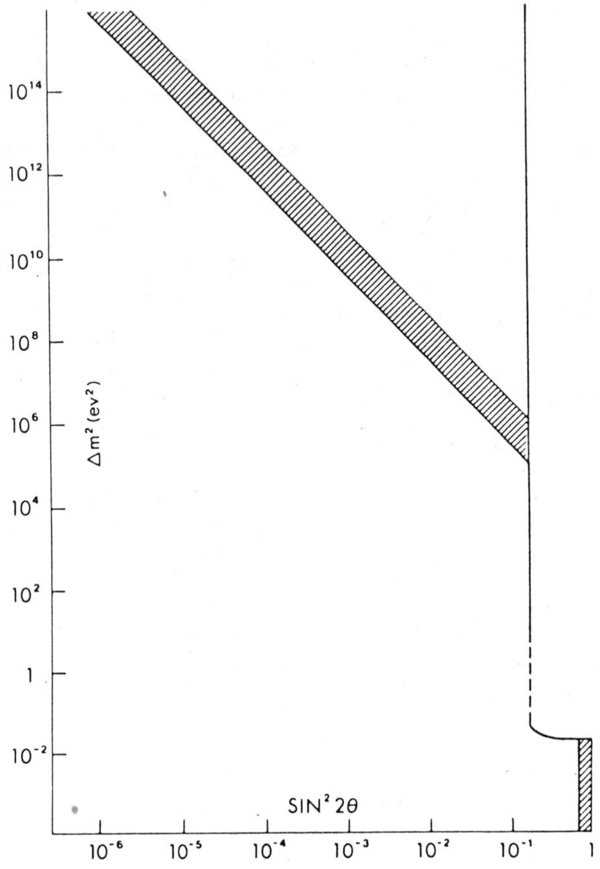

Fig. 2. Constraints relating $(m_2^2 - m_1^2)$ and mixing angle θ. The region to the right of and above the curve is forbidden by neutrino oscillation data. (The dashed line replaces a wiggly line in the original analysis of the data; on the scale I have chosen these are relatively insignificant.) Only the dashed regions are allowed when the sample data on the H^3 end-point spectrum and double beta-decay are used. The lower shaded region (solution B) extends indefinitely downward and the upper (solution A) indefinitely upward. For all allowed regions $m_1 \gtrsim 30$ ev.

REFERENCES

1. For a discussion of possibilities see N. Cabibbo, Phys. Lett. 92B, 333 (1978).
2. L. Wolfenstein, Phys. Lett. 107B, 77 (1981); P. B. Pal and L. Wolfenstein, Phys. Rev. D25, 766 (1982).
3. R. E. Shrock, Phys. Rev. D24, 1232 (1981).
4. Vuilleumier, et al., Phys. Lett. 114B, 298 (1982).
5. V. A. Lubimov, et al., Phys. Lett. 94B, 266 (1980).
6. W. C. Haxton, et al., Phys. Rev. Lett. 47, 153 (1981). Larger and smaller limits on "$m(\nu_e)$" can be found in papers given at this conference.
7. L. Wolfenstein, Nucl. Phys. B186, 147 (1981).
8. S. Petcov, "On Pseudo-Dirac Neutrinos, Neutrino Oscillations, and Neutrinoless Double Beta-Decay", to be published.
9. A. Halprin, et al., Phys. Rev. D13, 2567 (1976).
10. See M. Turner in Neutrino-81 Proceedings Vol. I, p. 95 (Cence, Ma, Roberts, eds., University of Hawaii).
11. This is illustrated in Fig. 4 of L. Wolfenstein in Neutrino-81, p. 329.
12. J. J. Simpson, Phys. Rev. D24, 2971 (1981).
13. D. A. Bryman, TRIUMF preprint PP-81-63, October 1981.
14. L. Wolfenstein in Neutrino Mass Mini-Conference, eds. V. Barger and D. Cline, University of Wisconsin, Report 186 (1980).
15. Gell-Mann, Ramond, and Slansky, private communication (1977); in Supergravity, P. Van Nieuwenhuizen and D. Freedman (eds.) North Holland (1979), p. 317.
16. D. Wyler and L. Wolfenstein, to be published.
17. A. Zee, Phys. Lett. 93B, 389 (1980); L. Wolfenstein, Nucl. Phys. B175, 93 (1980).
18. D. Chang and P. B. Pal, Phys. Rev., to be published.

CAN RELIC NEUTRINOS BE OBSERVED?

T. Weiler*
Physics Department
University of California, San Diego, La Jolla, CA 92093

ABSTRACT

Within the framework of big-bang cosmology and the standard electroweak model, it appears that the only possibly detectable interaction of relic neutrinos is their annihilation with cosmic ray neutrinos on the Z resonance. However, measurement will require the existence of a large density of $z \geqslant 5$ red-shifted sources with intense neutrino emission in the energy region 10^{21} eV^2/m_ν.

INTRODUCTION

The discovery in 1965 of an isotropic 2.7°K photon gas permeating the universe marked the major triumph for big-bang cosmology. Big-bang theory had predicted an isotropic Planck-distribution of photons released from equilibrium when hydrogen formed at a temperature of 4000°K (1/2 eV) and a time of roughly 10^5 years after the initial singularity. The present photon temperature reflects the red-shifting due to the universe's expansion over the past 10 to 20 billion years.

Standard big bang cosmology also predicts a 1.9°K gas of neutrinos and antineutrinos. Due to the weakness of their interaction, these neutrinos last interacted when the universe's temperature was 10^{10}°K (1 MeV) and its age was but one second! Thus the properties of the relic neutrinos offer a glimpse of our embryonic universe at an age twelve orders of magnitude earlier than that afforded by the photon background. Yet, seventeen years after the discovery of the photon gas, the neutrino ether remains elusive. The reason is simple to divine: the very weakness of interaction which enabled neutrinos to decouple so early in cosmological evolution also enables them to defy detection today.

Current ideas for relic neutrino detection may be classified into one of three categories:

Flux detection proposes to measure directly the relic flux incident on earth. The predicted mean momentum of the neutrinos (assuming zero chemical potential) is 5×10^{-4} eV; the predicted

*Work supported by the Department of Energy, contract DE-AT03-81ER-40029.

number density of neutrino plus antineutrino is 100 cm^{-3} per flavor, down by a factor of four from the photon density. Given the weakness of the neutrino interaction strength, $G_F = 4 \times 10^{-33}$ cm^2, the feebleness of the relic momentum, and the moderate number density, it is clear that attempts to measure the relic flux must trigger on $\mathcal{O}(G_F)$ effects rather than $\mathcal{O}(G_F^2)$ cross sections. One is led to forward scattering effects.[1] However, $\mathcal{O}(G_F)$ forward scattering effects have recently been proven to vanish, or (for the single non-vanishing effect, torque transfer from relics with a nonzero chemical potential to a polarized target) be negligibly small.[2] In target detection the existence of the relic neutrinos is deduced from their interaction effects on another flux[3,4] (e.g. an accelerator beam, or a cosmic ray flux). We will argue in section three that the interaction need be an $\mathcal{O}(G_F)$ resonance production. In deductive detection, the indirect effects of the relics upon terrestrial phenomena are interpreted as evidence for the relics. Examples of indirect effects include alteration of end-point spectra in neutrino-emitting weak decays due to the existence of a Fermi sea of relic neutrinos,[5] and alteration of K_L - K_S mass splitting or regeneration due to interactions with a nonzero net relic neutrino number (CP asymmetric).[6] Unless the bounds on relic densities to be derived in the next section are a gross underestimate, proposed indirect effects are completely immeasurable. Thus we need consider further only target detection.

In the next section the following assertion will be proven: Given big bang cosmology and the standard electroweak model of Glashow-Salam-Weinberg, the annihilation of a cosmic ray ν with a relic $\bar{\nu}$ (or vice versa) on the Z resonance is the unique process having sensitivity to the relic neutrino density. The magnitude of absorption of cosmic ray neutrinos from a single source is then derived in the following section. The absorption energy is $\mathcal{O}(10^{21}\, eV^2/m)$ for a neutrino of mass m. An anticipated paucity of flux at this energy then suggests a convolution of absorption per source over all sources. This is done for a realistic source distribution in the final section. Our conclusion is: detection of relic neutrinos appears possible only if there exist astrophysical objects with very intense neutrino emissions in the energy region $10^{21}\, eV^2/m$ and very large cosmological red-shifts. The content of this conclusion is not encouraging.

COSMOLOGICAL CONSIDERATIONS

In this section the properties of relic neutrinos expected in big bang cosmology[7] are summarized and used to prove the assertion in the preceding section. As mentioned in the introduction, the temperature T_d at which neutrinos decoupled from thermal equilibrium is approximately 1 MeV. Neutrinos with mass negligible on this temperature scale then approximately satisfy a massless Fermi-Dirac momentum distribution. After decoupling, momenta are red-shifted as $p \sim 1/R$, where R is the scale size of the Friedmann-Robertson-Walker universe. Hence the present relic number density for neutrino species i is

$$n_{\nu_i}(\xi_i) = (2\pi)^{-3} \int d^3\vec{p}\, f_{\nu_i}(p) \;,$$

where (1)

$$f_{\nu_i(\bar{\nu}_i)}(p) = \frac{1}{e^{\beta p \mp \xi_i} + 1} \;,$$

$\beta^{-1} \equiv T_d R_d / R = 1.9°K \times (T_\gamma / 2.7\,°K)$, and ξ_i is the degeneracy parameter (chemical potential multiplied by β) characterizing a possible neutrino number asymmetry.[8]

$$n_{\nu_i} - n_{\bar{\nu}_i} = 96\,(\xi_i + \xi_i^3/\pi^2)\ \mathrm{cm}^{-3} \times (T_\gamma/2.7\,°K)^3 \;. \quad (2)$$

T_γ is the present value of the relic photon temperature and R is the present scale size. From the integration measure, d^3p, it is clear that the number density falls as R^{-3} as expected. The present symmetric density value is $n_{\nu_i}(0) = n_{\bar{\nu}_i}(0) = 53 \times (T_\gamma/2.7\,°K)^3$ cm^{-3}. For $\xi_i \gg 1$,

$$n_{\nu_i} \simeq 0.18\,|\xi_i|^3\, n_{\nu_i}(0) \text{ and } n_{\bar{\nu}_i} \simeq 1.1\, e^{-|\xi_i|}\, n_{\bar{\nu}}(0) \;. \quad (3)$$

For $\xi_i \ll -1$, exchange ν_i and $\bar{\nu}_i$.

ξ and m are not known. A cosmological bound on $\xi(m)$ or $m(\xi)$ results from requiring that the neutrino energy density ρ_ν (a monotonically, increasing function of m and ξ) not exceed the total energy density of the universe,[7] $\rho_o \leqslant 4 \times 10^{-29}$ g/cm^3, i.e.

$$\sum_i \int \sqrt{p^2 + m_i^2}\, \frac{d^3\vec{p}}{(2\pi)^3}\, [f_{\nu_i}(p) + f_{\bar{\nu}_i}(p)] \leq \rho_o \;. \quad (4)$$

The sum is over light species. This gives $|\xi_i| \leqslant 60$ for $m_i \ll \beta^{-1}$, and $|\xi_i| \leqslant 10\,(m_i/\mathrm{eV})^{-1/3}$ for neutrino masses in the electron volt range (as suggested by grand unified models). Setting $\xi_i = 0$ yields a bound on the neutrino masses, viz. $\Sigma_i m_i \leqslant 200$ eV.

A more complicated and probably less trustworthy argument relates the observed He^4/H abundance ratio to the n/p ratio when nucleons decoupled, and leads to a more restrictive bound[9, 8] on the electron neutrino degeneracy parameter, $|\xi_{\nu_e}| \leq 2$. Families other than ν_e are not bounded by the He^4/H ratio. Even a degeneracy parameter as small as 1 leads to a large neutrino asymmetry (in units of photon number), $[n_\nu(1) - n_{\bar{\nu}}(1)]/n_\gamma \sim 1/4$, as compared to the known baryon asymmetry, $\Delta B/n_\gamma \sim 10^{-10}$.

Armed with these results from standard cosmology, we now prove that the annihilation of a cosmic ray ν with a relic $\bar{\nu}$ (or vice versa) on the Z resonance is the unique particle-relic interaction sensitive to a reasonable density.[4] Consider first scattering of cosmic rays by relic neutrinos. The mean free path (mfp) of a particle through the relic neutrinos is roughly $1/n_\nu \sigma_W$. The weak cross section is $\sigma_W \simeq (G_F^2/\pi)[s/(1+s/M_W^2)] \leqslant (G_F^2/\pi)s$. M_W is the W-boson mass and $\sqrt{s}$ the center-of-mass energy. For a cosmic ray of energy E, impinging on a relic neutrino with mean energy $\langle \epsilon \rangle$, $\sigma_W \leqslant (2G_F^2/\pi) E \langle \epsilon \rangle$. But $\langle \epsilon \rangle n_\nu$ is just the neutrino energy density, certainly less than the total energy density ρ_o. Therefore $\lambda_{mfp} > \pi/2G_F^2 E\rho_o$. For the scattering rate to be significant, the mfp must be less than or of order of the Hubble radius $H_o^{-1} = h^{-1} \times 10^{28}$ cm, with $1 \leq h^{-1} \leq 2$ the observational uncertainty. Thus one requires $E > \pi/2G_F^2 \rho_o H_o^{-1}$, which in turn is $\geqslant 2 \times 10^{14}$ GeV. But the universe is opaque to electrons, nucleons and photons at such energies: radio and thermal photon backgrounds degrade electrons via inverse Compton scattering and e^+e^- pair creation,[10] the nucleons via photomeson production,[11] and absorb primary photons via e^+e^- production.[12] In addition, for the electron such energies are also disallowed due to synchrotron losses occurring inside or outside the galaxy, or even in the earth's magnetosphere.[13, 10] The universe (but not the earth) is transparent to cosmic ray neutrinos, but at energies in excess of 10^{14} GeV, the flux may well be negligible. Thus the scattering of cosmic rays by relic neutrinos is infinitesimal, except possibly for primary neutrinos with energy exceeding 10^{14} GeV. We may also use[14] $\sigma_W \leqslant (G_F^2/\pi) M_W^2$ to deduce $\lambda_{mfp}/H_o^{-1} \geqslant \pi/G_F^2 M_W^2 n_\nu H_o^{-1} \geqslant 10^4/(n_\nu/50\ \text{cm}^{-3})$, which says that regardless of incident energy, cosmic ray scattering on relic neutrinos is negligible unless the relic density is several orders of magnitude larger than the big-bang value predicted for $\xi = 0$.

Primary photons and electrons at extreme energy may undergo absorption by the relic neutrinos via $\gamma \bar{\nu}_\ell \to \ell W$ and $e \nu_e \to \gamma W$.

The charged lepton exchange graphs imply a cross section comparable to the Compton value of $(2\pi\alpha^2/s)\ln(s/m_\ell^2)$; one then finds $\lambda_{mfp} \gtrsim M_W^2/n_\nu\, 2\pi\alpha^2 \ln(M_W^2/m_e^2) \gtrsim 10^3\, H_o^{-1}/(n_\nu/50\ \mathrm{cm}^{-3})$. A neutrino degeneracy could in fact make these processes significant, but again the universe is opaque to electrons and photons in the relevant energy range $E \gtrsim M_W^2/2\langle\epsilon\rangle$.

Now, consider resonant absorption of a cosmic ray lepton by a relic neutrino. Integration over the relic momenta or over the universe's expansion is equivalent, by a change of variable, to integration over the resonance width. Thus the relevant weighted cross section for a Breit-Wigner form is $\bar{\sigma} \equiv \int ds\, \sigma(s)/M_R^2 = 16\pi^2 S\, \Gamma(R \to \ell\nu)/M_R^3$. S is the ratio of resonance spin states to incident lepton spin states. This time, the condition $\lambda_{mfp} \lesssim H_o^{-1}$ becomes

$$\frac{\Gamma(R \to \ell\nu)}{M_R} \gtrsim \frac{G_F M_R^2}{(n_\nu/50\ \mathrm{cm}^{-3})\, S} \quad . \tag{5}$$

We have replaced $10^{-5}\ \mathrm{GeV}^{-2}$ with G_F to make it clear that unless the relic neutrino density is several orders of magnitude larger than the value expected for $\xi = 0$, $\Gamma(R \to \ell\nu)/M_R$ must be of order $G_F M_R^2$ rather than $(G_F M_R^2)^2$. This leaves the $W^\pm$ and Z as the only significant resonance candidates. Since the universe is opaque to electrons near the resonant energy $E_R \sim M_R^2/2\langle\epsilon\rangle$, the Z is the only resonant candidate.

Finally, consider an accelerator beam of energy E and current j (number per unit time) impinging on a length ℓ of relic density. The number of scatterings per unit time is $n_\nu \sigma_W \ell j$. For $\sigma_W \lesssim G_F^2 s/\pi$ this is $\lesssim (2G_F^2/\pi)\rho_\nu E\ell j \lesssim 10^{-38}\,(E/\mathrm{GeV})(\ell/\mathrm{meter})j$. Even a one amp current of TeV particles over a distance of 100 meters yields only at most one scattering per 10^7 years! With massive neutrinos there exists the possibility of gravitational clustering of relic neutrinos around galaxies or galactic clusters. A density enhancement of 10^6 is possible, increasing the accelerator beam scattering rate by the same factor, but still leaving it negligible. Moreover, resonant Z or W production cannot help us here since an accelerator cannot attain the resonant energy. Thus we dismiss accelerator beam-relic interactions and our assertion is proven.

We now turn to a detailed calculation of cosmic ray neutrino absorption. Since the neutrino mean free path for annihilation on the Z resonance is comparable to the Hubble radius, the effects of an expanding universe must be included in the calculation.[15]

Let $dn_\nu(E,t)$ be the number of primary neutrinos per unit volume, at time t, with energy in the range E to E+dE. Assuming relativistic velocities for the cosmic ray neutrinos, the rate of density loss due to absorption is

$$\frac{d}{dt}(dn_\nu) = -\left[\int \frac{d^3\vec{p}}{(2\pi)^3} f_{\bar{\nu}}(p)\,\sigma_Z(1 - v_{\bar{\nu}}\cos\theta)\right] dn_\nu \qquad (6)$$

σ_Z is the annihilation cross section, $v_{\bar{\nu}}$ is the relic antineutrino velocity in units of c, and θ is the incident angle of collision. From Eq. (6) the present $(t = 0)$ neutrino density is

$$dn_{\nu_o}(E_o) = e^{-\tau} dn_\nu(E(t),t), \quad \tau \equiv \int_0^t dt \int \frac{d^3\vec{p}}{(2\pi)^3} f_{\bar{\nu}}(p)\,\sigma_Z(1 - v_{\bar{\nu}}\cos\theta) \qquad (7)$$

Here and hereafter a subscript zero will denote a present time value.

In an expanding universe, densities are diluted and momenta are red-shifted independent of the interaction. Introducing the red-shift variable for cosmological expansion from time t to present, $\omega(t) \equiv R_o/R(t) - 1$, one thus has

$$dn_{\nu_o}(E_o) = (\omega + 1)^{-3} e^{-\tau} dn_\nu((1+\omega)E_o, t) \ . \qquad (8)$$

Since the cosmological red-shift is a monotonic function of time, one may parameterize time by the red-shift. The relevant change of variable is $d\omega = -(\omega+1)H\,dt$, where $H \equiv \dot{R}/R$ is the Hubble parameter, itself a function of time. The Einstein equations for a matter-dominated (pressure $p \ll \rho$) era relate the Hubble parameter at time t to its present value (we assume zero cosmological constant): $H(t) = H_o(\omega+1)\sqrt{1+\Omega_o\omega}$. Ω_o is the present energy density of the universe in units of the critical value for closure. The bounds from observational astronomy are $0.02 \leqslant \Omega_o \leqslant 2$. Thus we have

$$d\omega = -(\omega+1)^2 \sqrt{1+\Omega_o\omega}\; H_o\, dt \ . \qquad (9)$$

Let us concentrate on the calculation of τ, the neutrino depth, in the two limits of highly relativistic and nonrelativistic relic neutrinos. Since the relic mean momentum is $\langle p_{\nu_i}\rangle \sim \beta^{-1}(1+\xi_i)$, these limits correspond to $m_i << \beta^{-1}(1+\xi_i)$ and $m >> \beta^{-1}(1+\xi_i)$ respectively. For the Z resonance, we assume a Breit-Wigner form and a standard electroweak model coupling; thus $\int ds\, \sigma_Z(s) = 2\pi\sqrt{2}\, G_F M_Z^2$. For massless neutrinos, Eq. (7) becomes

$$\tau(E,\xi,z) = \frac{G_F M_Z^4}{4\pi\sqrt{2}HE^2}\int_0^z \frac{d\omega}{(\omega+1)^3\sqrt{1+\Omega\omega}}\int_0^\infty dp\, f_\nu(p)\,\Theta(p(\omega+1)^2 - M_Z^2/4E). \tag{10}$$

Subscript zeros have been dropped since every variable in (10) is either a present time variable or an integration variable. The maximum red-shift value, z, is the cosmological red-shift of the extra-galactic neutrino source. The values of detected energy E for which some absorption may occur extend over the entire range from zero to infinity since for any cosmic ray energy there exists a value of relic energy that will guarantee the resonance CMS energy. Unfortunately, unless the degeneracy parameter is large, this smearing eliminates the possibility of an absorption dip. Furthermore, it is clear from the theta function of Eq. (10) that absorption reaches a maximum at energies of order $M_Z^2/(1+z)^2\langle p_\nu\rangle \gtrsim M_Z^2\beta/[(1+\xi)(1+z)^2]$, i.e. $E > 10^{13}$ GeV for $z \leqslant 4$ and $\xi \leqslant 50$. Transmission probabilities $e^{-\tau}$ are shown in Fig. 1. Prospects for detection of relativistic relic neutrinos appear poor.

In grand unified models, neutrino masses in the eV range arise naturally.[16] Furthermore, 30 eV neutrinos are a panacea for unresolved questions concerning the large scale structure of the universe.[17] Relic neutrinos with an eV mass satisfy the nonrelativistic criterion. The nonrelativistic limit of Eq. (7) is

$$\tau(\tilde{E},z) = \frac{\tau(1)\,\Theta(1-\tilde{E})\,\Theta(\tilde{E}(1+z)-1)}{\tilde{E}^{3/2}\sqrt{\Omega_0 + \tilde{E}(1-\Omega_0)}} \tag{11}$$

Here $\tilde{E} = 2mE_0/M_Z^2$ is the energy incident at earth in units of the resonant energy $M_Z^2/2m$, and $\tau(1) = 2\pi\sqrt{2}\,G_F\, n_{\nu o} H_o^{-1} = 0.017\ h^{-1}$ $(n_{\nu o}/50\ \text{cm}^{-3})$. The allowed range of $\tilde{E}$ has a simple interpretation: a neutrino received with energy E_o left its red-shifted source with energy $E_o(1+z)$, and was a candidate for annihilation only if the resonance energy, $E_o/\tilde{E}$, lay within this range. Equivalently, $M_Z^2/2m(1+z) \le E_o \le M_Z^2/2m$.

The transmission probability for nonrelativistic relics is plotted as a function of received energy in Fig. 2. An absorption dip is apparent beginning at an energy of 1/1+z. It is clear that an absorption dip of 15% to 50% can be expected for neutrinos

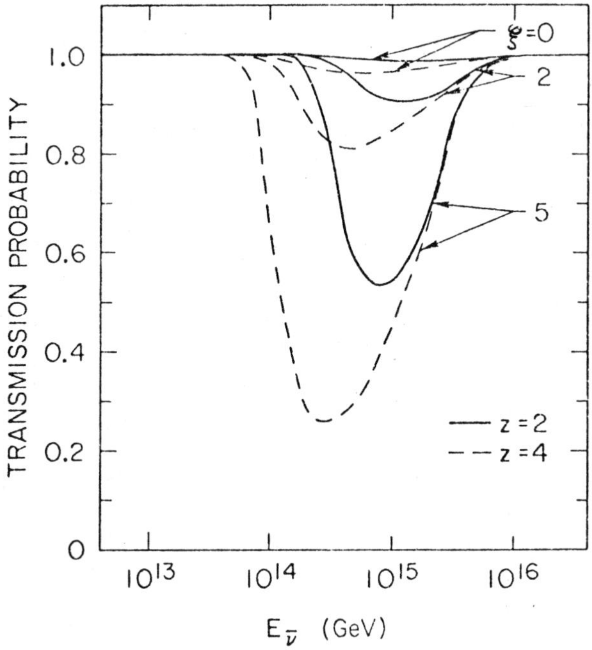

Fig. 1. Transmission probability for massless ($m \ll \beta^{-1}(1+\xi)$) cosmic ray antineutrinos from a single source (red-shift z) as a function of their energy. The assumed values of ($h^{-1}, \Omega, \xi, T_\gamma$) are (2, 1, 0, 2.7 °K) unless stated otherwise in the figure. $M_Z = 90$ GeV is assumed. Transmission probabilities for incident neutrinos are obtained by taking $\xi \to -\xi$.

from a $z = 3.5$ quasar source. If the degeneracy parameter is non-zero, the dip will be much larger. For a 90 GeV Z-boson mass and a $z = 3.5$ source, the dip energy is 10^{21} eV^2/m.

We have shown that a sizeable absorption dip at $\sim 10^{20}$ eV is unambiguously predicted for neutrinos emanating from a highly red-shifted source and having mass in the electron volt range. Detection feasibility depends on the magnitude of neutrino flux emitted by the source. Quasars, active galactic nuclei, pulsars, supernovae and accreting black holes are suggested sources.[18] Although their ultrahigh energy emission spectrum is unknown, it is very unlikely that a single source spectrum can be measured at the energies here required. Consider as a measurement criterion, detection of 10^3 events per year in the 10^{20} eV energy region by a 100% efficient DUMAND-type detector (one cubic kilometer of H_2O target). Since the ultrahigh energy neutrino-nucleon cross section is approximately[14]

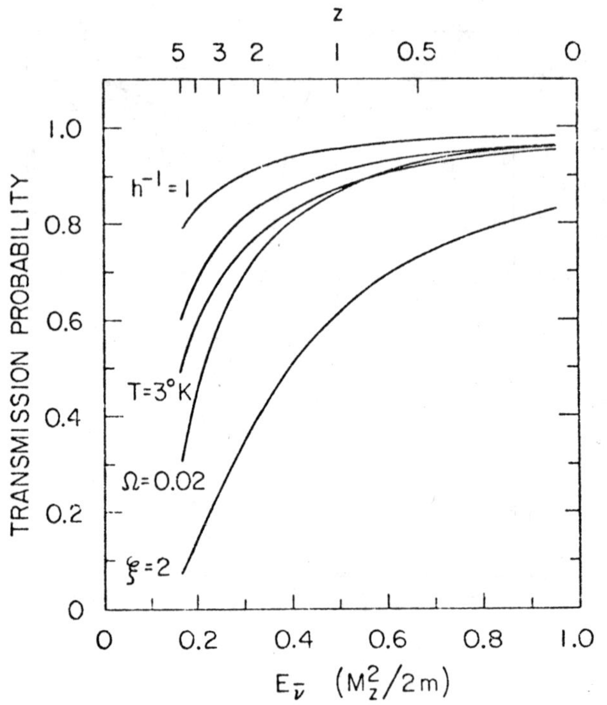

Fig. 2. Same as Fig. 1 but for massive $[m \gg (\xi+1)\beta^{-1}]$ neutrinos. The positions of the absorption dip for various z values are shown at the top of the figure.

$G_F^2 M_W^2 \ln(s/M_Z^2)/\pi \sim 10^{-33}$ cm^2 (the $\ln$ arises in QCD), an incident neutrino flux of $F \sim 10^{-10}$/cm^2/s is required at 10^{20} eV. The distance to a highly red shifted sourced is a large fraction of the Hubble radius, so the power generating this minimum flux is $P \sim F \cdot E_\nu \cdot 4\pi H_0^{-2} \sim 10^{55}$ erg/s. This may be contrasted with a 10^{45} to 10^{49} erg/s optical emission from quasars, the most powerful sources observed to date. If the mass of a quasar is of order of that of a typical galaxy, the required neutrino flux at $E \sim 10^{20}$ eV represents a power conversion of 1% per year and an upper bound on the lifetime of such a source of a century. Thus the existence of a single source of sufficient flux to enable measurement of the dip appears untenable.

SUM OVER SOURCES

In order to enhance the flux of 10^{20} eV neutrinos incident on earth, a sum over sources will be carried out. Taking the energy density of the universe to be $\sim 10^{-29}$ g/cm, the necessary flux value results if 10^{-4} of the energy has been converted to 10^{20} eV neutrinos. This is possible, and we proceed to calculate the absorption of the neutrino flux integrated over sources. The summation over a distribution of source red-shifts will tend to smear the absorption dip, but it will now be shown that depending on the history of the sources, detectable absorption may survive.

Let the number of neutrinos emitted per unit volume in the time interval dt_e and in the energy increment dE be

$$d^2 n_\nu(E,z) = \left[\sum_s \rho_s(E,z)\, n_s(z)\right] dE\, dt_e \tag{12}$$

where $n_s(z)$ is the source density at "time" z, $\rho_s(E,z)$ is the source power (neutrino number emitted per unit time per unit energy) at energy E and time z, and the sum is over the various types of neutrino sources. Then the relativistic neutrino flux incident at earth (neutrino number per unit area, time, steradian and energy) is

$$\mathfrak{F}_o(E_o) = \frac{1}{4\pi}\int dt_e \frac{d^2 n_{\nu o}(E_o, z)}{dE_o\, dt_e} \ . \tag{13}$$

Substituting from Eqs. (8) and (9) (with t replaced by t_e and ω replaced by z), and (12), gives for Eq. (13)

$$\mathfrak{F}_o(E_o) = \frac{1}{4\pi H_o}\sum_s \int_0^\infty dz \frac{\rho_s((1+z)E_o, z)\, n_s(z)\, e^{-\tau(E_o, z)}}{(1+z)^4\sqrt{1+\Omega_o z}} \ . \tag{14}$$

Then writing Eq. (11) as

$$e^{-\tau} = 1 + \Theta(1-\tilde{E})\,\Theta((1+z)\tilde{E}-1)\left\{\exp\left[\frac{-\tau(1)}{\tilde{E}^{3/2}\sqrt{\tilde{E}(1-\Omega_o)+\Omega_o}}\right] - 1\right\} \tag{15}$$

the flux incident at earth becomes

$$\mathfrak{F}_o(E_o) = \frac{1}{4\pi H_o}\sum_s \int_0^\infty dz \frac{\rho_s(1+z)E_o, z)\, n_s(z)}{(1+z)^4\sqrt{1+\Omega_o z}} - \delta\mathfrak{F}_o(E_o) \ , \tag{16}$$

where the flux loss due to absorption is

$$\delta \mathfrak{F}_o(\tilde{E}) = \frac{\Theta(1-\tilde{E})}{4\pi H_o}$$

$$\times \left\{1 - \exp\left[\frac{-\tau(1)}{\tilde{E}^{3/2}\sqrt{\tilde{E}(1-\Omega_o)+\Omega_o}}\right]\right\} \sum_s \int_{1/\tilde{E}-1}^{\infty} dz \frac{\mathcal{P}_s((1+z)E_o, z)\, n_s(z)}{(1+z)^4\sqrt{1+\Omega_o z}} . \qquad (17)$$

The relative flux loss is simply the ratio $\delta\mathfrak{F}_o/\mathfrak{F}_o = \delta \ln \mathfrak{F}_o$, and the transmission probability is $1 - \delta\mathfrak{F}_o/\mathfrak{F}_o$.

For further calculation, a source density distribution $n_s(z)$ and a source power distribution $\mathcal{P}_s(E, z)$ are needed. Since extragalactic neutrino astronomy has not yet begun as an observational science, nothing certain is known about ultraenergy neutrino sources. However, the skies seem to favor power law spectra for galactic fluxes. Radio waves have a spectral index of ~ 0.7, while charged particles have an index of 2 to 3 (consistent with the synchrotron relation, $\alpha_{charge} = 2\alpha_{radio} + 1$). Power laws are scale invariant, since $A(E/\hat{E})^{-\alpha} = A' E^{-\alpha}$. Thus, an assumption of a power law for the neutrino emission spectrum leads to a factorizing power distribution,

$$\mathcal{P}_s(E, z) = A_s E^{-\alpha_s} f_s(z) = A_s E_o^{-\alpha_s} (1+z)^{-\alpha_s} f_s(z) . \qquad (18)$$

It is clear that luminosity evolution, characterized by nonconstant $f(z)$, and source creation or destruction, characterized by nonconstant $(1+z)^{-3} n(z)$, mimic each other. We may therefore fix $(1+z)^{-3} n(z)$ to be constant, and parameterize all source evolution by $f_s(z)$. A useful and simple parameterization is

$$f_s(z) = (1+z)^{\beta_s} \Theta(z_s - z) . \qquad (19)$$

The theta function defines the critical time, z_s, at which the source began emitting neutrinos, and $\beta \neq 0$ characterizes source evolution. Employing Eqs. (18) and (19), one has

$$\mathfrak{F}_o(E_o) = \sum_s \frac{n_{so} A_s E_o^{-\alpha_s}}{4\pi H_o} \int_0^{z_s} dz \frac{(1+z)^{\beta_s - \alpha_s - 1}}{\sqrt{1+\Omega_o z}} - \delta\mathfrak{F}_o(E_o) \qquad (20)$$

and

$$\delta \mathfrak{F}_o(\tilde{E}) = \Theta(1-\tilde{E})$$

$$\times \left\{ 1 - \exp\left[\frac{-\tau(1)}{\tilde{E}^{3/2}\sqrt{\tilde{E}(1-\Omega_o)+\Omega_o}} \right] \right\} \sum_s \frac{n_{so} E_o^{-\alpha_s} A_s}{4\pi H_o} \int_{1/\tilde{E}=1}^{z_s} dz \frac{(1+z)^{\beta_s - \alpha_s - 1}}{\sqrt{1+\Omega_o z}}$$

(21)

The values $\Omega_o = 0$ and 1 give an integrand probably bracketing the true value and lead to especially simple results:

$$\mathfrak{F}_o(E_o) = \sum_s \frac{n_{so} A_s E_o^{-\alpha_s}}{4\pi H_o} \left[\frac{(1+z_s)^{\Gamma_s} - 1}{\Gamma_s} \right] - \delta \mathfrak{F}_o(E_o) \; , \tag{22}$$

$$\delta \mathfrak{F}_o(\tilde{E}) = \Theta(1-\tilde{E})$$

$$\times \left\{ 1 - \exp\left(\frac{-\tau(1)}{\tilde{E}^{3/2}\sqrt{\tilde{E}(1-\Omega_o)+\Omega_o}} \right) \right\} \sum_s \frac{n_{so} A_s E_o^{-\alpha_s}}{4\pi H_o} \left[\frac{(1+z_s)^{\Gamma_s} - (1/\tilde{E})^{\Gamma_s}}{\Gamma_s} \right]$$

(23)

with $\Gamma_s = \beta_s - \alpha_s - (0, 1/2)$ for $\Omega_o = (0, 1)$ in the integrand. For $\Gamma_s = 0$ replace the square brackets with $\ln(1+z_s)$ and $\ln[\tilde{E}(1+z_s)]$ respectively. The values in front of the square brackets are just what one gets by summing all sources out to the Hubble radius H_o^{-1}, ignoring red-shifting and source evolution. It is clear that such a naive sum is qualitatively correct. Significantly larger fluxes result only if evolution is important, and there exist sources at large red-shift (i.e. $z_s \gg 1$).

We will now assume that in the energy region of absorption ($\tilde{E} \sim 1$), one source type dominates. Then the sum on sources and the source subscripts may be dropped with the understanding that A, n_{so}, β, α and Γ refer to the dominant source. The relative flux loss is simply

$$\delta \ln \mathfrak{F}_o(\tilde{E}) = \Theta((1+z_c)\tilde{E}-1)\,\Theta(1-\tilde{E})\,\mathcal{U} \left\{ 1 - \exp\left(\frac{-\tau(1)}{\tilde{E}^{3/2}\sqrt{\tilde{E}(1-\Omega_o)+\Omega_o}} \right) \right\}$$

(24)

with

$$\mathscr{W}(\tilde{E}, z_c, \Gamma) = \left[\frac{1 - [\tilde{E}(1+z_c)]^{-\Gamma}}{1 - (1+z_c)^{-\Gamma}} \right], \cdot \Gamma \neq 0 \qquad (25)$$

$$= \ln[\tilde{E}(1+z_c)]/\ln[1+z_c], \ \Gamma = 0.$$

z_c is the critical time at which the dominant source began neutrino emission. Note that except for the factor $\mathscr{W}$, Eq. (24) has the form of the familiar single source result with $z = z_c$ (c.f. Eq. (15)). Thus the whole effect of summing over source red-shifts is to reduce the single source result by the "wash-out" factor $\mathscr{W}$. For all choices of Γ and z_c, and $\tilde{E}$ consistent with the theta functions, $0 \leq \mathscr{W} \leq 1$. $\mathscr{W}$ rises from zero at $\tilde{E} = 1/1+z_c$, where only the most red-shifted source may contribute, to unity at $\tilde{E} = 1$, where all sources contribute. $\mathscr{W}$ increases with increasing Γ or increasing z_c. The infinite evolution limit, $\Gamma \to \infty$, pushes all sources back to $z = z_c$, thereby setting $\mathscr{W} = 1$ and restoring the single source result for relative flux loss.

The simple functional form of $\delta \ln \mathscr{F}_o$ belies the fact that it depends on six ill-known parameters: Ω_o, T, ξ, h, z_c, Γ. $\mathscr{F}_o$ depends on A, n_{s_o} and α in addition to these. Predictions for the transmission probability of the neutrino flux summed over source red-shifts, $1 - \delta \ln \mathscr{F}_o(\tilde{E})$, are given in Fig. 3 for the standard set of (Ω, T, ξ, h^{-1}) values. Results for other values may be inferred by referring to Fig. 2.

The inference to be drawn from Fig. 3 is that unless the history of neutrino sources shows strong evolution (large Γ) and early emission (large z_c), the relative flux loss is negligible. What values of Γ and z_c might one expect? It is common belief that quasars emit ultraenergy neutrinos. The emission mechanism is presumed to be the collision of hadrons, accelerated by enormous magnetic fields, yielding pions with decay products $\nu_\mu e \nu_e$. Neutrinos produced in the quasar interior do not suffer the strong absorption and energy degradation in the high density environment that other quanta endure, and therefore escape with their ultraenergy intact. Quasar counts as a function of z show definite evolutionary effects, at least for $z \lesssim 2$, corresponding to $\beta \simeq 3.5$.[19] The apparent quasar density peaks around $z \sim 2$ (perhaps a selection effect), and vanishes abruptly above $z \sim 3.5$. This cut-off at $z \sim 3.5$ may be interpreted as signifying the age of quasar production, or alternatively, as signaling the end of a radio, optical and x-ray dense era. With the latter hypothesis, quasar birth occurred even earlier than $z = 3.5$, and neutrino astronomy will be sensitive to even larger z values. Also with the latter hypothesis,

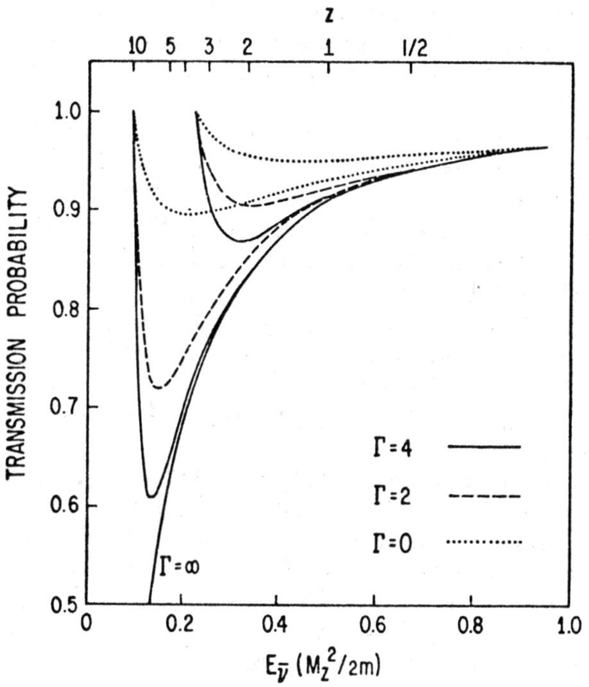

Fig. 3. Same as Fig. 2 but for a neutrino flux integrated over sources with red shift $\leq z_c$. Γ parameterizes source evolution and spectral index of neutrino emission (see text). Sets of curves for $z_c = 3.5$ and $z_c = 10$ are shown.

relic detection is perhaps possible if evolution continues to large z and the neutrino emission spectrum in the 10^{20} eV range is intense and not too steep, such that $\Gamma \geqslant 2$. Otherwise, the only presently viable hope for relic detection, cosmic ray ν-relic $\bar{\nu}$ annihilation on the Z resonance, joins the prior proposals as well-intentioned immeasureabilia.

I wish to thank E. M. Burbidge, M. Schmidt and H. E. Smith for useful conversations.

REFERENCES

1. J. Royer, Phys. Rev. 174, 1719 (1968); L. Stodolsky, Phys. Rev. Lett. 34, 110 (1975); R. Opher, Astron. Astrophys. 37, 135 (1974); R. R. Lewis, Phys. Rev. D21, 683 (1980); R. Opher, Astron. Astrophys. 108, 1 (1982).

2. N. Cabibbo and L. Maiani, Phys. Lett. 114B, 115 (1982); P. Langacker, J. P. Leveille and J. Shieman, Michigan preprint UM HE 82-28 (1982); J. P. Leveille, talk at this conference.
3. First considered by J. Bernstein, M. Ruderman, G. Feinberg, Phys. Rev. 132, 1227 (1963); B. P. Konstantinov and G. E. Kocharov, JETP 19, 992 (1964); R. Cowsik, Y. Pal and S. N. Tandon, Phys. Lett. 13, 265 (1964); T. Hara and H. Sato, Prog. Theor. Phys. 64, 1089 (1980); 65, 477 (1981).
4. T. Weiler, Phys. Rev. Lett. 49, 234 (1982).
5. S. Weinberg, Phys. Rev. 128, 1457 (1962).
6. S. Wolfram, unpublished.
7. See, e.g., S. Weinberg, Gravitation and Cosmology, Wiley New York (1972).
8. G. Beaudet and P. Goret, Astron. Astrophys. 49, 415 (1976).
9. R. V. Wagoner, W. A. Fowler and F. Hoyle, Astrophys. J. 148, 3 (1967); A. Yahil and G. Beaudet, Astrophys. J. 206, 26 (1976); S. Dimopoulos and G. Feinberg, Phys. Rev. D20, 1283 (1979); A. D. Linde, Phys. Lett. 83B, 311 (1979).
10. V. L. Ginzburg and S. I. Syrovatskii, The Origin of Cosmic Rays, Pergamon Press Ltd., Oxford (1964).
11. K. Greisen, Phys. Rev. Lett. 16, 748 (1966); G. T. Zatsepin and V. A. Kuzmin, JETP Lett. 4, 78 (1966); F. W. Stecker, Phys. Rev. Lett. 21, 1016 (1968).
12. A. I. Nikishov, JETP 14, 393 (1962); P. Goldreich and P. Morrison, JETP 45, 344 (1963); R. J. Gould and G. Schreder, Phys. Rev. Lett. 16, 252 (1966); Phys. Rev. 155, 1404, 1408 (1967); J. V. Jelly, Phys. Rev. Lett. 16, 479 (1966).
13. G. Khristiansen, G. Kulikov and J. Fomin, Cosmic Rays of Superhigh Energies, Verlag Karl Thiemig, Munich (1980).
14. For high energy neutrino-nucleon collisions, the weak cross section is slightly enhanced by the factor $\frac{1}{2} F_2(x = 0, Q^2 = M_W^2) \ln(1 + s/M_W^2)$; see V. S. Berezinskii and A. Z. Gazizov, Yad. Fiz. 29, 1589 (1979).
15. Our derivation is a generalization of that found in Ref. 7 to a background gas with arbitrary momentum distribution.
16. P. Ramond, Proc. of the First Workshop on Grand Unification, eds., P. Frampton, S. Glashow, A. Yildiz, U.N.H. (1980); E. Witten, ibid.
17. J. R. Bond and A. S. Szalay, Proc. of Neutrino-'81, Honolulu, Hawaii; G. Steigman, ibid, and references therein.

18. S. Margolis, D. Schramm and R. Silberberg, Astrophys. J. 221, 990 (1978); D. Eichler, Astrophys. J. 222, 1109 (1978); 232, 106 (1979); also see Proc. DUMAND Summer Workshop, Vol. 2, Ultrahigh Energy Interactions & Neutrino Astronomy, ed. A. Roberts (1978).

19. M. Schmidt, Ann. Rev. Astron. and Astrophys. 7, 527 (1969); Ap. J., 151, 393 (1968); 162, 371 (1970).

THE LEPTON ASYMMETRY OF THE UNIVERSE

Paul Langacker, Gino Segrè, Sanjeev Soni
University of Pennsylvania, Philadelphia, Pa. 19104

THE LEPTON ASYMMETRY

The standard hot big bang cosmological model has been extremely successful in explaining the $T_\gamma \simeq 2.7^oK$ microwave radiation and the relative abundance of primordial helium and deuterium.[1] When combined with the ideas of grand unification[2] it may also give a plausible dynamical explanation of the small baryon asymmetry $B \equiv (n_B - n_{\bar{B}})/n_\gamma \simeq 10^{-10\pm1}$ observed in the present universe.

The hot big bang model predicts the existence of cosmological relic neutrinos analogous to the microwave radiation. The neutrinos would have stayed in equilibrium via the weak interactions until they decoupled at a temperature of about 1 MeV. The momentum distribution of the decoupled neutrinos would have been subsequently redshifted, so that at present the relic neutrinos would have a distribution of relativistic thermal form characterized by an effective temperature

$$T_\nu = \left(\frac{4}{11}\right)^{\frac{1}{3}} T_\gamma \simeq 1.9^oK \quad , \tag{1}$$

corresponding to a number density $n_{\nu_i} \simeq n_{\bar{\nu}_i} \simeq 50/cm^3$.

This scenario involves one largely untested assumption, however, viz. it assumes that the lepton asymmetry

$$L_i \equiv \frac{n_{e_i^-} - n_{e_i^+} + n_{\nu_i} - n_{\bar{\nu}_i}}{n_\gamma} \simeq \frac{n_{\nu_i} - n_{\bar{\nu}_i}}{n_\gamma} \tag{2}$$

of the ith lepton family is negligibly small. In fact the existing limits on L_i are very weak. It is useful to define the dimensionless parameters $\xi_i \equiv \mu_i/T_\nu$ where μ_i is the chemical potential of the ith species. (The ξ_i remain constant for an adiabatically expanding universe if there are no lepton number violating interactions). Then the asymmetry is

$$L_i = \frac{2}{3}\left(\frac{T_\nu}{T_\gamma}\right)^3 \left(\xi_i + \frac{1}{\pi^2}\xi_i^3\right) \tag{3}$$

and the energy density associated with ν_i and $\bar{\nu}_i$ is

$$\rho_{\nu_i} + \rho_{\bar{\nu}_i} = \frac{\pi^2}{15} T_\nu^4 \left[\frac{7}{8} + \frac{15}{4\pi^2}\xi_i^2 + \frac{15}{8\pi^4}\xi_i^4\right] \tag{4}$$

0094-243X/83/990076-05 $3.00 Copyright 1983 American Institute of Physics

The only significant direct limit on the ξ_i is that the neutrino energy density not exceed $\rho_o = 8 \times 10^{-29} gm/cm^3$, the upper limit on the total energy density of the universe[1,3]. For massless neutrinos this implies

$$[\sum_i \xi_i{}^4]^{1/4} < 80 \Rightarrow |L_i| < 4 \times 10^4$$

i.e. an enormous asymmetry is allowed.

For massive neutrinos the situation is somewhat improved, but large asymmetries are still allowed[4]. For $m_{\nu_i} \simeq 20$ eV, for example, one obtains $|\xi_i| < 6$ or $|L_i| < 20$.

Additional constraints are imposed by nucleosynthesis. In the standard model ($\xi_i = 0$) the abundances of primordial ^{4}He and D are successfully given[5] and the number of neutrino species limited; furthermore, the weak dependence on the baryonic density requires a low baryon density universe, $0.01 < \Omega_N < 0.10$, corresponding to $n_B/n_\gamma = (4\pm1) \times 10^{-10}$, where Ω_N is the ratio of baryon density to critical density.

It has been pointed out by many authors[6], however, that the situation is completely changed if $\xi_i \neq 0$. An asymmetry in the electron neutrino directly affects the equilibrium fraction of neutrons to nucleons

$$X_n \simeq \frac{\lambda(p\to n)}{\lambda(p\to n) + \lambda(n\to p)} = \frac{1}{1+\exp\left[\xi_{\nu_e} + \frac{M_n - M_p}{T}\right]} \tag{5}$$

where $\lambda(p\to n)$ is the reaction rate for processes where a proton is converted into a neutron. We see that $\xi_{\nu_e} > 0$ (<0) implies fewer (more) neutrons and hence less (more) ^{4}He. Asymmetries in ν_μ or ν_τ increase the expansion rate of the universe, implying a higher freezeout temperature for X_n and therefore more ^{4}He. The observed element abundance allows $|\xi_{\nu_e}| \lesssim 0.2$ or $|\xi_i| \lesssim 2$ ($i = \nu_\mu$ or ν_τ) as perturbations around the standard model. In addition, David and Reeves[7] have found a continuum of new solutions to nucleosynthesis in which ξ_{ν_e} increases from 0 to $\simeq 1.2$; the effect of ξ_{ν_e} is balanced by an increased expansion rate, which could be due to asymmetries in ν_μ or ν_τ (with $0<|\xi_i|<80$), additional neutrino species, anisotropic shear, magnetic monopoles, etc. The required Ω_N increases with ξ_{ν_e}.

Let us conclude this introduction by mentioning the theoretical expectations for the lepton asymmetry. A priori, any initial value of L_i at the time of the big bang is possible. If lepton number is conserved then L_i (or more accurately, the ratio of $n_{\nu_i} - n_{\bar{\nu}_i}$ to the entropy) is unchanged by an adiabatic expansion of the universe.

It is generally assumed[8] that the lepton number violating interactions in GUTs would dilute an initial large asymmetry to a negligibly small value ($|L_i| \simeq n_B/n_\gamma \simeq 10^{-10}$) long before nucleosynthesis.

However, Harvey and Kolb[3] have shown that $|L_i| \gg n_B/n_\gamma$ is possible even in a GUT provided there are (a) a large initial asymmetry in some quantum number and (b) approximately conserved global quantum numbers. They have constructed an explicit SO_{10} model with an arbitrarily large value of $|L_i|$. (In practice the requirement that this model lead to a realistic cosmology implies[9] $|L_i| < 0(1)$).

The theoretical and observational constraints on the lepton asymmetry are therefore very weak. Furthermore, large asymmetries could significantly alter the usual nucleosynthesis scenario and therefore the determination of n_B/n_γ and the limits on the number of neutrino species.

THE ROLE OF MAJORANA NEUTRINO MASSES

In this paper we will consider the possibility that for some (unexplained) reason the asymmetry in one or more neutrino species was large (ξ_i>1) subsequent to the epoch of grand unification (or at the big bang if one does not believe in GUTs). We consider the role of the lepton number violation associated with Majorana neutrino masses in the 10 eV range in reducing an initial large lepton asymmetry. We find[9] that in most models with explicit hard lepton number violation any initial lepton asymmetry would be reduced to an insignificant level long before nucleosynthesis. For models in which lepton number is violated spontaneously or relies on $SU_2 \times U_1$ breaking, we naively expect the symmetry to be restored at high temperatures and therefore the mechanism for lepton asymmetry erasure to be inoperative at high temperatures. Following an argument of Linde[10], we find this to not be the case, i.e. a large enough asymmetry $L = \sum_i L_i > L_c$ acts on the Higgs potential to prevent restoration of symmetry. In fact we expect, for L not too much larger than L_c, that the relevant vacuum expectation values (v.e.v.'s) v(T) behave as

$$\left[\frac{v(T)}{T}\right]^2 \simeq \beta\left[\frac{L^2(T)}{L_c^2} - 1\right] \tag{6}$$

where β is a (model dependent) constant of order unity; for $L < L_c$, $v(T) = 0$.

We consider starting with a large $|L_i|$ as in the SO_{10} model[3]. We then discuss the two currently popular models for spontaneously breaking lepton number; the first[11] introduces a Majorana neutrino N_R coupled to an SU_2 singlet field Φ (the so-called Majoron) which acquires a very large v.e.v. N_R also couples to ordinary neutrinos via the Higgs doublet field which has a v.e.v. of $v_\phi(T)$. The N-ν mixing leads to a neutrino mass $m_\nu(T) \sim \frac{h^2 v_\phi^2(T)}{M_N}$ where h is an appropriate Yukawa coupling. The second model[12] introduces a Higgs triplet χ with a direct coupling to lepton doublets. The neutrinos acquire Majorana masses proportional to $v_x(T)$ of the form

$$f_{ij} \bar{\nu}^c_{R,i}\, \nu_{L,j}\, v_x(T) + h.c. \tag{7}$$

so in this case $m_\nu(T) \sim f\, v_x(T)$.

The lepton asymmetry erasure mechanisms occur through scatterings such as $\nu + a \to \bar{\nu} + a'$, where a is the target. These reactions violate lepton number conservation and are therefore proportional to $m_\nu(T)$, the characteristic magnitude of the violation. Studying the Boltzmann equation for L, we find, for velocity u, density of target n_a and lepton number violating cross section σ that

$$\frac{1}{L}\frac{dL}{dT} \sim -2\, n_a(T) \langle u\sigma \rangle \qquad (8)$$

with $\sigma \propto m_\nu^2(T)/T^4$. Using equation (6) and the fact that $n_a \sim T^3$, we find for the Higgs singlet model that $\dot{L}/L \sim T^3$ and in the Higgs triplet model that $\dot{L}/L \sim T$. The difference is caused by the fact that $m_\nu \sim v_\phi^2(T)$ in the first case and like $v_x(T)$ in the second. We compare the resulting equation with the expansion rate of the universe

$$\frac{\dot{R}}{R} \sim 1.66\, g^* \frac{T^2}{M_p} \qquad (9)$$

where $M_p \simeq 1.2 \times 10^{19}$ Gev and g^* is the number of relativistic species of particles at temperature T (appropriately modified if there are large chemical potentials).

For both models, in a large class of initial conditions, we find that $L_i \to 0$ for all but the species with the lightest mass neutrino, which we call ν_1. L_1 tends to a fixed point value L_c which is of order unity. Intuitively this is due to the driving mechanism of a large lepton number violating interaction proportional to v(T), which however is turned off as $L(T) \to L_c \sim 0(1)$.

Our conclusion therefore is that, for a wide range of models and parameters, we arrive at neutrino asymmetries of the order of magnitude to significantly affect nucleosynthesis. We believe therefore that this possibility should be seriously entertained.

REFERENCES

1. S. Weinberg, "Gravitation and Cosmology" (John Wiley and Sons, Publishers, N.Y. 1972).
2. P. Langacker, Phys. Reports 72, 4, p. 185 (1981).
3. J. Harvey and E. W. Kolb, Phys. Rev. D24, 2090 (1981).
4. J. M. Cohen and P. Langacker, to be published.
5. For a recent paper on the subject see K. A. Olive, D. N. Schramm, G. Steigman, M. S. Turner and J. Yang, Astrophys. Jour. 246, 557 (1981).
6. R. V. Wagoner, W. A. Fowler and F. Hoyle, Astrophys. Jour. 148, 3 (1967); A. Yahil and G. Baudet, Astrophys. Jour. 206, 261 (1976); A. Linde, Phys. Lett. 83B, 311 (1979).
7. Y. David and H. Reeves, Phil. Trans. R. Soc. Lond. A296, 415 (1980).
8. S. Dimopoulos and G. Feinberg, Phys. Rev. D20, 1283 (1979); D. Schramm and G. Steigman, Phys. Lett. 87B, 141 (1979); D. V. Nanopoulos, D. Sutherland, and A. Yildiz, Lett. Nuovo

8. (Cont.) Cimento 28, 205 (1980); M. S. Turner, Phys. Lett. 98B, 145 (1981).
9. The details of these calculations will be presented elsewhere, P. Langacker, G. Segrè and S. Soni, Phys. Rev. D, to be published.
10. A. Linde, Rep. Prog. Phys. 42, 25 (1979).
11. Y. Chikashige, R. N. Mohapatra and R. Peccei, Phys. Lett. 98B, 265 (1981); Phys. Rev. Lett. 45, 1926 (1980).
12. G. B. Gelmini and M. Roncadelli, Phys. Lett. 99B, 411 (1981). H. M. Georgi, S. L. Glashow, and S. Nussinov, Nucl. Phys. B193, 297 (1981).

CORRECTIONS TO PRIMORDIAL NUCLEOSYNTHESIS

Duane A. Dicus,[1] Edward W. Kolb,[2] A. M. Gleeson,[1]
E.C.G. Sudarshan,[1] Vigdor L. Teplitz,[3] and Michael S. Turner[4]

[1]Center for Particle Theory
University of Texas
Austin, TX 78712

[2]Theoretical Division
Los Alamos National Laboratory
Los Alamos, NM 87545

[3]U.S. Arms Control and Disarmament Agency
Washington, DC 20451
and
Physics Department
University of Maryland
College Park, MD 20740

[4]Astronomy and Astrophysics Center
The University of Chicago
Chicago, IL 60637

ABSTRACT

The changes in primordial nucleosynthesis resulting from small corrections to rates for weak processes that connect neutrons and protons are discussed. The weak rates are corrected by improved treatment of Coulomb and radiative corrections, and by inclusion of plasma effects. The calculations lead to a systematic decrease in the predicted ^{4}He abundance of about $\Delta Y = 0.0025$. The relative changes in other primordial abundances are also 1-2%.

In this talk I will discuss improvements in the calculation of the primordial ^{4}He abundance. The corrections are related to improved treatment of the weak $n \leftrightarrow p$ rates. I will compare the new values of the ^{4}He abundance to the standard calculation of Wagoner.[1] For details of our calculations, the reader is referred to our original paper.[2]

The standard hot big-bang model of the universe seems to provide a reliable framework for understanding the origin and evolution of our universe.[3] One of the features of our present universe which is naturally explained in this model is the large abundance of ^{4}He. The success of primordial nucleosynthesis in predicting the large primordial abundance of ^{4}He, and the relatively large abundance of D, is usually considered to be the strongest evidence that the universe can be described by a Friedmann-Robertson-Walker cosmology at very early times. Because of this concordance, it is attractive to assume that the Friedmann-Robertson-Walker cosmology was applicable at the time of nucleosynthesis, and then demand that the resulting primordial abundances of the light elements be within bounds extrapolated from present observations.

This latter approach has been recently employed to limit the number of neutrinos at nucleosynthesis, and to put constraints on a number of exotic prticles, such as photinos, gravitinos, goldstinos, monopoles, quarks, stable leptons, heavy neutrinos, etc. The ^{4}He abundance has also been recently used to argue that the standard cosmological model does not fit the data.[4] The same phenomenon has been called by others the "crowning achievement of the standard model."

Because primordial nucleosynthesis provides such a powerful probe of the conditions in the universe at early times, it is important to have precise predictions of the primordial light element abundances, particularly that of ^{4}He. In this talk, I will consider modifications to the calculation of the ^{4}He abundance due to: 1) use of an explicit numerical integration of the rates for n-p transitions rather than fits to the numerical rates; 2) correct treatment of Coulomb corrections; 3) inclusion of radiative corrections--both the usual radiative corrections, and the finite temperature and finite density radiative corrections that depend on the presence of a plasma; 4) inclusion of the effect of the plasma on the mass of the electron; and 5) heating of electron neutrinos by e^+e^- annihilations. We find that the above five effects result in a systematic decrease in the ^{4}He abundance of about 0.003, or about a one percent relative decrease. (Addition of a light neutrino species leads to a ΔY of $\Delta Y \simeq 0.01$.) There are similar 1-2% changes in the abundances of the other light elements which are produced (D, ^{3}He, ^{7}Li, etc.).

Although we find that the corrections to the weak rates are in general temperature dependent, we can get a rough estimate of the sensitivity of the primordial ^{4}He abundance, Y, on the weak rates by considering the change in Y resulting from a change in the neutron half-life.[5] For instance, a decrease in the neutron half-life from 10.6 minutes to 10.4 minutes ($\Delta\lambda/\lambda \cong +0.02$ where $\lambda \equiv \tau_{\frac{1}{2}}^{-1}$) results in a decrease in the ^{4}He abundance of about 0.004. This decrease in the neutron half-life is equivalent to an increase in G_F, the Fermi constant, hence a temperature independent increase in all the weak rates. This suggests that the dependence of Y on $\Delta\lambda$ may be approximated as $\Delta Y \cong -0.2\ \Delta\lambda/\lambda$.

It should be emphasized that corrections 1-5 above are universal in the sense that they must be applied regardless of the values of the neutron lifetime, the number of neutrinos, or the ratio of baryons to photons.

The six weak reactions which interconvert neutrons and protons are $n \leftrightarrow pe\bar{\nu}$, $n\bar{e} \leftrightarrow p\bar{\nu}$, and $n\nu \leftrightarrow pe$. The rates for these processes were first calculated in the context of the early universe by Alpher, Follin, and Herman,[5] and we will adopt their notation (see also ref. 1). The rates depend on $Q = m_n - m_p$, the photon temperature T, the neutrino temperature T_ν, and the electron energy E_e. The rates depend on these quantities through the dimensionless variables $z = m_e/T$, $z_\nu = m_e/T_\nu$, $q = Q/m_e$, and $\varepsilon = E_e/m_e$. The weak rates may be written in terms of these variables. For instance, the rate for neutron decay is

$$\lambda_{n\to pe\nu} = (\tau\lambda_0)^{-1} \int_1^q d\varepsilon \frac{\varepsilon(\varepsilon - q)^2(\varepsilon^2 - 1)^{\frac{1}{2}}}{[1 + \exp(-\varepsilon z)]\{1 + \exp[(\varepsilon - q)\, z_\nu]\}} , \qquad (1)$$

where τ in the neutron lifetime and λ_0 is defined as $\lambda_0 \equiv \int_1^q d\varepsilon\, \varepsilon(\varepsilon - q)^2(\varepsilon^2 - 1)^{\frac{1}{2}} = 1.63615$. The factor $(\tau\lambda_0)^{-1}$ fixes the nuclear matrix element to be used.

In the computer code developed by Wagoner,[1] the rates (2.7) are fit by an analytic function of the photon temperature. The fit to λ_n is good to an accuracy of about 1.4% over the temperature range $T_9 = 100$ to $T_9 = 0.3$ (T_9 is the temperature in units of 10^9 K). The fit to λ_p is good to about 1.4% over a range $T_9 = 100$ to $T_9 = 3$. This fit is rather remarkable considering that λ_n and λ_p change by more than seven orders of magnitude over the above temperature range. However, based on our approximate formula for ΔY given in Eq. (1.1), a 1.4% error in the weak rates lead to a potential ΔY of $|\Delta Y| = (0.2)(1.4\%) \cong .003$.

We have modified Wagoner's code to evaluate the rates numerically for every time step. We evaluate the integrals to an accuracy of better than 0.005%, which should introduce an error in ΔY of $|\Delta Y| \lesssim 0.0001$. The change in Y is given in Table I.

Our next change is to correctly treat the Coulomb corrections. Coulomb corrections are taken into account by multiplying the phase space density for $n \leftrightarrow pe\nu$ and $n\nu \leftrightarrow pe$ by F_+, where F_+ is the Fermi function given in terms of the electron velocity, β, as

$$F_+(\beta) \cong \frac{2\pi\alpha/\beta}{1 - e^{-2\pi\alpha/\beta}} \quad . \tag{2}$$

It is also necessary to include F_+ in the phase space for λ_0, which changes λ_0 from 1.63615 to about 1.69. Wagoner accounted for Coulomb corrections by decreasing λ_0 by 2%. The effect of correctly treating the Coulomb corrections is shown in Table I.

Our next correction is to include the radiative corrections. Radiative corrections have been considered by many authors;[6] the effect is to multiply the phase space for all processes by $[1 + \frac{\alpha}{2\pi} C(\beta,y)]$ where β is the electron's velocity, and y is the neutrino energy divided by m_e (y is different for the different weak processes). The function C is given by[6]

$$C(\beta,y) \cong 40 + 4(R - 1)(y/3\varepsilon - 3/2 + \ln 2y)$$

$$+ R[2(1 + \beta^2) + y^2/6\varepsilon^2 - 4\beta R]$$

$$- 4[2 + 11\beta + 25\beta^2 + 30\beta^3 + 20\beta^4 + 8\beta^5]/(1 + \beta)^6. \quad (3)$$

In Eq. (3) the last term in the brackets is from expansions of Spence functions, and R is defined to be

$$R \equiv \beta^{-1} \tanh^{-1} \beta \quad . \quad (4)$$

Including radiative corrections changes λ_0 to 1.75321. Wagoner ignored radiative corrections. The effect of including the radiative corrections is given in Table I.

We next correct the rates by doing the radiative corrections at finite temperature. These corrections were also done independently by Cambier, Primack, and Sher.[7] The real-time formalism[8] is particularly useful for this purpose, since there is a clean separation between temperature dependent and the temperature independent parts of the radiative corrections.

The finite temperature radiative corrections involve double integrals which must be done numerically. The integrals are too tedious to write here; they are given elsewhere.[2,7] The finite temperature radiative corrections cause the rates to be multiplied by an additional factor of the form

$$1 + \frac{\alpha}{\pi} C_i'(T) \quad , \quad (5)$$

where the C_i' differ for different processes.

It turns out that the finite temperature corrections depend strongly on the temperature and on the individual reaction considered. It is also turns out that the rate for neutron decay receives the largest correction. This large correction is mostly due to the effect of a photon in the initial state: $\gamma n \to pe\nu$. The effect of the photon can be thought of as an effective increase in the Q value for $n \to pe\nu$, $Q \to Q + \omega$, where ω is the energy of the photon, $\langle\omega\rangle \cong 2.7T$. Since the neutron decay rate is a more sensitive function of Q than the 2 → 2 weak processes, this particular consequence of finite temperature is larger for $\lambda_{n\to pe\nu}$. However, at the time of the freeze out of the neutron-proton ratio, $\lambda_{n\to pe\nu}$ makes only a small contribution to the total rate. The effect of the finite temperature radiative corrections is shown in Table I.

The next effect we consider is the mass shift of the electron due to finite temperature effects. The electron mass at finite temperature is found by calculating the electron self-energy diagram with the temperature dependent real time propagators. If one defines mass renormalization at zero temperature by the addition of a counter term $\delta m = -\mathrm{Re}\ \Sigma(p^2 = m_e^2)$ with $\Sigma(p)$ the electron self-energy, then finite temperature effects add to the electron mass a term δm_T defined by[9]

$$\delta m_T = \delta m + \mathrm{Re}\ \Sigma_T(p^2 = m_e^2) \quad , \tag{6}$$

where Σ_T is the electron self-energy evaluated with the finite temperature photon and electron propagators. This results in a temperature dependent effective electron mass, m_T, of the form[9]

$$\begin{aligned} m_T &= m_e + \delta m_T \\ &= m_e + B\alpha T^2/m_e \quad , \quad (\alpha T^2/m_e^2 \leq 1) \quad , \end{aligned} \tag{7}$$

where m_e is the zero temperature electron mass (0.511 MeV), and B is a slowly varying function of the temperature with a value between 1 and 2.

It is easy to include this correction in the rates by the substitution of m_T for m_e in the rates and by multiplying all the rates by $(m_T/m_e)^5$, as the cancellation of the m_e^5 from the numerator with the m_e^5 from λ_0 is true only at zero temperature. (We did not explicitly include the factor of m_e^5 in the definitions of the rates or in the definition of λ_0.)

The modification of the electron mass also has the effect of changing the relationship between the neutrino and photon temperatures. The neutrino temperature differs from the photon temperature because, to a good approximation, the neutrinos are decoupled when the entropy in the e^+e^- gas is released as e^+e^- annihilate. Therefore, the entropy released heats the photons, but not the neutrinos. (Below, we discuss the validity of the approximation that the neutrinos are not heated by e^+e^- annihilations, and the effect on Y of relaxing this assumption.)

An additional effect of the change in the electron mass that leads to an even smaller effect on ΔY, is that as the electron mass changes, the electron's contribution to the total energy density, and hence to the expansion rate, changes. The changes in the primordial ^{4}He abundance due to finite temperature effects on the electron mass is given in Table I.

A final small effect we have considered is neutrino heating due to e^+e^- annihilation. Part of the "early Universe" lore is that neutrinos decouple at a temperature $T_d \cong 1$ MeV and therefore do not share in the entropy release from $e^\pm$ annihilations which occur at temperatures $\lesssim 0.5$ MeV. As a consequence, neutrinos are predicted to have a lower temperature today than the photons, $T_\nu = (4/11)^{1/3} T_\gamma$. The assumption of complete neutrino decoupling is explicitly incorporated into Wagoner's code. However, since the weak rates are extremely sensitive to the neutrino temperature, we have investigated the validity of this assumption. We have found that the usual approximation of $T_\nu = (4/11)^{1/3} T_\gamma$ is valid to 2×10^{-3} for ν_e, and 9×10^{-4} for ν_μ and ν_τ. The slightly warmer neutrinos change the weak rates, and change the contribution to the total energy density due to neutrinos. The effects go in opposite directions, and tend to cancel. The size of the effect is listed in Table I.

The changes in the predicted ^{4}He mass fraction due to all the effects discussed here are summarized in Table I. For a wide range of input parameters ($\tau_{1/2}$ = 10.1 - 11.1 min, N_ν = 2 - 10, and $\eta = 3 \times 10^{-11} - 3 \times 10^{-9}$), the sum of all these corrections results in an approximately constant, systematic decrease in Y, $\Delta Y \cong -0.0025$ (~ 1% relative change). The changes in the predicted abundances of the other light elements (D, ^{3}He, and ^{7}Li) are also in the range of a few percent. However, their present abundances are known to much less precision than the present ^{4}He abundance, typically to only within a factor of 2. In Fig. 2 we show the predicted primordial ^{4}He abundance with and without the corrections we have discussed in this paper, as a function of η for $\tau_{1/2}$ = 10.6 min and N_ν = 2, 3, and 4.

The net ΔY we find is about the size of Wagoner's estimated uncertainty, and just slightly less than the uncertainty due to a 1 σ change ($\pm$ 0.16 min) in $\tau_{1/2}$. Nevertheless, it is a systematic decrease in the predicted primordial abundance of ^{4}He. Within a few years experiments with confined neutrons should significantly improve the determination of $\tau_{\frac{1}{2}}$. Recent studies of extragalactic, very metal-poor objects (in which the stellar contribution to the ^{4}He abundance should be small) have led to more accurate and reliable determination of the <u>primordial</u> mass fraction of ^{4}He.[10] A recent result typical of these studies is $Y_p = 0.24 \pm 0.01$.[10] As more objects are studied the uncertainty in Y_p should continue to decrease. It has been argued on the basis of the abundances of D and ^{3}He that η must be greater than $(2 - 3) \times 10^{-10}$.[11] For $N_\nu = 3$ the standard model did predict $Y_p \gtrsim 0.240 - 0.246$ - leaving little room for concordance. We now predict $Y_p \gtrsim 0.237 - 0.243$ - easing the situation a bit. In any case, since primordial nucleosynthesis is the most vigorous test of the standard, hot big bang model, and is also our most powerful probe of the early universe, it is important to continue to sharpen and to reexamine its predictions as the uncertainties in the input parameters decrease.

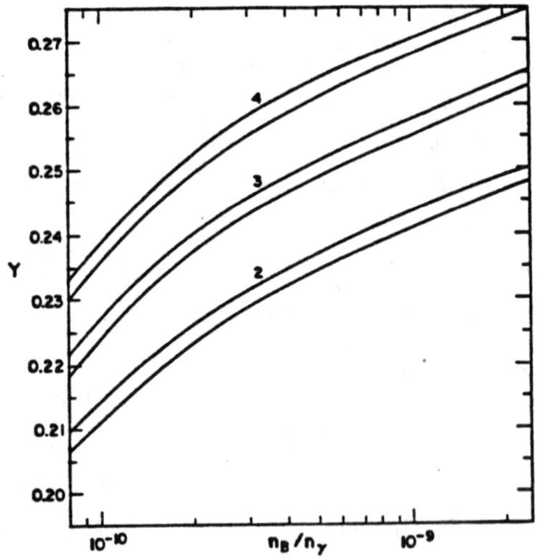

Fig. 1.

The primordial ^{4}He mass fraction as a function of the baryon-to-photon ratio with 2, 3, and 4 light neutrino families assuming a neutron half-life of 10.61 minutes. The upper curve in each set shows the result of Wagoner's calcualtion, and the lower curve shows the results of including all the corrections to the weak rates.

TABLE I

SENSITIVITY OF THE PRIMORDIAL ^{4}He ABUNDANCE TO THE CORRECTIONS DISCUSSED IN THE TEXT
(The values quoted are for $n_B/n_\gamma = 3\times 10^{-10}$, 3 light neutrinos, and $\tau_{1/2} = 10.6$ minutes.)

	Wagoner 1973	$Y_0 = 0.2456$		
1.	Wagoner 73 with numerical evaluation of rates	$Y_1 = 0.2443$	$Y_0 - Y_1 = 0.0013$	
2.	Above corrections plus Coulomb corrections	$Y_2 = 0.2434$	$Y_0 - Y_2 = 0.0022$	$Y_1 - Y_2 = 0.0009$
3.	Above corrections plus zero temperature radiative corrections	$Y_3 = 0.2439$	$Y_0 - Y_3 = 0.0017$	$Y_2 - Y_3 = -0.0005$
4.	Above corrections plus finite temperature radiative corrections	$Y_4 = 0.2435$	$Y_0 - Y_4 = 0.0021$	$Y_3 - Y_4 = 0.0004$
5.	Above corrections plus corrections to electron mass	$Y_5 = 0.2436$	$Y_0 - Y_5 = 0.0020$	$Y_4 - Y_5 = -0.0001$
6.	Above corrections plus e^+e^- heating of ν_e	$Y_6 = 0.2434$	$Y_0 - Y_6 = 0.0022$	$Y_5 - Y_6 = 0.0002$

REFERENCES

1. The version of the code we compare to is described in R. V. Wagoner, Astrophys. J. 1979, 343 (1973).

2. D. A. Dicus, E. W. Kolb, A. M. Gleeson, E. C. G. Sudarshan, V. L. Teplitz, and M. S. Turner, Phys. Rev. D., to be published.

3. For a review of the standard big-bang cosmology, see, e.g., S. Weinberg, Gravitation and Cosmology (Wiley, New York, 1972), Chap. 15.

4. Needless to say, we do not agree with this interpretation.

5. R. A. Alpher, J. W. Follin, and R. C. Herman, Phys. Rev. 92, 1347 (1953).

6. E. S. Abers, D. A. Dicus, R. E. Norton, and H. R. Quinn, Phys. Rev. 167, 1461 (1968); D. A. Dicus and R. E. Norton, Phys. Rev. D1, 1360 (1970); M. A. B. Bég, J. Bernstein, and A. Sirlin, Phys. Rev. D6, 2597 (1972); T. W. Appelquist, J. R. Primack, and H. R. Quinn, Phys. Rev. D6, 2998 (1972); and A. Sirlin, Rev. Mod. Phys. 50, 573 (1978).

7. J.-L. Cambier, J. R. Primack, and M. Sher, unpublished.

8. See, e.g., E. S. Fradkin, Zh. Eksp. Teor. Fiz. 36, 1286 (1961) [Sov. Phys. JETP 9, 912 (1959)]; V. N. Tsytovich, Zh. Eksp. Teor. Fiz. 40, 1775 (1961) (Sov. Phys. JETP 13, 1249 (1961)]; A. Bechler, Ann. of Phys. 135, 19 (1981).

9. S. L. Glashow, E. C. G. Sudarshan, and A. Yildiz, unpublished; also see G. Peressutti and B.-S. Skagerstam, Phys. Lett. 110B, 406 (1982).

10. Some recent ^{4}He determinations for metal-poor objects include: H. B. French, Ap. J. 240, 41 (1980); J. Lequeuz, M. Peimbert, J. F. Rays, A. Serrano, and S. Torres-Piembert, Astro. Astrophys. 80, 155 (1979); D. L. Talent, Ph.D. Thesis, Rice University (1980); J. F. Rays, M. Piembert, and S. Torres-Piembert, Ap. j. 255, 1 (1982).

11. J. Yang, M. S. Turner, G. Steigman, D. N. Schramm, and K. A. Olive, in preparation (1982).

Neutrino Oscillations in the Early Universe II.

Bruce H. J. McKellar
Theoretical Division, Los Alamos National Laboratory
Los Alamos, NM 87545 USA
and
School of Physics, University of Melbourne, Parkville Vic

and

Henry Granek
School of Physics, University of Melbourne, Parkville Vic
Australia 3052

ABSTRACT

We show that it is possible for neutrino oscillations to create a state of the universe in which the net lepton numbers are appreciable and different for each species at the time of nucleosynthesis. Under these conditions mixing induced by the oscillations can influence the abundances of the light elements.

INTRODUCTION

At the first Telemark Neutrino Conference we presented preliminary results on the influence of neutrino oscillations and mixing on the cosmological production of ^{4}He.[1] We promised further results on the cosmological production of the light elements and also a study of the effects of neutrino oscillations during the non-equilibrium period when the oscillation length is of the order of the size of the universe. Now we make good on these promises.

To understand the possible effects neutrino oscillations may have during the early stages of cosmology, it is important to understand the four characteristic times associated with the neutrinos. These are the inverse of the rate of expansion of the universe

$$t \sim \frac{1}{\sqrt{G_N} T^2} \quad , \tag{1}$$

the mean time between left-handed neutrino interactions,

$$\tau_L \sim \frac{1}{G_F^2 T^5} \sim \frac{(\sqrt{G_N} t)^{5/2}}{G_F^2} \quad , \tag{2}$$

the corresponding time for right-handed neutrino interaction τ_R, which, since $m_{w_R} \gg m_{w_L}$, is much greater that τ_L, and the oscillation time[2]

$$t_0 \sim \frac{T}{\Delta m^2} \sim \frac{1}{\Delta m^2(\sqrt{G_N}t)^{\frac{1}{2}}} \quad . \tag{3}$$

As the universe cools down, τ, τ_2 and τ_R all increase, whereas t_0 decreases, as is shown in figure 1. At small enough t, $t_0 >> t$ and the oscillations have no time to develop and thus exert no cosmological influence. There is always a time t_0^* at which $t_0 = t$, but whether the oscillations have any influence depends on whether t_0^* is greater or less that τ_L^*, the left-handed neutrino decoupling time at which $\tau_L = t$. If $t_0^* > \tau_L^*$, then by the time that the neutrino oscillations are able to develop fully the neutrinos have already decoupled and the oscillations cannot influence the evolution of the universe. However if $t_0^* < \tau_L^*$ as illustrated in figure 1, then there is a stage at which the oscillation time is of the order of the expansion times while being greater than τ_L. At this stage the oscillations are a non-equilibrium perturbation on the time evolution of the universe. If we introduce a general neutrino mass matrix, and include CP non-conservation in the lepton section we have the classic requirements for generating a lepton excess - a point which was made by Kholopov and Petrov.[3]

We have investigated the non-equilibrium problem in more detail, and find that the effect discussed by Kholopov and Petrov[3] in fact allows a different lepton excess for each species of lepton. This implies that it is possible for the different species of neutrinos to have different but appreciable chemical potentials. It was this circumstance we showed had significant effects on the ^4He generation in our earlier paper, provided that $t_0 << \tau$ near $t \sim \tau^*$, as is the case in figure 1.

The condition that $t_0^* < \tau_L^*$ for oscillations to be potentially interesting requires

$$\Delta m^2 \geq G_N/G_F^2 \sim 10^{-9} eV^2 \tag{4}$$

showing that cosmology is able to probe very small mass differences between neutrinos.

In this paper we will first set up the formalism, and then describe in detail the non-equilibrium processes which generate these interesting conditions. Finally we present our results on nucleosynthesis of light elements when these conditions prevail near the decoupling time.

II. FORMALISM

We introduce a set of weak eigenstates $\nu_{\alpha L}$ which participate in the usual left-handed weak interactions, appearing in the charged current as $\bar{\ell}_\alpha \gamma_\mu (1+\gamma_5)\nu_{\alpha L}$ where ℓ_α is the charged lepton and $\alpha = e,\mu,\tau,\ldots,\ell_{n_f}$. In addition we allow for right-handed neutrinos

$\nu'_{\alpha R}$ which participate in the much weaker right-handed weak interactions. The mass matrix is then of the general Majorana + Dirac form

$$= \sum_{\alpha,\beta} \Big[M_{\alpha\beta} \overline{(\nu_{\alpha L})} \nu^c_{\beta L} + m_{\alpha\beta} \overline{(\nu'_{\alpha R})}^c \nu'_{\beta R}$$

$$+ D_{\alpha\beta} \Big\{ \overline{(\nu_{\alpha L})} \nu'_{\beta R} + \overline{(\nu'_{\beta R})}^c \nu^c_{\alpha L} \Big\} + \text{h.c.} \Big] \quad . \qquad (5)$$

The diagonalization of mixes $\nu_{\alpha L}$ and $\nu_{\beta L}$, and also $\nu_{\alpha L}$ and $(\nu'_{\beta R})^c$. In the usual terminology not only are the left-handed neutrinos mixed, but the left-handed neutrinos are also mixed with left-handed antineutrinos.

The mass eigenstates are Majorana particles. If we designate these eigenstates by N_i where $i = 1,2,\ldots,2n_f$, and set

$$\nu_{(\alpha+n_f)L} = (\nu'_{\alpha R})^c \quad , \qquad (6)$$

then we can regard the ν_γ as obtained from the mass eigenstates by a unitary transformation

$$\nu_{\gamma L} = U_{\gamma i} N_i \quad . \qquad (7)$$

Oscillations, of both the first and second classes, may then be obtained in the usual way.[2]

It is, however, more convenient to introduce the density matrix $\underline{\underline{f}}(t)$, which is such that $\langle\alpha|\underline{\underline{f}}(t)|\alpha\rangle$ is the number of neutrinos in the state α at time t, given the inital number of neutrinos $\langle\alpha|\underline{\underline{f}}(0)|\alpha\rangle$ of each species. To allow for expansion of the universe, and for collisions we replace the usual evolution equation for $\underline{\underline{f}}$

$$\frac{d\underline{\underline{f}}}{dt} = i[\underline{\underline{f}},\underline{\underline{H}}_0] \quad , \qquad (9)$$

(where $\underline{\underline{H}}_0$ is the Hamiltonian for free neutrinos) with the Boltzman type of equation

$$\frac{d\underline{\underline{f}}}{dt} - \frac{3\dot{R}}{R} \underline{\underline{f}} = i[\underline{\underline{f}},\underline{\underline{H}}] + \underline{\underline{C}}[\underline{\underline{f}}] \quad , \qquad (10)$$

where the matrix $\underline{\underline{C}}$ describes the collision terms, which tend to maintain $\underline{\underline{f}}$ in the equilibrium distribution. Our attempts to construct detailed interesting solutions to (10) are still in progress (perhaps to be reported at Telemark III?), but we believe that some qualitative statements can be made, following Kholopov and Petrov.[3] This we now do.

III. THE NON-EQUILIBRIUM PHASE

We first note that the right-handed neutrinos interact much more weakly than the left-handed neutrinos, and so decouple at an earlier time τ_R^*. As a result they do not participate in any subsequent heating of the universe that occurs because of annihilation of particles and antiparticles of mass m as T drops below m. Thus T_R, the temperature of the right-handed neutrinos will be related to T_L by

$$\frac{T_R}{T_L} = \left(\frac{4}{11}\right)^{s/3}$$

where s is the number of species which have annihilated between τ_R^* and t.

If $\tau_R^* << \tau_L^*$ then $T_R << T_L$ and the number density of right-handed neutrinos (or left-handed antineutrinos) will be much less than that of the left-handed neutrinos. This is the driving force which, coupled with oscillations, produces a net neutrino number for the universe.

In general, the neutrino densities are given by

$$n[\nu_{\alpha L}] = \sum_{\gamma=1}^{2n_f} \langle n[\nu_{\gamma L}] P(\nu_{\gamma L} \to \nu_{\alpha L}) \rangle_E \tag{11}$$

where the average is over the energy distribution of the neutrinos and $P(\nu_{\gamma L} \to \nu_{\alpha L})$ is the transition probability which strictly should be obtained from the Boltzman equation (10), or an equivalent master equation. Since we are assuming $T_R << T_L$,

$$n[\nu_{\alpha+n_f,L}] << n[\nu_{\alpha,L}] \tag{12}$$

and we may truncate the sum in (12) at n_f, obtaining

$$n[\nu_{\alpha,L}] = \sum_{\gamma=1}^{n_f} \langle n[\nu_{\gamma,L}]\, P(\nu_{\gamma L} \to \nu_{\alpha L}) \rangle_E \tag{13}$$

However, unitarity requires

$$\sum_{\gamma=1}^{2n_f} P(\nu_{\gamma L}\to\nu_{\alpha L}) = 1 \tag{14}$$

which allows

$$\sum_{\gamma=1}^{n_f} P(\nu_{\gamma L}\to\nu_{\alpha L}) \neq 1 \tag{15}$$

in the presence of second class oscillations. Thus even if we start with $n[\nu_{\gamma L}] = n[\nu_{\alpha L}]$ for $\alpha,\gamma, \leq n_f$, equation (13) does not preserve this equality. We thus obtain our first lemma:

Lemma 1: The number densities of the different neutrino species will in general be different in the presence of neutrino oscillations of the second class.

The antiparticle densities $n[(\nu_{\alpha L})^c]$ satisfy an equation analogous to (13),

$$n[(\nu_{\alpha L})^c] = \sum_{\gamma=1}^{n_f} \langle n[(\nu_{\gamma L})^c]\; P\;((\nu_{\gamma L})^c\to(\nu_{\alpha L})^c)\rangle_E \tag{16}$$

In general the initial conditions will be such that $n[(\nu_{\gamma L})^c] = n[(\nu_{\gamma L})]$; there is no net lepton number of the universe (see however, Harvey and Kolb[4] for a scenario where this is not true in general).

If however, we have CP violation as well as second class oscillations then

$$P(\nu_{\gamma L}\to\nu_{\alpha L}) \neq P((\nu_{\gamma L})^c\to(\nu_{\alpha L})^c) \tag{17}$$

Let us introduce the net lepton density of the α species, L_α, defined by

$$L_\alpha = n[\nu_{\alpha L}] - n[(\nu_{\alpha L})^c] \tag{18}$$

which then satisfies the equation

$$L_\alpha = \langle \sum_{\gamma=1}^{n_f} L_\gamma P(\nu_{\gamma L}\rightarrow\nu_{\alpha L})$$

$$+ n[(\nu_{\gamma L})^c]\cdot[P(\nu_{\gamma L}\rightarrow\nu_{\alpha L}) - P((\nu_{\gamma L})^c\rightarrow(\nu_{\alpha L})^c)]\rangle_E \quad . \tag{19}$$

So even if, initially, $L_\gamma = 0$ and $n[(\nu_{\gamma L})^c] = n[(\nu_{\alpha L})^c]$ for γ and $\alpha \leq n_f$, the second class oscillations can develop $L_\gamma \neq 0$, and $L_\gamma - L_\alpha \neq 0$.

Thus we arrive at our second and third lemmas:

Lemma 2: In the presence of second class oscillations and CP violations, generations of appreciable net lepton numbers L_α is possible.

Lemma 3: In the presence of second class oscillations and CP violation, the lepton numbers of different neutrino species need not be equal.

Lemma 2 is the result of Khopolov and Petrov, but lemmas 1 and 3 are, to our knowledge, new.

We proceed to remark that if these oscillations occur while the left-handed neutrinos are still in thermal equilibrium with the universe, then these number distributions will be thermalized, and will thus be characterized by the equilibrium temperature T, and a chemical potential μ_α which will in general be nonzero and different for different species.

We note that the experimental limit on μ is

$$\xi_\alpha = \frac{\mu_\alpha}{T} \leq 60$$

as long as the neutrinos are light (of rest mass less than 1 meV). This limit applies to all species and is simply a consequence of requiring that the density of background neutrinos not exceed the critical density. Values of ξ_α or order 1 are not excluded by observational evidence.

After this phase $t \sim t_0^* > \tau_L$, the oscillation period decreases so that $t_0 \ll t, \tau_L$. At this stage the oscillations simply develop an equilibrium distribution characterized by

$$[\underline{\underline{f}},\underline{\underline{H}}] = 0 \ , \tag{20}$$

to eliminate term of order t_0^{-1} which would otherwise dominate the Boltzman equation, and

$$\underline{\underline{C}}[\underline{\underline{f}}] = 0 \tag{21}$$

which eliminates the term in τ_L^{-1}.

(20) requires that $\underline{\underline{f}}$ the diagonal in the mass basis, and (21) that $\underline{\underline{f}}$ be an equilibrium distribution. We examine the form and consequences of this equilibrium distribution in the next section.

IV. THE EQUILIBRIUM SITUATION

We have already described the circumstances in which $\underline{\underline{f}}$ takes on the equilibrium form for the left-handed neutrinos. In general $\underline{\underline{f}}$ is diagonal in the mass basis, and

$$f_{ij} = f_i \delta_{ij} \tag{22}$$

with $$f_i \propto \frac{1}{e^{(E_i-\mu_i)/kT}+1} \tag{23}$$

with $\mu_i \neq \mu_j$.

It is then easy to compute the reaction rate, e.g., for $\nu_e n \to ep$, for which the T matrix element is T_{ee}, where the subscripts refer to lepton states in the weak interaction basis. Then

$$\Gamma(\nu_e n \to ep) = (T\underline{\underline{f}}T)_{ee} \tag{24}$$

$$= \sum_i |T_{ee}|^2 |U_{ei}|^2 f_i \quad . \tag{25}$$

If either the μ_i are different for different i, or there are second class oscillations, or both, then rates computed through (25) will differ from those without neutrino mixing.

We have set up the network equations up to ^{4}He and solved them for the light element abundances for a simple two neutrino species model, say ν_e and ν_μ, ignoring for this stage of the calculation the right-handed neutrinos. In this case there are three additional parameters in the problem and it should be no surprise that we can obtain good fits to the ^{4}He, ^{3}He and d abundances, even if the limit $N_\nu \leq 4$ is relaxed. For example, with $\Theta = 30°$, we obtain $Y_{He} \sim 0.21$ for $N_\nu = 6$ at $\eta = 3 \times 10^{-10}$ for $\xi_1 = 0.3$, $\xi_2 = 0.5$. These same parameters then give

$$\frac{X_d}{X_H} \sim 3 \times 10^{-5} \text{ and } \frac{X_{^3He}}{X_H} \sim 4 \times 10^{-5} ,$$

which are in good agreement with observations. Detailed results will be published elsewhere, but it should be obvious that there is now more than enough freedom in the parameters to fit the observed abundances without serious restriction on N_ν. Indeed, for nucleon to photon ratios smaller that 10^{-9} the sensitivity of X_d and $X_{^3He}$ to ξ_1 and ξ_2 is weak enough that any pair of values of ξ_1 and ξ_2 which fit $Y_{^4He}$ will also fit X_d and $X_{^3He}$.

V. CONCLUSIONS

We have shown that if second class oscillations and CP violations are introduced in the neutrino sector it is possible to generate a situation where the lepton species numbers L_α are appreciable and distinct at the time of nucleosynthesis. These are precisely the conditions under which neutrino mixing induced by neutrino oscillations can influence the nucleosynthesis process. These circumstances introduce many new parameters into the nucleosynthesis problem enabling one, for example, to fit the observed light element abundances without constraining the number of neutrino species.

REFERENCES

1. B. H. J. McKellar and H. Granek, Proc. Neutrino Mass Mini-Conference, Telemark 1980 (University of Wisconsin Report #186, edited by V. Barger and D. Cline) p. 165.

2. see, e.g., J. Leveille, ibid p. 133.

3. M. Yu Kholopov and S. T. Petcov, Phys. Lett. 99B, 117 (1981), erratum 100B, 520 (1981).

4. J. Harvey and E. W. Kolb, Phys. Rev. D24, 2090 (1981).

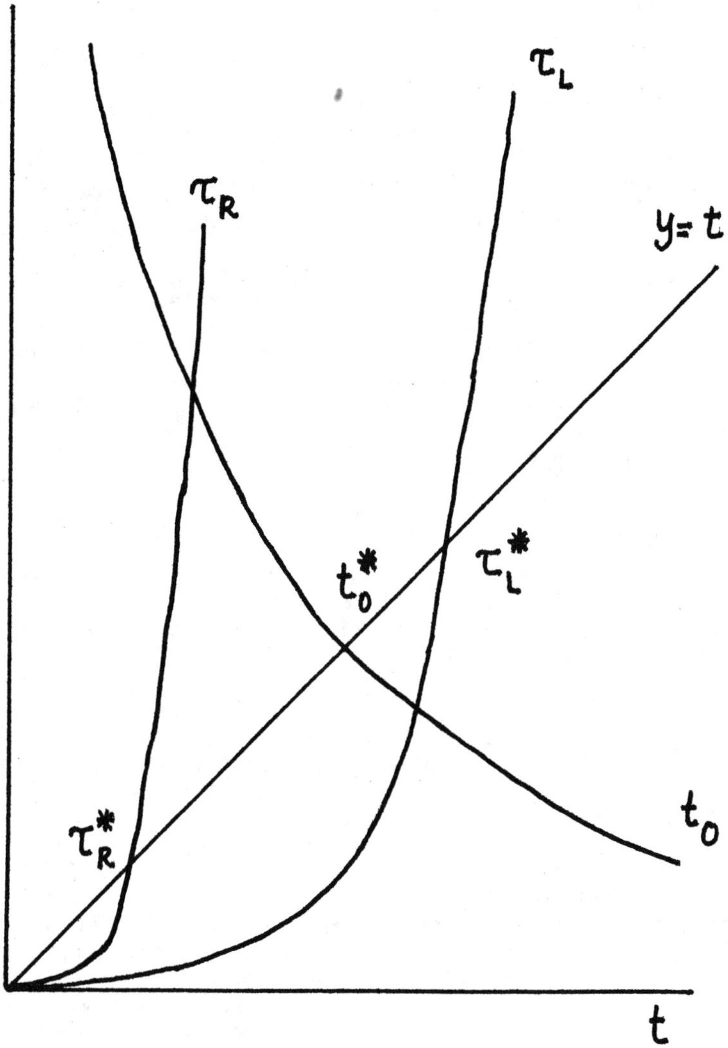

Fig. 1: Development of the characteristic times associated with neutrinos with the evolution of the universe.

RESULTS FROM A FERMILAB NEUTRINO BEAM DUMP EXPERIMENT

R.C. Ball, C.T. Coffin, H.R. Gustafson, L.W. Jones, M.J. Longo,
T.J. Roberts[a], B.P. Roe, E. Wang
University of Michigan, Ann Arbor, MI 48109

C. Castoldi, G. Conforto
INFN, Florence, Italy

M.B. Crisler, J.S. Hoftun, T.Y. Ling, T.A. Romanowski, J.T. Volk
Ohio State University, Columbus, OH 43210

S. Childress[b]
University of Washington, Seattle, WA 98195

M.E. Duffy, G.K. Fanourakis, R.J. Loveless, D.D. Reeder,
D.L. Schumann, E.S. Smith
University of Wisconsin, Madison, WI 53706

Presented by: Robert C. Ball

ABSTRACT

The flux of prompt neutrinos from a beam dump has been measured in an experiment at the Fermi National Accelerator Laboratory (E613). Assuming that the charm production has a linear dependence on atomic number and varies as $(1-|x|)^5\ e^{-2m_T}$, a model dependent cross section of 27±5μb/nucleon can be derived. For neutrino energies greater than 20 GeV, the flux of electron neutrinos with respect to muon neutrinos is 0.78±0.19. For neutrinos with energy greater than 30 GeV and $p_\perp$ greater than 0.2, the flux of $\bar{\nu}_\mu$ compared to ν_μ is 0.96±0.22.

INTRODUCTION

Prompt neutrinos are those neutrinos produced in the creation and subsequent semi-leptonic decay of charmed particles. In producing prompt neutrinos, there is a background from non-prompt neutrinos, those from the decay of pions and kaons also produced in the target. The goal of this experiment was to maximize the prompt neutrino flux with respect to the background, and measure it as a function of the neutrino energy E_ν, target atomic number A, and neutrino transverse momentum p_T. The relative flux of electron and muon neutrinos was also measured.

EXPERIMENTAL ARRANGEMENT

The layout of the experiment is shown schematically in FIgure 1. A beam of 400 GeV protons was incident on a (minimum) 3 interaction

[a] Presently with Bell Labs, Naperville, IL
[b] Presently with Fermilab, Batavia, IL

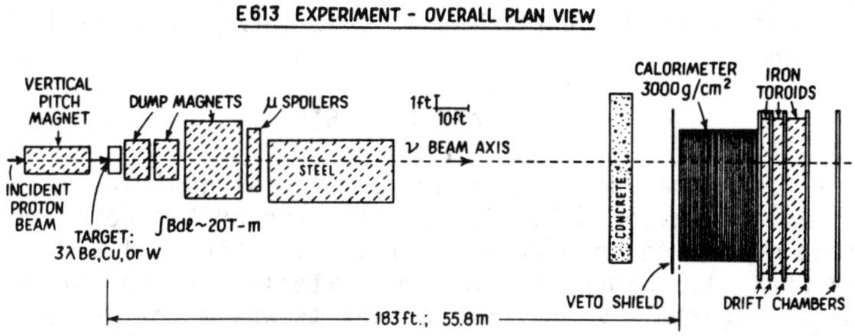

Fig. 1. Schematic layout of the experiment.

length tungsten target. The target was followed immediately by an 11 m magnetized iron shield which absorbed remaining strongly interacting particles and ranged out or swept aside muons aimed at the detector, which was 60 m from the target. The muon flux was further reduced to 3×10^5 per 2×10^{12} incident protons by 11 m of passive iron shield.

The detector consisted of a 3000 g/cm^2 calorimeter followed by a muon spectrometer of iron toroids interspersed with drift chamber planes. Data collection was triggered by interactions in the calorimeter which deposited sufficient energy to exceed established thresholds. The calorimeter was segmented into 30 modules longitudinally, and each module was followed by x and y proportional tubes read out in the proportional mode. The modules, lead and liquid scintillator sandwiches, were split vertically into 5 cells viewed by phototubes at each end. The signals from summed, overlapping sections of phototubes were used to form the trigger. The fiducial volume was confined to modules 3-25 with a transverse window 2.6 m wide by 1.0 m high. The calorimeter center was offset from beam center by 0.75 m in the long transverse dimension.

Earlier experiments at CERN[1-3] were performed at a distance of about 900 m from their copper production target, and subtended a maximum angle of 2 mrad. This experiment was at a distance of 60 m, and subtended angles up to 37 mrad. Since the non-prompt neutrinos were concentrated at smaller angles than the prompt, this helped to increase the ratio of prompt to non-prompt events over the earlier experiments' rates. Additionally we used a tungsten target which had a higher density than copper and a consequently smaller non-prompt background.

To control the neutrino background from upstream sources, a system of more than 30 beam line monitors was installed and maintained. They were calibrated by varying the beam pipe vacuum and by inserting known amounts of material into the beam. In this way, the beam line related background was determined to be less than 2.0% of the full density tungsten prompt signal.

Background from material near the target, such as air, vacuum windows, and proton monitors, was more serious but calculable. For our 1981 data period reported here, this background was 17% of the prompt signal.

DATA ANALYSIS

The technique used to extract the prompt neutrino signal was that of extrapolation. Data were collected on tunsgen targets of two different densities, nominally full density (full ρ) and 1/3-full density (1/3 ρ). Since for the same material the non-prompt signal increases linearally with inverse density while the prompt signal remains unchanged, extrapolation to $1/\rho = 0$ gives the prompt signal. The near-target background correction was made by correcting the nominal inverse density ratio from 3:1 to 2.49:1. This is shown in Figure 2 where the total number of events normalized by the number of incident protons is plotted for the two targets.

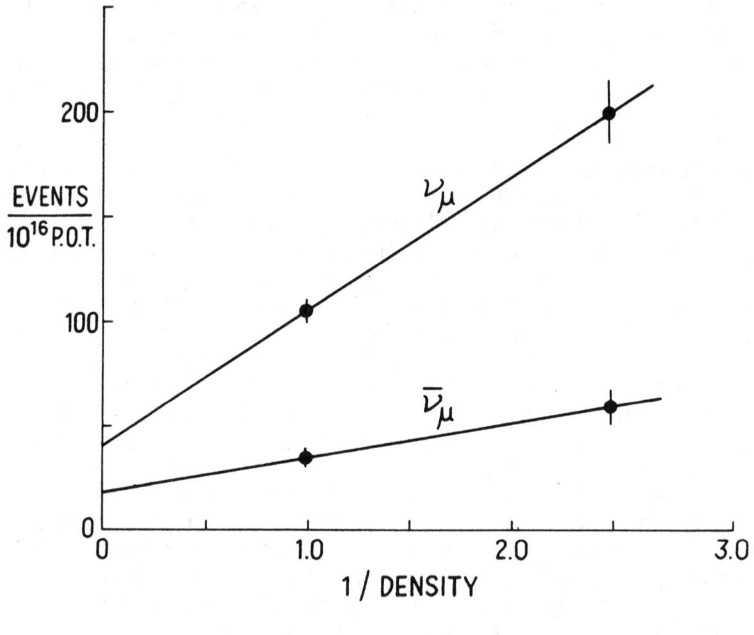

Fig. 2. Extrapolation to infinite density for ν_μ and $\bar{\nu}_\mu$ events.

During the run period, a total of 1.6×10^{17} protons were targeted on the two densities of tungsten. After correcting for the experiment live time (~70%) and discarding beam spills in which beam line backgrounds were high or the beam was mis-steered, a total of 6.6×10^{16} protons incident for the full density tungsten and 1.6×10^{16} protons for the partial density tungsten targets were kept. A total of approximately 300,000 triggers were written to magnetic tape during this time. Approximately 1/3 of these triggers resulted from cosmic rays, and the remainder were beam related, mostly due to muons which interacted in the floor or concrete roof shielding and showered into the calorimeter. Effects of the cosmic rays were monitored by triggering with no beam on target for a time equal to that in which there was beam. These "beam-off" triggers were treated by the analysis in the same way as all other triggers. The total number of ν_μ charged current interactions found in the fiducial volume and successfully momentum fit was 854. The number of 0μ events, which includes ν_e charged current (CC) events along with ν_e and ν_μ neutral current (NC) events, based on a slightly smaller sample of protons, was 752.

RESULTS

Three results were derived from this data, the cross section for production of $D\overline{D}$ pairs, the ratio of $\overline{\nu}_\mu$ to ν_μ fluxes, and the ratio of ν_e to ν_μ charged current interactions. The first result is very model dependent, the latter two are not.

To derive the cross section, the full and partial density events were divided into bins in E_ν and θ_ν so that the prompt signal could be extracted and the acceptance of the detector could be calculated for each bin. It was then assumed[4,5] that the neutrino production varied as $(1-|x|)^n e^{-ar_\perp}$ where x is Feynman x, n is an integer, $r_\perp$ is either $p_\perp$ or $m_\perp=(p_\perp^2+M_D^2)^{1/2}$, and a is a variable describing the $r_\perp$ dependence. In addition, cascading of the beam protons in the target is described using a mean proton elasticity (ε) of 0.3 with energy dependence of s^k where $s=(2M_pE_{LAB})^{1/2}$. J. Leveille[5] suggests the best value of k is 1.3 and urges the use of $r_\perp=m_\perp$. The semileptonic branching ratio ($D\to\mu$) is taken as the average of the D^+ and $D°$ branching fractions, 8.2%[6,7]. Table I shows the results reported for the various experiments[1,2,8,9] along with their model assumptions and then corrects all results to a common model using n=5, a=2, ε=0.3, k=1.3 and $r_\perp=m_\perp$, which we have found gives a better fit to our data and gives a cross section for $D\overline{D}$ production of 27±5μb. Figures 3 and 4 show the data plotted as functions of E_ν and $p_\perp(\nu)$ for E_ν>20 GeV and θ_ν<37 mrad. The data have been corrected for trigger efficiency, muon acceptance by the toroids, and the incomplete azimuthal acceptance for neutrinos. The flux of antineutrinos compared to the flux of neutrinos from the target, restricted to E_ν>30 GeV/c and $p_\perp$>0.2 GeV where systematics are less severe, is found to be 0.96±0.22.

TABLE I
Cross Sections for $D\overline{D}$ Production Quoted by Various Experiments

Group	$\sigma(D\overline{D})$	Model Parameters n	a	k	ε	$\sigma(D\overline{D})$ n=5, a=2, k=1.3, ε=0.3, $m_\perp$
BEBC	17±4 (ν_e) 30±10 (ν_μ)	3	2	0.5	2/3	46±17 81±27
CHARM	19±6	4	2	--	--	29±9
CCFRS	13±1	3	2	1.3	0.3	25±2 (n=6, a=2.5, $p_\perp$)
E-613	27±5	5	2($m_\perp$)	1.3	0.3	27±5

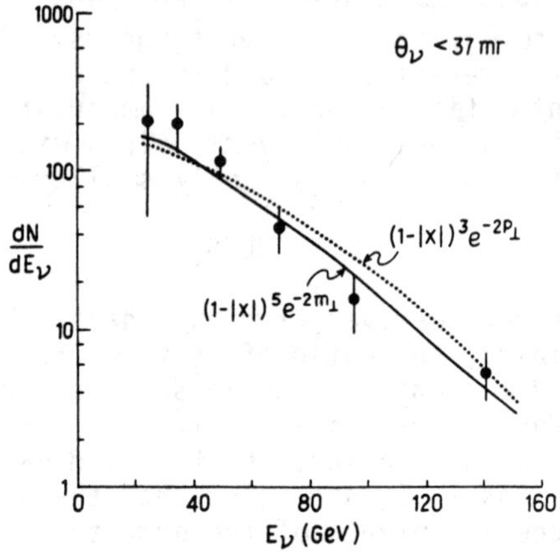

Fig. 3. Number of prompt neutrino events per 10 GeV plotted against E_ν, with the predictions of two models superimposed. The data are corrected for trigger efficiency, muon acceptance, and incomplete azimuth.

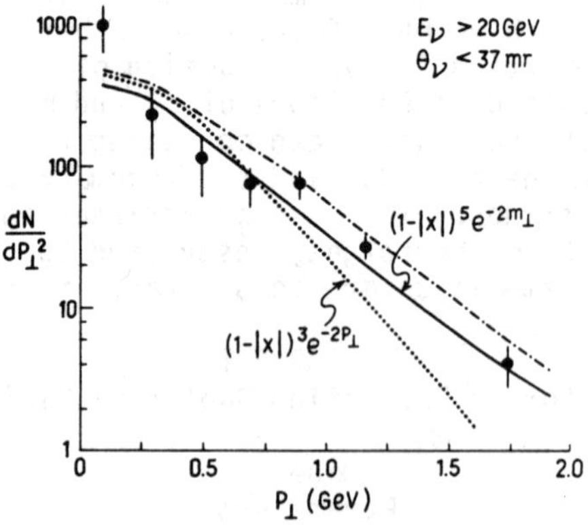

Fig. 4. Number of prompt neutrino events per 0.2 GeV/c plotted vs. $p_\perp$. The dot-dashed line indicates the model prediction (n=5, a=2) with no energy or angular restrictions imposed.

The method used to determine the ratio $R=\nu_e(CC)/\nu_\mu(CC)$ took advantage of the large difference in size between electromagnetic showers and hadronic showers in our calorimeter which was due to the use of Pb plates rather than iron or marble. Each module was 14.4 radiation lengths but only 0.5 interaction lengths. In the future, this feature will be used to make a direct separation of $\nu_e(CC)$ events from the neutral current contamination. For now it has been used to make a probabilistic determination of the neutrino energy of the 0μ events (those events with no visible final state muon). Cosmic rays were then subtracted, the prompt signal was extracted, and the result was normalized to the number of incident protons. The resulting distribution included $\nu_e(CC)$ events as well as $\nu_e(NC)$ and ν_μ (NC) events. These latter two were subtracted by using the normalized hadronic energy distribution of the $\nu_\mu(CC)$ events. The direct separation method described above gives results which agree with the method used here.

The result of this analysis is that for $E_\nu > 20$ GeV, the ratio $R=\nu_e(CC)/\nu_\mu(CC)=0.78\pm0.19$. For $E_\nu>20$ GeV the value $R=1.0\pm0.3$, and for $E_\nu>40$ GeV, $R=1.1\pm0.4$. This result is shown in Figure 5. The region where the ratio R significantly deviates from unity is also the region in which the systematics, such as event finding and muon reconstruction, are most severe and can most easily distort the results.

ACKNOWLEDGEMENTS

We wish to acknowledge the efforts of the Fermilab staff and in particular the Meson Lab personnel for their efforts in our behalf. The support of the technical staff at our various institutions has been invaluable. This work was supported in part by the U.S. National Science Foundation and by the U.S. Department of Energy.

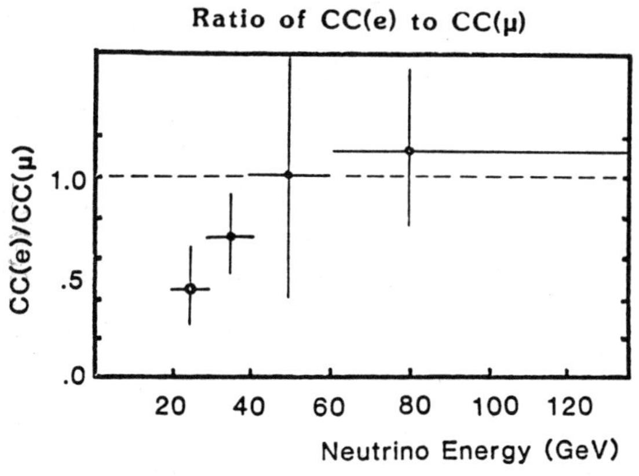

Fig. 5. The ratio of prompt $\nu_e(CC)$ over $\nu_\mu(CC)$ events vs. energy (preliminary).

REFERENCES

1. P. Fritze et al., Phys. Lett. 96B, 427 (1980).
2. M. Jonker et al., Phys. Lett. 96B, 435 (1980).
3. H. Abramowicz et al., Z. Physik C 13, 179 (1982).
4. C. Michael, Proceedings of the Fourteenth Recontre de Moriond, Les Arcs (Savoie), France, March, 1979, Vol. I Editions Frontieres, Dreux, France (1979), edited by J. Tran Than Van.
5. J. Leveille, University of Michigan Preprint, UM HE 82-20.
6. J.M. Feller et al., Phys. Rev. Letters 40, 274 (1978).
7. W. Bacino et al., Phys. Rev. Letters 43, 1073 (1979).
8. A. Bodek, Proceedings of Neutrino '82, Balatonfüred, Hungary, June, 1982, edited by A. Frenkel and L. Jenik, p. 109.
9. A. Bodek et al., talk presented by J.L. Ritchie at the XIII International Symposium on Multiparticle Dynamics, Volendam, Netherlands, June, 1982.

PLANS FOR A HIGH-RESOLUTION MEASUREMENT OF THE TRITIUM BETA-SPECTRUM END POINT TO DETERMINE THE NEUTRINO MASS*

R.L. Graham, M.A. Lone, H.R. Andrews, J.S. Geiger
J.L. Gallant, J.W. Knowles, H.C. Lee and G.E. Lee-Whiting

Chalk River Nuclear Laboratories
Chalk River, Ontario, Canada, KOJ 1JO

ABSTRACT

The Chalk River $\pi\sqrt{2}$ iron-free beta spectrometer is being recommissioned and upgraded for a precise measurement of the shape of the tritium spectrum near the end point. With a multiple strip source and 60-element detector array an overall energy resolution of $\leq$ 19 eV FWHM is expected. Computer simulations of the expected experimental Kurie plots are presented for various anti-neutrino mass assumptions.

The report of a non-zero mass for the electron anti-neutrino by Lubimov et al.[1] has stirred interest in many laboratories throughout the world including ours at Chalk River. Since the world's largest iron-free beta spectrometer[2] is at our disposal, an experiment to remeasure the shape of the tritium beta spectrum is planned.

Our first step was to reassess the data of Lubimov et al. A computer program has been developed[3] to simulate experimental spectrum shapes, Kurie plots etc., under various assumptions e.g. resolution setting, response function, source strength, background counting rate, neutrino mass and the degrees of excition of various final ^{3}He atomic states. With this program, using the response function shown in ref. 4, and the other parameters cited by Lubimov et al.[1], it was found that the data in their paper are consistent with a neutrino rest mass of about 30 eV assuming that the ^{3}He atoms are excited by 43 eV in 30% of the decays e.g. the atomic approximation of Bergkvist[5]. A more precise analysis to establish error limits was not attempted since it is difficul to extract accurate data from their published Kurie plot and there is som doubt as to the validity of an atomic approximation for ^{3}H atoms bound to valine as indicated by the calculations of Ching and Ho[6].

For our experiment we are recommissioning the Chalk River spectrometer[2] which is located in the non-magnetic building, shown in Fig. 1. Since this type of spectrometer has a rather small transmission at high resolution, a number of improvements are planned:-

a) the single proportional counter shown in Fig. 2 will be replaced by a 60 - element proportional counter array having coincidence and anti-concidence features to minimize the ambient background counting rate,

b) a multi-strip source similar to that used by Bergkvist[5]. The initial plan is for 180 strips 10 cm long by 1 mm wide of titanium hydride (≈ 1 μg/cm^2) on quartz backing plates, and

c) computer control of spectrometer settings and data acqusition.

Calculations are in progress to determine the optimum geometry for the source array, detector array and baffle arrangement. With 1 mm wide defining apertures for each counter element we should realize an overall momentum resolution of dp/p = 0.05%, i.e. a full width at half maximum of 18.6 eV at 18.6 keV. The properties of our expected spectrometer performance for the neutrino experiment are compared in Table 1 with those of others used for this experiment.

It is clear that the limiting factor in our experiment will be source thickness. We hope to be able to produce titanium tritide source layers as thin as 1 μg/cm^2. Experiments to determine feasibility are in progress.

A major concern is, of course, the energy shifts in the beta spectrum components due to the excitation of the ^{3}He atom. While the situation for the free tritium atom can be calculated with some accuracy[5], that for tritium atoms bound in solids is less clear[6]. Also constancy of source quality during the experiment is essential; one must know the energy loss contribution to the energy response function to be able to analyse the resultant experimental data in a quantitative way.

We are advised by New England Nuclear that tritiated organic compounds such as valine, are unstable under beta bombardment and are not recommended for sources of high specific activity.

We have used the simulation program to give us some idea of how different degrees of excitation of the ^3He atom will affect the resultant spectrum shape for different assumed neutrino masses. The Kurie plots shown in Fig. 3 have arbitrarily been normalized (intensity and slope) at 18.05 keV. These simulations are for the experimental conditions we expect to achieve in practice. The upper part of Fig. 3 shows Kurie plots for zero neutrino mass and excitation of a 40 eV level in ^3He of 0%, 20% and 40%. In the lower part the change in normalized shape is shown assuming no ^3He excitation and several assumptions of neutrino mass.

To make comparison easier to visualize, we now take the zero mass zero excitation as our reference and plot differences, delta K, for several mass excitation assumptions in Fig. 4. One notes that if nature is kind enough to give us only one excited state in ^3He at 40 eV then there is hope of being able to determine both the degree of excitation and neutrino mass given good statistics near the end point. The error bars shown are for a source strength of 370 MBq (10 MCi) an assumed background rate of only 0.1 cpm and a counting period of one day.

These simulations show that it would be advantageous to have a priori knowledge of the energies and degree of population of the final ^3He atomic states before attempting to extract the neutrino mass from the experimental data. Moreover it brings home the need for minimizing counter background both from cosmic rays and from scattered beta rays.

REFERENCES

1. V.A. Lubimov et al., Phys. Lett. 94B (1980) 266.
2. R.L, Graham, G.T. Ewan and J.S. Geiger, Nucl. Instr. and Methods 9 (1960) 245.
3. R.L. Graham, AECL-7683 section 2.21, 1982 July, to be published.
4. E.F. Tretyakov et al., Izv. Akad, Nauk SSSR 40, 2026.
5. K.E. Bergkvist - Nucl. Phys. B39 (1972) 317.
6. Ching Cheng-rui and Ho Tso-hsia, Preprint ASITP-81-008.
7. T. Okshima, Tokyo report INS-Rep. 406, March 1981, K. Kageyama et al., CYRIC Annual Report 1981.

TABLE 1

COMPARISON OF BETA SPECTROMETERS

PROPERTY	STOCKHOLM[5)]	MOSCOW[1)]	TOKYO[7)]	CHALK RIVER
Type	$\pi\sqrt{2}$	Toroidal	$\pi\sqrt{2}$	$\pi\sqrt{2}$
Radius (cm)	50	21.3	75	100
Momentum res. (%)	0.11	0.12	0.03	0.05
Source elements	85	18	?	180
Element height(cm)	10	1.8	?	10
width (cm)	0.17	0.26	?	0.1
Total Area (cm^2)	145	8.4	2	180
Transmission (%)	0.5	0.83	0.10	0.10
Dispersion (mm)	2000	3900	3000	4000
Det. slit width (mm)	2.0	3.0	≈ 2[(a)]	1.0
No. det. channels	1	3	≈ 60[(a)]	60
Overall res. FWHM(ev)	55	45	≈ 25[(a)]	19
Det. range (ev)	55	3x45	1450[(a)]	1120

(a) using a single wire position sensitive proportional counter. The effective slit width of 2 mm is an estimate[7)].

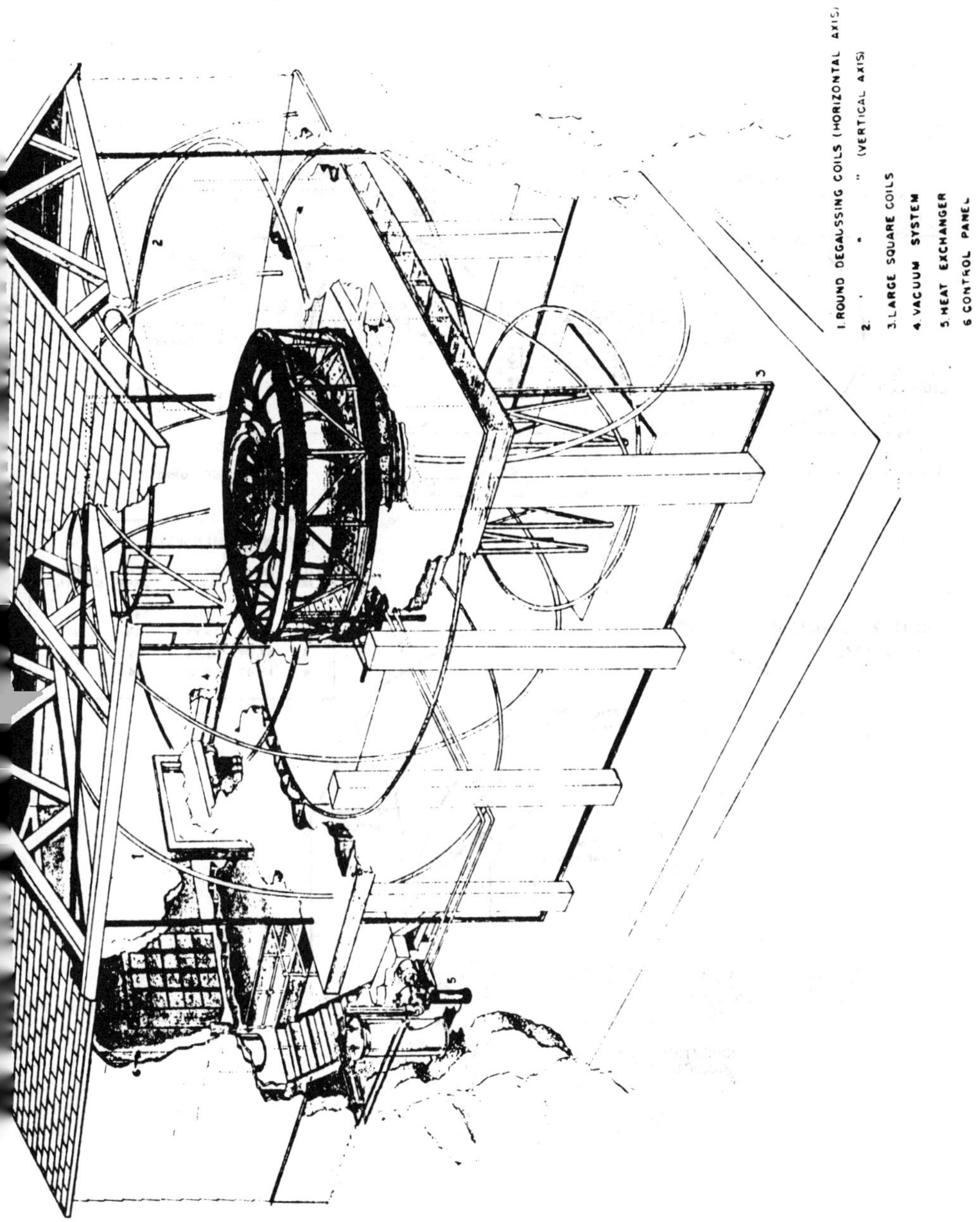

Fig. 1 Cutaway sketch of the iron-free building housing the Chalk River $\pi\sqrt{2}$ beta-ray spectrometer. The largest degaussing coils are 8.4 m in diameter.

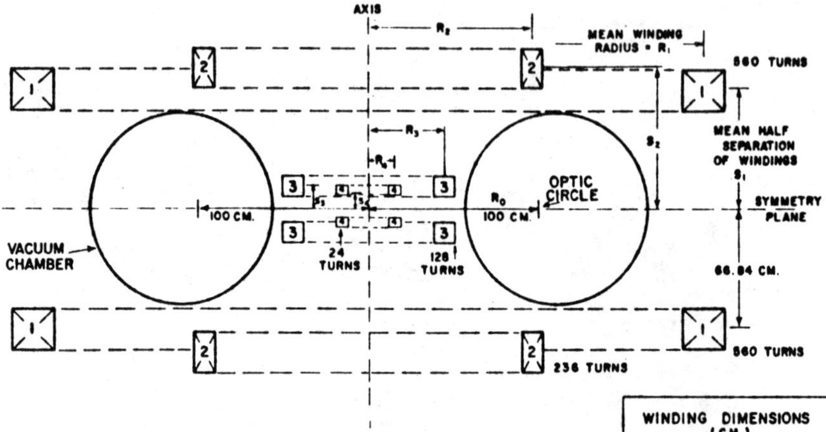

WINDING DIMENSIONS (CM.)

COIL NO	MEAN RADIUS R	HALF SEPN. S	NO. OF TURNS	CURRENT DIRECTION
1	197.55	66.84	560	+
2	96.67	79.77	236	+
3	45.06	12.79	128	−
4	15.47	8.88	24	−

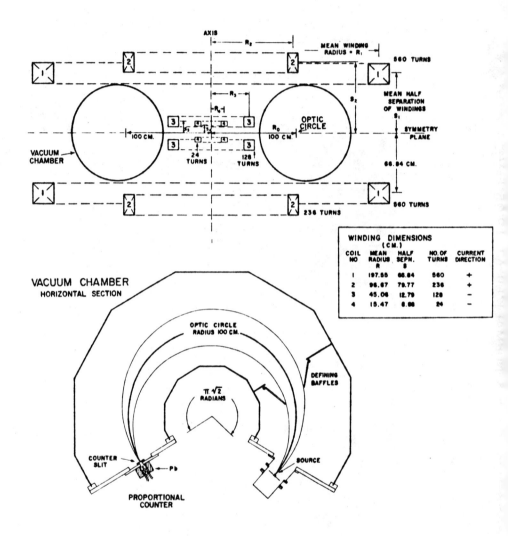

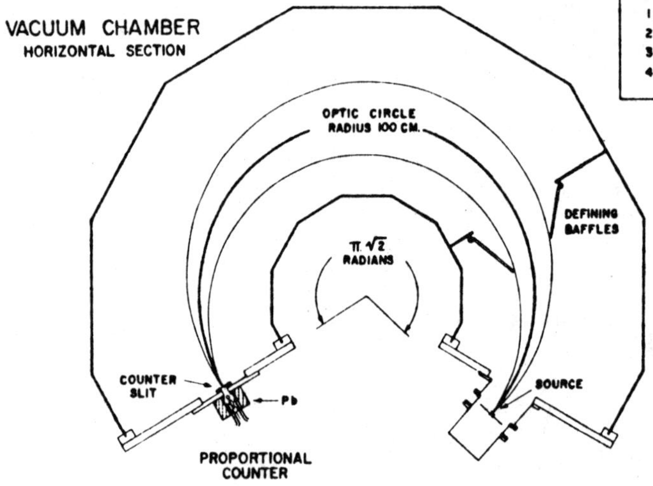

Fig. 2 The upper part of this figure shows a vertical section of the spectrometer coils and vacuum chamber. The lower part shows a plan section of the vacuum chamber. The single strip source and single counter will be replaced by a multi-strip source and a multi-element counter array.

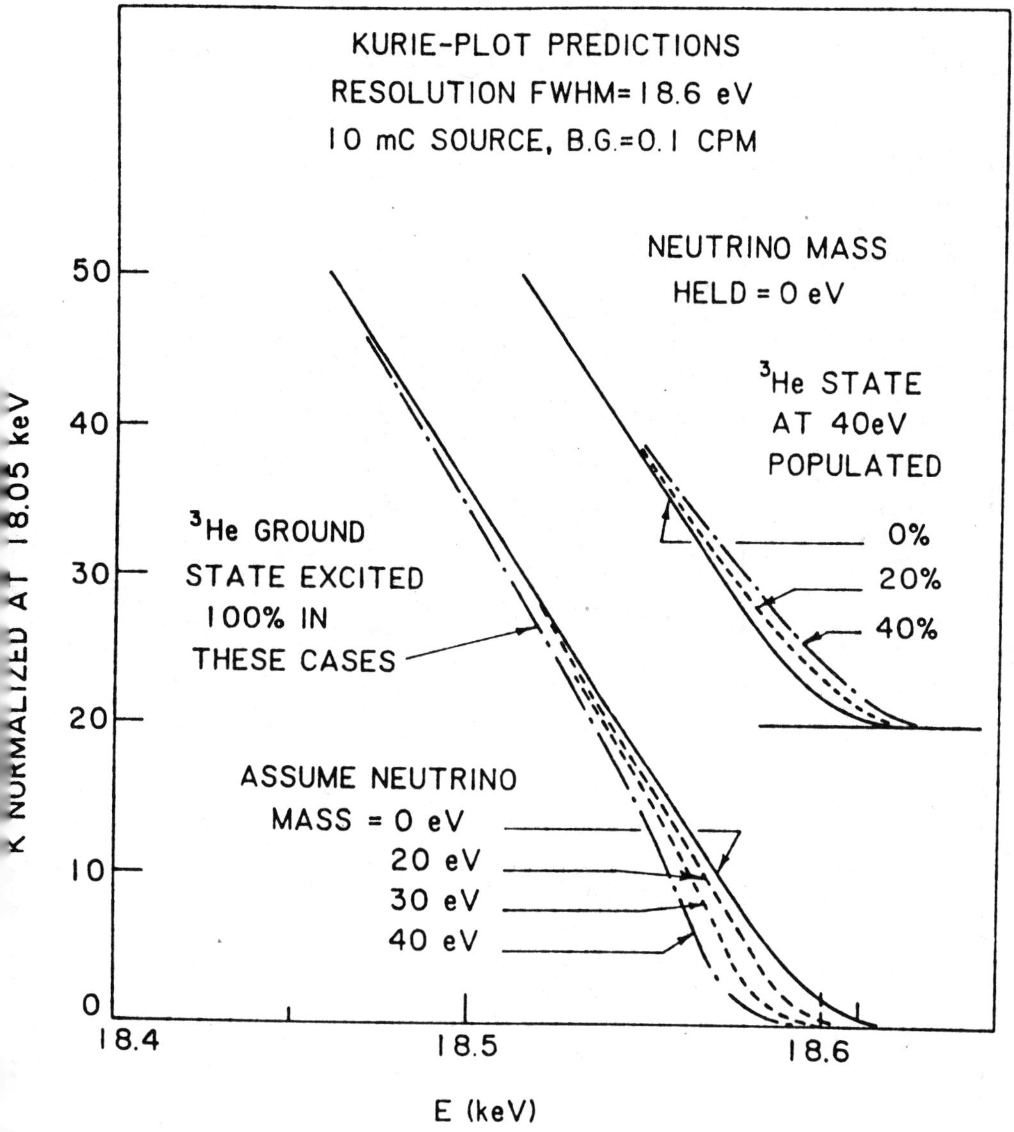

Fig. 3 Predictions of the shape of the Tritium beta-spectrum Kurie's plot near the end point for the various assumptions discussed in the text.

Fig. 4

In these plots we show the differences in shape of Kurie plots near the endpoint under various assumptions. In each case the horizontal straight line corresponds to the zero neutrino-mass 100% ground-state excitation case shown in Fig. 3. The ΔK scale here is in the same units as for the K scale in Fig. 3.

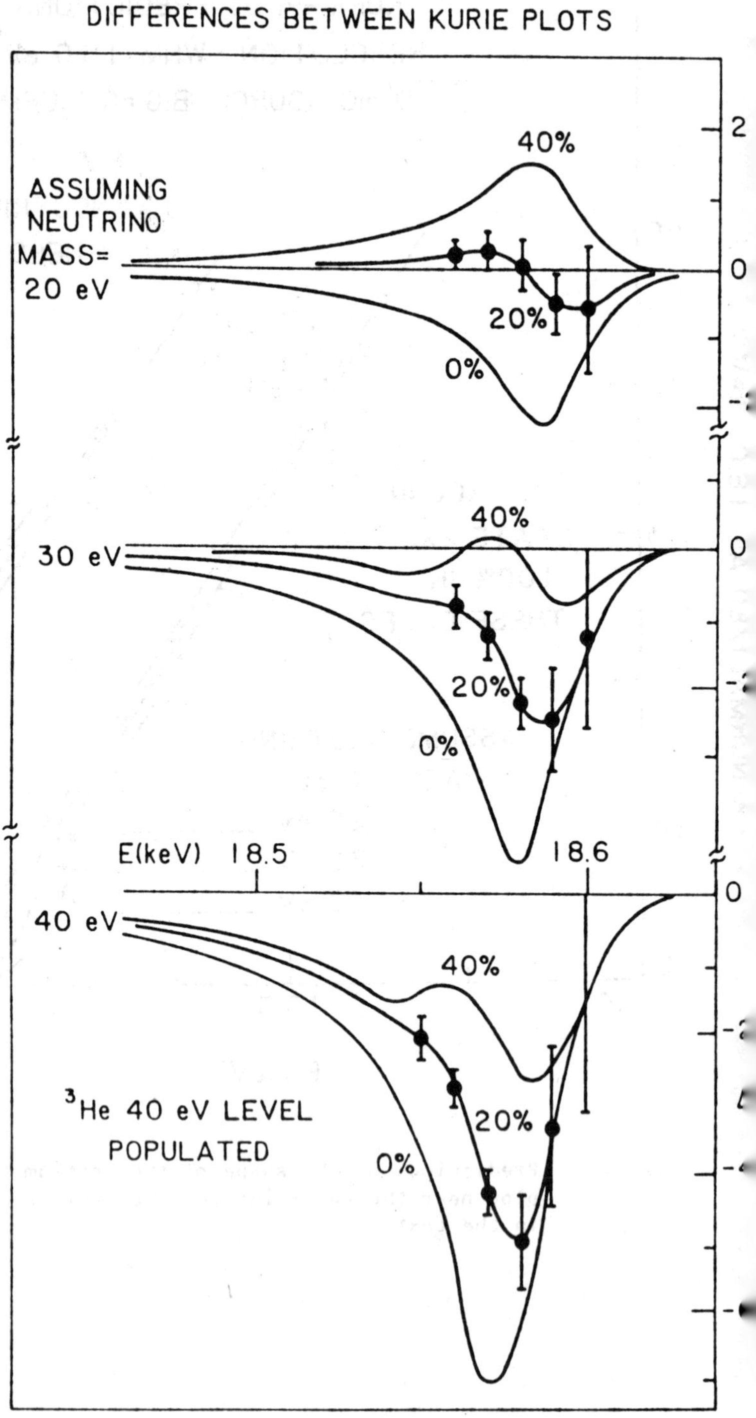

MATTER - ANTIMATTER OSCILLATIONS AND NEUTRINO MASS

Goran Senjanović
Brookhaven National Laboratory, Upton, New York 11973, U.S.A.

ABSTRACT

A discussion of neutron-antineutron (n-$\bar{n}$) and hydrogen-antihydrogen (H-$\bar{H}$) transitions is presented. An $SU(2)_L \times U(1) \times SU(3)_c$ model with spontaneously broken global B-L symmetry is shown to predict the interesting connection between oscillation times $T_{n-\bar{n}}$, $T_{H-\bar{H}}$, neutrino mass and the mass of a doubly charged Higgs scalar. A case of B-L as a gauge symmetry is discussed in the context of $SU(2)_L \times SU(2)_R \times U(1)_{B-L} \times SU(3)_c$ gauge model, with the emphasis on matter oscillations. Finally, an analysis of Higgs mass scales in GUTS and their impact on such processes is offered.

I. INTRODUCTION

Now that the idea of grand unification[1] has become so well established, one of the main questions that is poses is which kind of baryon number (B) and lepton number (L) violating processes is to be observed in the near future. A few years ago Weinberg and Wilczek and Zee have offered[2] a model independent way of analyzing that question based on the short distance expansion of the leading effective baryon number violating operators. If an assumption of a single mass scale beyond M_W is made, the leading operators are those of the lowest dimension and they take the necessary form which conserves B-L

$$O = \frac{1}{M_X^2} (qqq\ell) \tag{1}$$

where q stands for quarks and ℓ for leptons and M_X is the large scale beyond M_W. As is well known, these operators correspond to the $\Delta B = 1$, B-L conserving nucleon decay $n \to \pi^- + e^+$, $p \to \pi^0 + e^+$ and the limits on nucleon stability[3] $\tau_p \gtrsim 10^{31}$ yr imply $M_X \gtrsim 10^{14}$-10^{15} GeV. Using Georgi, Quinn and Weinberg program[4] in GUTS with a single scale above M_W, such as SU(5), M_X is predicted from unification constraints to be between 10^{14}-10^{15} GeV; a spectacular prediction to be tested in the near future.

If the above is true, we will encounter the unprecedented situation in the history of physics: an enormous "desert" in energies between 100-10^{14} GeV is predicted in which no new physics is to be found! No matter how exciting the prediction of proton decay is, this would certainly make the future of particle physics rather grim. Various scenarios that suggest filling the desert with one or more oasis, such as technicolor[5] or supersymmetric GUTS[6] have come forward, however, yet without full success. A more modest approach would be to search for an oasis within the context of conventional grand unification, an approach that I will follow here in my talk and an approach more suitable to the character of this conference.

To understand better the emergence of a desert and the relevance of B-L symmetry, we will start with the discussion of B and L nonconservation in the standard $SU(2)_L \times U(1) \times SU(3)_c$ model. Although perturbatively the above quantum numbers are absolutely conserved, 't Hooft has demonstrated that the nonperturbative, instanton induced effects, lead to the violation of B and L, but still conserve B-L global symmetry. Again, B-L emerges as the relevant symmetry! It appears then natural to study the question of an oasis in the desert through the properties of B-L opertor and this study comprises the main part of this paper.

We start first with the discussion of B-L conservation in the standard model, as presented in Section II. If, as it is natural in gauge theories, B-L is broken spontaneously, an interesting possibility of hopefully observable n-$\bar{n}$ oscillations (which violate B-L) and H-$\bar{H}$ oscillations (which conserve B-L) appear in the context of the extended $SU(2)_L \times U(1)$ model. The relevant transition times, although not predicted in this model, become related to the neutrino mass and the mass of the doubly charged Higgs scalar. This model is easily extended to simple GUTS, such as SU(5). Next, in Section III, a different possibility, in which B-L becomes a part of a guage electroweak group $SU(2)_L \times SU(2)_R \times U(1)_{B-L}$ is briefly reviewed. These left-right symmetryic models have been extensively analyized in the past and we cover them here from the point of view of B-L breaking, with the emphasis on n-$\bar{n}$ and H-$\bar{H}$ transitions. In Section IV we discuss the question of oscillation times expected in simple GUTS. This issue is intimately tied up to the question of neutral Higgs mass scales in GUTS. It is agreed that a proper treatment of Higgs boson masses renders these processes unobservable, at least in simple grand unified models. Finally, in Section V a summary of this peper is offered.

§II. THE STANDARD MODEL AND GLOBAL B-L SYMMETRY BREAKING

Let us first briefly recall the main features of the minimal $SU(2)_L \times U(1) \times SU(3)_c$ model which will also be useful in setting our notations,

Fermions

$$\psi_L = \binom{\nu}{e}_L \; (2, -1, 1_c) \quad ; \quad e_R \; (1, -2, 1_c) \tag{2.1}$$

$$Q_L = \binom{u}{d}_L \; (2, 1/3, 3_c) \quad ; \quad u_R \; (1, 4/3, 3_c), \quad d_R (1, -2/3, 3_c)$$

Higgs

$$\phi = \binom{\phi^+}{\phi^0} \quad (1, 2, 1_c) \tag{2.2}$$

The most general Yukawa interaction is

$$L_Y = h_u \bar{Q}_L i\tau_2 \phi^* u_R + h_d \bar{Q}_L \phi \, d_R + h_e \bar{Q}_L \phi \, e_R + \text{h.c.} \tag{2.3}$$

As encouraged by experiment, ν_R is assumed not to exist and neutrino is massless to all orders in perturbation theory. It is easy to see that (2.3) and the rest of the Lagrangian is invariant under $(U1)_B$ and $U(1)_L$ global symmetries, with $B\phi = L\phi = 0$. A minimal realistic theory of quarks and leptons appears to naturally conserve baryon and lepton numbers.

However, the above is not the whole story. As is well known, if a global current has an axial anomaly, the current conservation does not hold true, i.e. one gets[8]

$$\partial^\mu J_\mu = \frac{g^2}{16\pi^2} A F^{\mu\nu} F_{\mu\nu}^{d} \tag{2.4}$$

where $F_{\mu\nu}$ is an Abelian or non-Abelian field strength and $F_{\mu\nu}{}^d = \varepsilon_{\mu\nu\alpha\beta} F^{\alpha\beta}$ and A is the anomaly coefficient. From (2.4) it is easy to see that the charge $Q = \int d^3x J^0$ is not conserved in that case, i.e.

$$\frac{d}{dt} Q = \frac{g^2}{16\pi^2} A \int d^3x\, F^{\mu\nu} F_{\mu\nu}^{d} \tag{2.5}$$

Although the integrand on the right hand side is the total derivative, in the presence of instantons the integral does not vanish. Now, the anomaly coefficient $A \alpha\, TrQT_iT_j$ and so from, say

$$Tr\, B\, T_{3L}^{2} = \frac{1}{3}\frac{1}{2} \cdot 3 = \frac{1}{2}$$

$$Tr\, L\, T_{3L}^{2} = 1 \cdot \frac{1}{2} = \frac{1}{2} \tag{2.6}$$

B and L are not anaomaly free and so are not conserved by the nonperturbative effects[7]. Notice, however that the B-L combination is anomaly free. The breakdown of B and L is in this case only of academic interest, since the nonperturbative effects are enormously suppressed by $e - 4\pi\alpha/\sin^2\theta_W$, where α is the electromagnetic coupling. Nevertheless, the special character of B-L is again seen.

Now, to get around the desert picture, B-L must be broken. Since the explicit breaking lacks any predictive power, in what follows we shall the pursue the idea of the spontaneous breakdown of B-L symmetry[9].

Extended $SU(2)_L \times U(1) \times SU(3)_c$ model and the spontaneous breakdown of global B-L symmetry: This model has been developed in Ref. 9 and the phenomenological implications presented here were worked out by Mohapatra and myself[9]. Our main idea was to incorporate the notion of spontaneously broken global B-L symmetry into a study of neutron ($n-\bar{n}$) and hydrogen ($H-\bar{H}$)oscillations; the result of which, as we shall see is the connection between these and the neutrino Majorana mass.

Let us start by describing the model. Besides the usual particles given in (2.1) and (2.2), in order to break B-L spontaneously, one is forced into extending the Higgs sector with the following $SU(2)_L$ triplets[9].

$$\Delta_\ell(3,\ 2,\ 1_c)\ \ ;\ \ \Delta_q(3,\ -\tfrac{2}{3},\ \bar{6}_c) \tag{2.7}$$

The existence of Δ_ℓ, suggested in Ref. 10, is sufficient to break B-L; but Δ_q enables the preservation of quark-lepton symmetry of the standard model, and furthermore will lead to n-$\bar{n}$ and H-$\bar{\text{H}}$ transitions.

We assign $B_{\Delta_q} = -2/3$, $L_{\Delta_q} = 0$, and $B_{\Delta_\ell} = -2$, and demand the full Lagrangian to possess global $U(1)_{B-L}$ symmetry. The most general Yukawa interaction and Higgs potential are then given by ($\Delta \equiv 1/\sqrt{2}\ \vec{\tau}\cdot\vec{\Delta}$)

$$L_Y = h_u\ \bar{Q}_L(i\tau_2\ \phi^*)u_R + h_d\ \bar{Q}_L\phi d_R + h_e\bar{\psi}_L\phi e_R$$

$$+ f_q\ Q_L^T C^{-1}\tau_2\Delta_q Q_L + f_e\psi_L^T C^{-1}\tau_2\Delta_\ell\psi_L + \text{h.c.} \tag{2.8}$$

and

$$V = -\mu_\phi^2\ \phi^+\phi + \lambda_\phi(\phi^+\phi)^2$$

$$-\mu_\ell^2\ \text{Tr}\Delta_\ell^+\Delta_\ell + \lambda_\ell(\text{Tr}\Delta_\ell^+\Delta_\ell)^2 + \lambda_\ell'\ \text{Tr}\Delta_\ell^+\Delta_\ell\Delta_\ell^+\Delta_\ell$$

$$-u_q^2\ \text{Tr}\Delta_q^+\Delta_q + \lambda_q(\text{Tr}\Delta_q^+\Delta_q)^2 + \lambda'_q\text{Tr}\Delta_q^+\Delta_q\Delta_q^+\Delta_q$$

$$+\alpha\ \text{Tr}\Delta_\ell^+\Delta_\ell\ \text{Tr}\Delta_q^+\Delta_q + \alpha'\ \text{Tr}\Delta_q^+\Delta_q\Delta_\ell^+\Delta_\ell$$

$$+\phi^+\phi(\beta_\ell\ \text{Tr}\Delta_\ell^+\Delta_\ell + \beta_q\ \text{Tr}\Delta_q^+\Delta_q) + \phi^+(\gamma_\ell\Delta_\ell^+\Delta_\ell + \gamma_q\Delta_q^+\Delta_q)\phi$$

$$\lambda\ \varepsilon_{ikm}\ \varepsilon_{j\ell n}(\text{Tr}\Delta_q^{ij}\Delta_q^{k\ell})(\text{Tr}\Delta_q^{mn}\Delta_\ell) + \text{h.c.} \tag{2.9}$$

where we have explicitly included the only nontrivial colour indices in the last term of (2.9). That term is very important: in its absence the Lagrangian admits two separate U(1) symmetries corresponding to baryon and lepton numbers; however, $\lambda \neq 0$ leaves only $U(1)_{B-L}$ broken.

Incidentally, the theory as a by-product possesses a discrete symetry D_B, with all the fields having the transformation property: (field) $\rightarrow e^{i\pi B}$ (field). In particular, under D_B, $p \rightarrow -p$, $n \rightarrow -n$, but $e \rightarrow e$ and mesons $\rightarrow$ mesons. This symmetry, as we shall see below, plays a crucial role in guaranteeing the stability of the proton.

The potential in Eq. (2.9) is minimized for the following values of the fields, which break the global B-L symmetry[10]:

$$\langle\phi\rangle = \begin{pmatrix} 0 \\ V/\sqrt{2} \end{pmatrix} , \quad \Delta_\ell = \begin{pmatrix} 0 & 0 \\ \kappa/\sqrt{2} & 0 \end{pmatrix} \tag{2.10}$$

leading to the usual expressions for quark and electron masses and a Majorana mass for the neutrino

$$m_\nu = \sqrt{2}\, f_e\, \kappa \tag{2.11}$$

The leptonic aspect of this model has been discussed at length before[10-12]. The spontaneous breakdown of global B-L symmetry leads to an existence of a zero mass Goldstone boson, the Majorana[13,10], whose presence requires re-doing some of the usual weak interaction phenomenology. The only constraints of importance for us is $f_e < 7 \times 10^{-3}$ from π and K meson decays[12] and $\kappa < 100$ keV from the astrophysical analysis of energy loss from red giants[11,14]. This results in $m_{\nu_\ell} < 700$ eV; obviously not a very useful bound.

Matter-antimatter transitions: We now turn our attention to $\Delta B = 2$, $\Delta L = 0$ n-$\bar{n}$ transitions[15] and $\Delta B = \Delta L = 2$ H-$\bar{\text{H}}$ transitions[16]. Notice that, since a discrete symmetry D_B remains unbroken, proton is absolutely stable. That means that we have no reason to assume the existence of new large mass scales in the theory; in what follows we shall make the most natural assumption of a single mass scale, i.e. that of weak interactions ~ 100 GeV.

H-$\bar{\text{H}}$ oscillations: This process proceeds in the manner shown in Fig. 1. Its strength is easily estimated to be

$$G_{H-\bar{H}} \simeq \frac{\lambda\, f_q^3\, f_e}{m_{\Delta_q}^6\, m_{\Delta^{++}}^2} \tag{2.12}$$

This in turn leads to an effective hydrogen-antihydrogen mixing of the strength

$$H_{eff} \simeq G_{H-\bar{H}} \left|\psi_N(0)\right|^4 (p_L^T C^{-1} p_L)(e_L^T C^{-1} e_L) + \text{h.c.} \tag{2.13}$$

where $\left|\psi_N(0)\right|^4$ is the quark wave function inside the nucleus, and e and p are electron and proton fields. From Ref. 16 we can then estimate the mixing time for H-$\bar{\text{H}}$ oscillation

$$T^{-1}_{H-\bar{H}} \simeq G_{H-\bar{H}} \left|\psi_N(0)\right|^4 \frac{(m_e \alpha)^3}{\pi} \tag{2.14}$$

In what follows we will take $m_{\Delta_q} \sim 100$ GeV and $\lambda \sim 1$ as natural values. Since $f_e < 7 \times 10^{-3}$, $m_{\Delta^{++}} > 15$ GeV,[11] by assuming $f_q < 10^{-3}$ and taking[17] $|\psi_N(0)|^4 \simeq 10^{-5}$ (GeV)^6 we get from (2.12) and (2.14)

$$T_{H-\bar{H}} \gtrsim 3 \times 10^{15} \text{ yr} \tag{2.15}$$

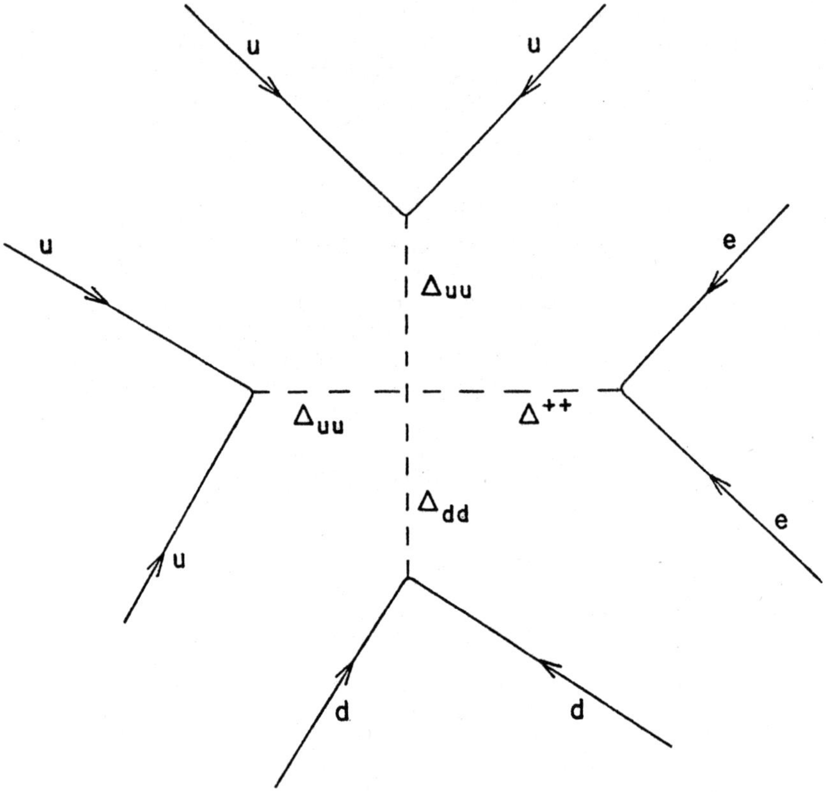

Fig. 1 Example of a quark diagram involving $H \to \bar{H}$ process. Δ_{uu} Δ_{dd} are charge −4/3 and +2/3 members of the Δ_q triplet, and Δ^{++} is the doubly charged Higgs scalar from Δ_ℓ.

We wish to add that, as is usual with $\Delta B \neq 0$ processes, our prediction depends sensitively on the choice of Higgs particle masses; however, our estimate is fairly reasonable. We eliminate a large degree of uncertainty when we make comparison of $T_{H-\bar{H}}$ with (possibly) observable $T_{n-\bar{n}}$.

n-$\bar{n}$ oscillations: It is easy to estimate $T_{n-\bar{n}}$ from Fig. 2, in exactly the same manner as for H-$\bar{H}$ transitions

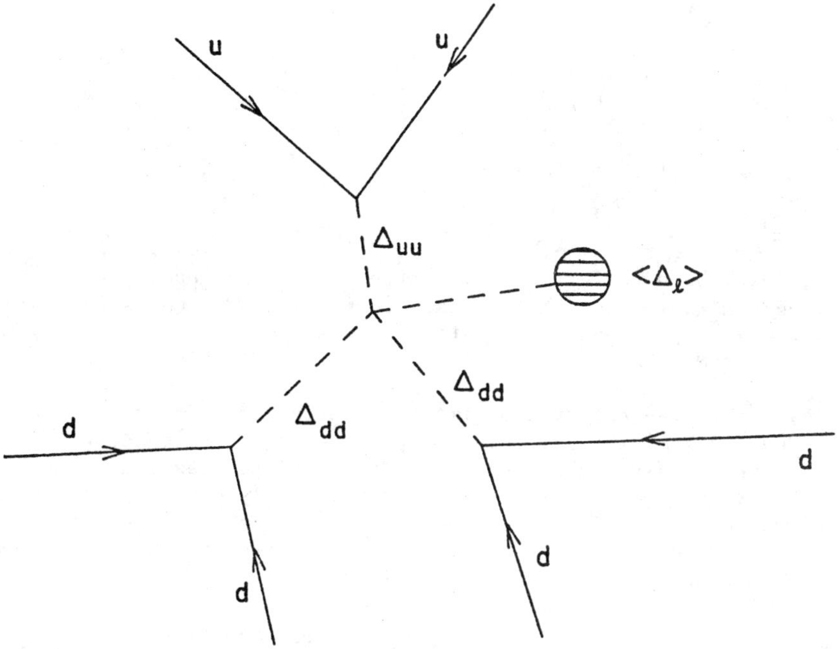

Fig. 2 A basic quark diagram depicting n-$\bar{n}$ transitions.

$$T^{-1}_{n-\bar{n}} \simeq \frac{\lambda f_q^3}{m_{\Delta_q}^6} \frac{m_\nu}{\sqrt{2} f_e} \left|\psi_N(0)\right|^4 \quad (2.16)$$

where we have used (2.11). For the same values of parameters as before, we obtain

$$T_{n-\bar{n}} \simeq 10^{22} m_\nu^{-1} \quad (2.17)$$

The experimental limit[18] $T_{n-\bar{n}} > 10^7 - 10^8$ sec requires then $m_\nu < 10^{-1}$ eV (of course, the smaller couplings and/or larger value of m_{Δ_q} would suppress the above process and allow for heavier neutrinos).

From (2.12), (2.14) and (2.16) we arrive at our main prediction

$$\frac{T_{H-\bar{H}}}{T_{n-\bar{n}}} = \frac{\pi\, m^2_{\Delta^{++}} m_\nu}{f_e^2 (\alpha\, m_e)^3} \tag{2.18}$$

which is free from most of the uncertainties that plagued $T_{n-\bar{n}}$ and $T_{H-\bar{H}}$. For example, if we choose $m_{\Delta^{++}} \simeq$ 10-100 GeV, we expect

$$T_{H-\bar{H}} \simeq 5 \times 10^{20}\, m_\nu (\mathrm{GeV}) T_{n-\bar{n}} \tag{2.19}$$

For $T_{n-\bar{n}} \simeq 1$ yr, $m_\nu \simeq$ (1-10) eV, $T_{H-\bar{H}} \simeq 10^{13}$ yr, which is close to the lower bound inferred from astrophysics by Feinberg et al.[16]. The important point to note is that acceptable and interesting baryon non-conserving processes can be predicted in theories without any mass scale beyong $M \simeq M_W$.

As far as laboratory experimental searches for H-$\bar{\text{H}}$ mixing is concerned, long mixing times would appear to make it inaccessible to observation in existing set ups. However, the processes such as $p + n \to e^+ + \nu$ and $n \to p + e^+ + \nu$, and especially double proton decay described below, are free from the suppression due to atomic wave functions and so could be searched for in the same set ups that look for proton decays.

Double proton decay: $p + p \to e^+ + e^+$: The process proceeds through the same diagram in Fig. 1 that describes H-$\bar{\text{H}}$ transition and can be characterized by the same effective strength $G_{H-\bar{H}}$ defined in (2.12). For the same values of the parameters as before, we estimate in our model $G_{H-H}|\psi_N(0)|^4 < 10^{-29}$ (GeV)$^{-2}$, or in other words $\tau_{pp} > 10^{31}$ yr, close to the experimental bound.

For values of $T_{H-\bar{H}}$ that we predict, there also exists the possibility of observing H-$\bar{\text{H}}$ mixing effects in astrophysics by looking for energetic interstellar γ-rays. As for other distinguishable features of this model, we note that:

i) doubly charged Higgs bosons (with $m_{\Delta^{++}}$ = 15 GeV or so) should be visible in existing high energy machines in the not too distant future;

ii) electron neutrino should weight about 10^{-1} eV, if both n-$\bar{\text{n}}$ and H-$\bar{\text{H}}$ processes are to have detectable rates;

iii) if there exist direct $\Delta S = 2$ tree level amplitude through the exchange of a diquark Δ_{dd}, with the strength $f_d f_s / m^2_{\Delta_q}$ For $f_d < 10^{-3}$ and $m_{\Delta_q} \simeq 100$ GeV, we need $f_s/f_d < 10^{-3}$ in order to guarantee the smallness of K_L-K_S mass difference. This would predict the large suppression of possible Λ-$\bar{\Lambda}$ transition, compared to n-n process

In closing we should mention that this model provides the simplest framework which realizes the general possibility of mutually related $T_{n-\bar{n}}$ and $T_{H-\bar{H}}$ when only left-handed operators contribute. The connection is necessary since we work in the context of purely left-handed, standard model.

SU(5) model with spontaneously broken global B-L symmetry: We have shown in the previous section how a fairly simple extension of the Higgs sector in the standard model can lead to mutually connected and possibly observable $n-\bar{n}$ and $H-\bar{H}$ oscillations. Here, we shall discuss its grand unified version.

Gauge group: $G = SU(5) \times U(1)_G$ with $U(1)_G$ a global symmetry specified below; we shall call its generator G.

Fermions: As in the minimal SU95) model[19], we have

$$\underline{5} = F_R = \begin{pmatrix} d_1 \\ d_2 \\ d_3 \\ e^+ \\ \nu^c \end{pmatrix}_R$$

$$\underline{10} = T_L = \begin{pmatrix} 0 & u_3^c & -u_2^c & -u_1 & -d_1 \\ -u_3^c & 0 & u_1^c & -u_2 & -d_2 \\ u_2^c & -u_1^c & 0 & -u_3 & -d_3 \\ u_1 & u_2 & u_3 & 0 & e^+ \\ d_1 & d_2 & d_3 & -e^+ & 0 \end{pmatrix}_L$$

Higgs sector: Apart from 5 dimensional representation H and 24 dimensional in representation Σ of the minimal model, we include the following additional Higgs multiplets to implement our idea of spontaneous B-L symmetry breakdown[20]

$$\underline{15} \; : \; \Delta_{pp}$$

$$\underline{50} \; : \; \chi^{pq}_{rst} \tag{2.21}$$

where Δ_{pq} is symmetric: $\Delta_{pq} = \Delta_{qp}$ and χ^{pq}_{rst} is totally antisymmetric upper and lower indices, respectively and is traceless: $\chi^{pq}_{pst} = 0$.

The most general Yukawa interaction consistent with SU(5) gauge symmetry is

$$L_Y = h_d\, F^p_R\, H^q\, T_{Lpq}$$

$$+ h_u\, \varepsilon^{pqrst}\, T^T_{Lpq}\, H_r\, C^{-1} T_{Lst} + h_\nu\, F^T_{Rp}\, C^{-1} \Delta^{pq} F_{Rq}$$

$$+ f\, \varepsilon^{pqrst}\, T^T_{Lpp_1}\, C^{-1} \chi^{p_1 q_1}_{rst}\, T_{Lqq_1} + \text{h.c.} \tag{2.22}$$

It is easy to see that the above interaction is invariant under a $U(1)_G$ global symmetry $U = e^{i\theta G}$, where

$$G\, T_L = T_L, \quad G\, F_R = 3\, F_R$$

$$G\, H = -\,2\, H, \quad G\, \Delta = G\, \chi = -\,2\, \chi \tag{2.23}$$

all other fields of the theory having no G charge. The invariance of the rest of the Lagrangian under this symmetry is not automatic. It requires omitting various renormalizable couplings such as $H\, \Delta^+\, H$, $\Sigma\, \Delta^+\, \Sigma$, etc. Therefore, we demand $U(1)_G$ to be a symmetry of the full Lagrangian.

Now, $\langle\Sigma\rangle \simeq M_X$ which is responsible for breaking of SU(5) down to $SU(2)_L \times U(1) \times SU(3)_c$ will not break $U(1)_G$. On the other hand, $\langle H\rangle \simeq M_W$ which is responsible for the second stage of the symmetry breaking, breaks $U(1)_G$, since $G\,\langle H\rangle = -\,2\,\langle H\rangle$ but preserves $(Y + G/2) \cdot \langle H\rangle = \langle H\rangle$. A simple computation shows[18] that this is nothing else but B-L symmetry when it acts on fermions

$$B\text{-}L = \frac{2}{5}\left(Y + \frac{G}{2}\right) \tag{2.24}$$

The potential of the theory can be chosen in the conventional manner to produce a minimum at $\langle\Delta_{55}\rangle \neq 0$, which leads to the spontaneous breakdown of B-L symmetry manifest through a Majorana mass for the neutrino

$$m_\nu = h_\nu\, \langle\Delta\rangle \tag{2.25}$$

The associated Goldstone boson, the majoron is now given by

$$M = \frac{\langle H\rangle \ \mathrm{Im}\ \Delta_{55} - \langle\Delta\rangle \ \mathrm{Im}\ H_5}{(\langle H\rangle^2 + \langle\Delta\rangle^2)^{1/2}} \tag{2.26}$$

and its properties are essentially the same as in the $SU(2)_L \times U(1)$ model.

We now turn our attention to matter oscillations. These processes, as we show below, are mediated by particles in Δ and χ multiplets. Let us then first display the $SU(2)_L \times U(1) \times SU(3)_c$ representation contents of Δ and χ

$$\Delta = \begin{matrix} (3, 2, 1_c) & + (1, -4/3, 6_c) & + (2, 1/3, 3_c) \\ \Delta_{ab} = \Delta_1 & \Delta_{ij} = \Delta_2 & \Delta_{ia} = \Delta_3 \end{matrix} \tag{2.27}$$

where a,b, ... are $SU(2)_L$ indices and i,j,k, ... are $SU(3)_c$ indices. Similarly, for χ we get

$$\chi = \begin{matrix} (1, -4, 1_c) & + (1, -2/3, 3_c) & + (2, -7/3, 3_c) \\ \chi^{ab}_{ijk} = \chi_1 & \chi^{ab}_{cdi} = \chi_2 & \chi^{ab}_{cij} = \chi_3 \end{matrix}$$

$$+ \begin{matrix} (1, 8/3, 6_c) & + (3, -2/3, 6_c) & + (2, 1, 8_c) \\ \chi^{ij}_{kab} = \chi_4 & \chi^{ai}_{bjk} = \chi_5 & \chi^{ia}_{jbc} = \chi_6 \end{matrix} \tag{2.28}$$

Before we discuss any physical consequences due to these Higgs scalars, we would have to know their mass scales. If one assumes minimal fine tuning, .e. one makes no more fine tuning of the parameters beyond the one needed for the gauge hierarchy $M_W/M_X \simeq 10^{-12}$, then the surival hypothesis is operative: only in those particles which <u>have</u> to be light will remain light (see Section IV). In this case it is the $SU(2)_L$ doublet (H_4, H_5) from H and the $SU(2)_L$ triplet Δ_1 from Δ. Namely (H_4, H_5) contains Goldstone bosons longitudinal components of $W^{\pm}$ and Z) and the physical scalar η; whereas Δ_1 contains the majoron to prevent the rest of the multiplet from becoming superheavy. All other Higgs scalars would however become heavy. This of course, would imply the suppression of all other processes but proton decay and we somewhat uneasily give it up. We could imagine that some symmetry (maybe supersymmetry) could eventually keep other particles light, but for the moment we just assume it (it is always technically possible to do it by adjusting the parameters). Still, however, χ_2 has to be superheavy because it mediates proton decay and so $m_{\chi 2} \simeq M_X$. As far as other particles are concerned, we will keep them light and in particular we assume that χ_1, χ_4, χ_5 and Δ_1 submultiplets have masses of order 100 GeV.

It is work noting that χ_1 and Δ_2 are doubly charged Higgs bosons, where χ_1 couples to right-handed leptons and Δ_1 couples to the left-handed ones and is the analogue of the Δ_ℓ of Section III. Since this is somewhat conceptual, I will not puruse it here. Suffice it to say that both n-$\bar{n}$ and H-$\bar{H}$ transitions proceed in the manner described in this section. Furthermore, the model can be easily made consistent with constraints as $\sin^2\theta_W$ and τ_p. For the details, the interested reader should consider a paper[22] of Mohapatra and myself which discusses these questions at length.

III. B-L AS A GAUGE GENERATOR: LEFT-RIGHT SYMMETRY

In the previous section we have reviewed the possibility of B-L being a global symmetry, subsequently spontaneously broken by a Higgs mechanism. This led to a plethora of interesting new phenomena, in particular Majorana neutrino mass, existence of Goldstone bosons (Majorons) and the connection between m_ν and $T_{n-\bar{n}}$, $T_{H-\bar{H}}$. The other, equally interesting possibility is that B-L is a gauge symmetry, again spontaneously broken. This actually happens naturally in a class of gauge theories, characterized by left-right symmetry[23]. We offer a brief review of such models here, with the emphasis on B-L breaking.

The gauge groups which incorporate this idea is $SU(2)_L \times SU(2)_R \times U(1)_{B-L} \times SU(3)_c$ with the fermionic assignment[23]

$$Q_L = \begin{pmatrix} u \\ d \end{pmatrix}_L \quad (2, 1, \tfrac{1}{3}, 3_c) \quad ; \quad Q_R = \begin{pmatrix} u \\ d \end{pmatrix}_R \quad (1, 2, \tfrac{1}{3}, 3_c)$$

$$\psi_L = \begin{pmatrix} \nu \\ \ell \end{pmatrix}_L \quad (2, 1, -1, 1_c) \quad ; \quad \psi_R = \begin{pmatrix} \nu \\ \ell \end{pmatrix}_R \quad (1, 2, -1, 1_c) \tag{3.1}$$

which is completely left-right symmetric as is the rest of the theory. The electromagnetic charge in this model is seen to be[24] from (3.1)

$$Q_{em} = I_{3L} + I_{3R} + \frac{B-L}{2} \tag{3.2}$$

so that the U(1) generator is physical: it is B-L. The above relation indicates the clear connection between the breaking of parity (I_{3R}) and B-L. This was realized in the work of Mohapatra and the author[25], through the connection of the smallness of neutrino Majorana mass and the maximality of parity violation in weak interactions. In order to describe it we must introduce the minimal Higgs sector which incorporates[25] the breaking of parity and B-L

$$\Delta_L^\ell\ (3, 1, 2, 1_c) \quad ; \quad \Delta_R^\ell\ (1, 3, 2, 1_c) \quad ; \quad \phi\ (2, 2, 0, 1_c) \tag{3.3}$$

Subsequent symmetry breaking

$$\langle\Delta_L^\ell\rangle \simeq 0, \quad \langle\Delta_R^\ell\rangle \simeq g\, M_R, \quad \langle\phi\rangle \simeq g\, M_W \tag{3.4}$$

in the symbolic notation gives a realistic theory with broken parity, through $M_R > M_W$. What is more interesting to us, from the Yukawa interaction for leptons

$$L_Y = f_e(\psi_L^T\, \tau_2\, C\, \Delta_L\, \psi_L + \psi_R^T\, \tau_2\, C\, \Delta_R\, \psi_R) + h\, \bar{\psi}_L\, \phi\, \psi_R \tag{3.5}$$

where

$$\Delta_{L,R} \equiv \frac{1}{\sqrt{2}}\, \vec{\tau} \cdot \vec{\Delta}_{L,R},$$

right-handed neutrino gets a large Majorana mass

$$m_{\nu_R} \simeq f\, M_R \tag{3.6}$$

and in turn the left-handed neutrino becomes naturally small

$$m_{\nu_L} \simeq \frac{h^2 \langle p\rangle^2}{m_{\nu_R}} \tag{3.7}$$

In the V-A limit of the theory, (i.e. $M_R \to \infty$), $m_{\nu R} \to \infty$ and $m_{\nu L} \to 0$; hence, the smallness of neutrino mass gets tied up to the maximality of observed parity violation in weak interactions.

An interesting extension of the minimal model has been suggested by Marshak and Mohapatra[26], by making the model completely quark-lepton symmetric. They add, besides leptonic triplets Δ^ℓ, quark triplets which are sextets of colour

$$\Delta_L^q\ (3,\ 1,\ -\tfrac{2}{3},\ \bar{6}_c) \quad ; \quad \Delta_R^q\ (1,\ 3,\ -\tfrac{2}{3},\ \bar{6}_c) \tag{3.8}$$

This model is in some sense left-right symmetric extension of the $SU(2)_L \times U(1)$ model with B-L broken global symmetry, as discussed in the previous section. This, of course, will lead to the existence of n-$\bar{n}$ and H-$\bar{\text{H}}$ transitions. The main reason is the existence of an extra term in the Yukawa interaction

$$\Delta L_Y^q = f_e(Q_L^T\, C\, \tau_2\, \Delta_L^q\, Q_L + Q_R^T\, C\, \tau_2\, \Delta_R^q\, Q_R) + \text{h.c.} \tag{3.9}$$

n-$\bar{n}$ oscillators: For M_R which is not astronomically large (which is of phenomenological interest, anyway), the dominant contribution to this process comes from the exchange of the right-handed fields, in the manner depicted in Fig. 2. Assuming $m_{\Delta_R^q} \simeq M_R$ and using $\langle\Delta_R\rangle \simeq M_R$, one easily obtains

$$T^{-1}_{n-\bar{n}} = \frac{\lambda f_q^3}{M_R^6} \langle\Delta_R\rangle \left|\psi(0)\right|^4 \simeq \frac{\lambda f_q^3}{M_R^5} \left|\psi(0)\right|^4 \qquad (3.10)$$

If λ and f_q are not terribly small, say λ, $f_q \sim 10^{-2}$, we get

$$T_{n-\bar{n}} \simeq 10^{-13} M_R\ (\mathrm{GeV})^5\ \mathrm{sec} \qquad (3.11)$$

which would require $M_R \simeq 10^4$ GeV (or so), in order to make $n-\bar{n}$ transitions observable ($T_{n-\bar{n}} \simeq 10^7-10^8$ sec).

H-$\bar{\text{H}}$ transition: Here, clearly both $\Delta_L{}^q$ and $\Delta_R{}^q$ exchange leads to H-$\bar{\text{H}}$ transition amplitude. Without fine tuning the parameters, we expect (see Section IV)

$$m^q_{\Delta_L} \sim m^q_{\Delta_R} \sim m_{\Delta^\ell_L} \sim m_{\Delta^\ell_R} \simeq M_R .$$

$$T^{-1}_{H-\bar{H}} \simeq \frac{\lambda f_q^3 f_e}{M_R^8} \left|\psi(0)\right|^4 \frac{(m_e \alpha)^3}{\pi} \qquad (3.12)$$

We have a simple connection between $T_{n-\bar{n}}$ and $T_{H-\bar{H}}$

$$\frac{T_{H-\bar{H}}}{T_{n-\bar{n}}} = \frac{\pi M_R^3}{(m_e \alpha)^3 f_e} \qquad (3.13)$$

which for $M_R \simeq 10^6$ GeV, would make H-$\bar{\text{H}}$ transition hopelessly small: $T_{H-\bar{H}} > 10^{28} T_{n-\bar{n}}$. It is noteworthy that even if M_R is fairly low (~ 100 GeV or so), from (3.13) $T_{H-\bar{H}} > 10^{22} T_{n-\bar{n}}$ and so H-$\bar{\text{H}}$ transition would still be unobservable.

In conclusion, left-right symmetric models based on $SU(2)_L \times SU(2)_R \times U(1)_{B-L}$ provide a simple framework in which B-L becomes a dynamic variable, a part of a gauge group. Although the model predicts both n-$\bar{\text{n}}$ and H-H, the precise estimates of $T_{n-\bar{n}}$ and $T_{H-\bar{H}}$ are plagued by huge uncertainties. On the other hand, similarly as in the $SU(2)_L \times U(1)$ model with B-L a broken global symmetry, the ratio of $T_{H-\bar{H}}/T_{n-\bar{n}}$ becomes very much independent of all the unknown parameters. However unlike in the previous case, left-right model suggest possibly observable n-$\bar{\text{n}}$ oscillations, but not H-$\bar{\text{H}}$ transitions; the latter lifetime being hopelessly large: $T_{H-\bar{H}} > 10^{22}$ yr, from the existing limit $T_{n-\bar{n}} \gtrsim 1$ yr.

IV. GRAND UNIFICATION AND MATTER OSCILLATIONS: A LESSON ON SURVIVAL

As I emphasized in the introduction, minimal SU(5) theory, or any other theory that breaks in the same manner necessarily predicts a desert between M_W and $M_X \simeq 10^{14}$-10^{15} GeV. In particular it makes n-n and H-H transitions, if at all possible, hopelessly unobservable. In the last two sections we have shown how simple it is to find an oasis in this desert, if one is willing to restrict oneself to phenomenological theories such as $SU(2)_L \times U(1)$ or $SU(2)_L \times SU(2)_R \times U(1)_{B-L}$. Matter oscillations, especially in n-n, can easily be made observable and furthermore the oscillation times $T_{n-\bar{n}}$ and $T_{H-\bar{H}}$ become related to each other. The question is, could we expect the same in simple GUTS which are extended beyond SU(5)? The answer, as we shall see, appears to be no, if no fine tuning of the parameters of the theory is allowed. There may or may not be an oasis in the GUT desert, but matter oscillations are unlikely to be observed.

Now, in the simple models these processes are mediated by Higgs scalars, so that the relevant issue becomes whether such Higgs scalars can have naturally reasonably small masses, i.e. below $\sim 10^5$ GeV. I want to give a brief report here on the work done in collaboration with Rabi Mohapatra regarding Higgs mass scales, which completes the earlier work of Georgi[28] and del Aquila and Ibanez[29]. An interested reader can find more detailed discussion in Ref. 27 and in my earlier review[30].

To spare the impatient reader unnecessary boredom, let me announce the results before discussing them. Higgs bosons have trouble surviving from from becoming superheavy[27-29], and in particular the exotic scalars which radiate matter oscillations get the mass proportional to M_X. In other words, even when there is an oasis in the desert, they will never know about it. Instead of a general discussion, this will be presented through a simple example[27].

<u>Illustrative example: SU(5) with two 5's:</u> The model contains two 5's: H_1 and H_2 and a <u>24</u> dimensional multiplet Σ. Imagine the following discrete symmetries implemented for simplicity

$$D_1: \Sigma \to -\Sigma \qquad\qquad D_2: \Sigma \to \Sigma$$

$$H_2 \to H_1, \; H_2 \to H_2 \qquad\qquad H_1 \to H_2$$

The most general Higgs potential is then given by

$$V(\Sigma,H_1,H_2) = -\frac{1}{2}\mu_X^2 \,\mathrm{Tr}\Sigma^2 + \frac{1}{4} a(\mathrm{Tr}\Sigma^2)^2 + \frac{1}{2} b\, \mathrm{Tr}\Sigma^4$$

$$-\frac{1}{2}\mu_W^2(H_1^+H_1 + H_2^+H_2) + \frac{1}{4}\lambda_1[+(H_1^+H_1)^2 + (H_2^+H_2)^2] + \frac{1}{8}\lambda_3(H_1^+H_1)(H_2^+H_2)$$

$$+\frac{1}{4}\lambda_4(H_1^+H_2)(H_2^+H_1) + \frac{1}{2}\lambda_5[(H_1^+H_2)^2 + (H_2^+H_1)^2] + \frac{1}{8}\lambda_6(H_1^+H_1 + H_2^+H_2)\times(H_1^+H_2 + H_2^+H_1)$$

$$+\alpha(H_1^+H_1 + H_2^+H_2)\,\mathrm{Tr}\Sigma^2 + \beta(H_1^+\Sigma^2H_1 + H_2^+\Sigma^2H_2)$$

$$+\gamma(H_1^+\Sigma^2H_2 + H_2^+\Sigma^2H_1) + \delta(H_1^+H_2 + H_2^+H_1)\,\mathrm{Tr}\Sigma^2 \qquad (4.2)$$

The diagonalization of the potential leads then to the following values for Higgs boson masses. First, as expected, Σ is superheavy. Second, if we write

$$H_i = \begin{pmatrix} h_i \\ \phi_i \end{pmatrix}, \quad i = 1,2 \qquad (4.3)$$

where h_i are color triplets ($SU(2)_L$ singlets) and ϕ_i are the usual $SU(2)_L$ doublets (color singlets), then h_1 and h_2 become superheavy, too, as they should be for the sake of proton stability. This has not taught us much, since there was no freedom up to now. What about the doublets ϕ_i? Let us define

$$\eta = \sqrt{2}\,(\phi_1 + \phi_2) \quad ; \quad \rho = \sqrt{2}\,(\phi_1 - \phi_2) \qquad (4.4)$$

which become physical states

$$\eta = \frac{1}{\sqrt{2}}\begin{pmatrix} \sqrt{2}G_W^+ \\ \eta + V_W + iG_Z \end{pmatrix}, \qquad \rho = \frac{1}{\sqrt{2}}\begin{pmatrix} \rho^+ \\ \rho_1 + i\rho_2 \end{pmatrix} \qquad (4.5)$$

with

$$\langle\phi_1\rangle = \langle\phi_2\rangle = \frac{1}{\sqrt{2}}\begin{pmatrix} 0 \\ V_W \end{pmatrix}$$ and G_W^+ and G_Z are unphysical Higgs bosons

present in the renormalizable gauge which in the unitary gauge get eaten by W^+ and Z, respectively. Clearly then, η must be light ($\sim M_W$) to preserve $SU(2)_L$ invariance. A computation confirms it[2/]

$$m_\eta^2 = [\frac{1}{4}(\lambda + \lambda_3 + \lambda + \lambda_5 + \lambda_6) - \frac{6}{5}\frac{(10\alpha + 10\delta + 3\beta + 3\gamma)^2}{15a + 7b}$$

$$- \frac{g}{10}\frac{\beta + \gamma}{6}] v_W^2 \tag{4.6}$$

On the other hand, the other doublet ρ has no reason to be light, since $\langle\rho\rangle = 0$. Its mass is a test of a more general issue: what are the masses of the particles which are not phenomenologically forced to be superheavy? Now, no symmetry forbids ρ to receive a superlarge mass and so it does

$$m_\rho^2 \simeq -\frac{15}{2}(d + \frac{3}{10}\gamma) v_X^2 \tag{4.7}$$

We have analyzed[27] many other models with the same conclusion, that the survival principle is operative[28-29]: <u>if the minimal fine tuning is assumed</u>, then, <u>unless protected by some discrete, global or gauge symmetry, the Higgs bosons become superheavy</u>.

In particular, those scalars with quantum numbers of di-quarks and di-leptons have no protective symmetry, and so their mass is $\sim M_X$, independently of the existence of intermediate mass scales. To be more precise, at lease some of them become superheavy and they suppress matter oscillations.

Before concluding this section, let us, if only briefly discuss the possibilities of observable n-n in transitions in partially or grand unified models. Now that we know that the Higgs bosons tend to pick up the mass of the order M_X, the hope is to be able to find models with low unification scales. This, although not found in simple GUTS based on small groups is still possible in some, more complicated theories[31-32] such as $[SU(6)]^4$ or SU(16). Such models are too complex to be discussed here.

Another alternative would be a partially confined theory, such as the $SU(2)_L \times SU(2)_R \times SU(4)_c$ model of Pati and Salam[2]. Again, the scale of partial unification turns out to be too large: $M_C > 10^{10}$ GeV, from the $\sin^2\theta_W$ considerations[33]. Buras et al., suggest[33] extending the gauge group to $[SU(2)_L \times SU(2)_R]^2 \times SU(4)_c$ in order to bring M_c down; a solution I do not find particularly appealing, although it is certainly legitimate.

A possible hope arises when $SU(2)_L \times SU(2)_R \times SU(4)_c$ is embedded in SO(10). Namely, a possible chain of symmetry breaking[34]

$$O(10) \underset{M_X}{\rightarrow} SU(2)_L \times U(1)_R \times SU(4)_c \underset{M_c}{\rightarrow} SU(2)_L \times U(1)_R \times U(1)_{B-L} \times SU(3)_c$$

$$\underset{M_{BL}}{\rightarrow} SU(2)_L \times U(1) \times SU(3)_c \underset{M_W}{\rightarrow} U(1)_{em} \times SU(3)_c \tag{4.8}$$

allows a solution[35] $m_c \simeq 10^5$-10^6 GeV, close to its lower bound[36] ~ 10^5 GeV. However, still not all the relevant lepto-quarks remain at M_c and n-$\bar{n}$, H-$\bar{H}$ transitions are still suppressed by $1/M_X{}^2$. Namely, only those with $T_{3R} = 0$ have mass ~ M_c, but they are not sufficient to mediate matter oscillations (this can be easily check by inspection).

V. SUMMARY

In this talk, I have tried to go through the general discussion of matter oscillations in gauge theories. As we learned from the introduction and Section II, this question is intimately tied up to the analysis of B-L operator and its breaking. The special position that B-L holds is best illustrated by the facts that the nonperturbative effects in $SU(2)_L \times U(1)$ fail to break it and also that it becomes an automatic symmetry at the leading $\Delta B \neq 0$ operators in the case of a single mass scale beyond M_W.

On the phenomenological level, it is easy to construct models with observable n-$\bar{n}$ and H-$\bar{H}$ oscillations. Although by no means the only example, extended $SU(2)_L \times U(1) \times SU(3)_c$ model with B-L spontaneously broken provides an interesting link between $T_{n-\bar{n}}$ and $T_{H-\bar{H}}$ and neutrino mass; all possibly observable. When the model is made left-right symmetric, B-L becomes gauged; Majoran disappears and H-$\bar{H}$ oscillations become hopelessly suppressed.

Under the conditions of naturalness, the grand unification of these simple models fails. Namely, they can be easily embedded in SU(5) and SO(10), respectively, however the relevant Higgs scalars which mediate matter oscillations become superheavy. Of course, one may abandon the idea of minimal fine tuning, as has been done in most of the models[16,22], in which case one is not surprised to find observable n-$\bar{n}$ and H-$\bar{H}$ transitions[16,22,37].

The main question I have tried to answer here is whether we can hope to observe matter oscillations in the near future, i.e. whether there are single theories that predict $T_{n-\bar{n}}$ and $T_{H-\bar{H}}$. Unless counting on luck, my feeling, one which I have tried to portray in this talk is that the answer is no. Both n-$\bar{n}$ and H-$\bar{H}$ transitions are dramaticaly dependent on the relevant mass scale, usually Higgs mass scales and if anything, these mass scales are expected in simple GUTS to be very large. Still, on the pragmatic level one cannot yet rule out these processes and as usual we should let the experiment decide.

ACKNOWLEDGMENT

Most of the work described in this talk has been in collaboration with Rabi Mohapatra. I wish to thank him, Pavle Senjanović and Saško Sokorac for discussions.

The submitted manuscript has been authored under Contract NO. DE-AC02-76CH00016 with the U.S. Department of Energy.

REFERENCES AND FOOTNOTES

1. J.C. Pati and A. Salam, Phys. Rev. D10, 275 (1974); H. Georgi and S.L. Glashow, Phys. Rev. Lett. 32, 438 (1974).
2. S. Weinberg, Phys. Rev. Lett. 43, 1566 (1979); F. Wilczek and A. Zee, ibid, 43, 1571 (1979). See, also J. Nieves, Nucl. Phys. B189, 182 (1981).
3. M. Krishnaswamy, M. Menon, N. Mondal, V. Naraslimham, B. Sreekantan, N. Ito, S. Kawakami, Y. Hayashi and S. Miyake, Phys. Lett. 106B, 339 (1981).
4. H. Georgi, H. Quinn and S. Weinberg, Phys. Rev. Lett. 33, 451 (1974).
5. For a review and refernces of the technicolor idea, see E. Farhi and L. Susskind, Phys. Reports (1982).
6. See, e.g. reviews by C. Llewellyn Smith, Univ. of Oxford preprint (1982) and N. Sakai, KEK preprint (1982).
7. G. 't Hooft, Phys. Rev. Lett. 37, 8 (1976).
8. See, e.g. R. Jackiw, in Current Algebra and its Applications, Princeton Univ. Press (1972).
9. R. Barbieri and R.N. Mohapatra, Z. Phys. C11, 175 (1981); R.N. Mohapatra and G. Senjanović, Phys. Rev. Lett. 49, 7 (1982).
10. G.B. Gellmini and M. Roncadelli, Phys. Lett. 99B, 411 (1979).
11. H. Georgi, S.L. Glashow and S. Nussinov, Nucl. Phys. B (1982).
12. V. Barger, W.-Y. Keung and S. Pakvasa, Phys. Rev. D25, 907 (1982).
13. Y. Chikashige, R.N. Mohapatra and R.D. Peccei, Phys. Lett. 98B, 265 (1981).
14. M. Fukugita, S. Watamura and M. Yoshimura, KEK preprint (1982); J. Ellis and K. Olive, CERN preprint (1982).
15. For a review of the theory and phenomenology of $n-\bar{n}$ oscillations, see R.N. Mohapatra, talk at the Harvard Workshop on Neutron Oscillations, Harvard Univ. (1982) to appear in the proceedings; M. Baldo-Ceolin, Univ. of Pisa preprint (1982).
16. G. Feinberg, M. Goldhaber and G. Steigman, Phys. Rev. D18, 1602 (1982); Mohapatra and Senjanović, Ref. 9; L. Arnellos and W.J. Marciano, Phys. Rev. Lett. 48, 1708 (1982).
17. See, e.g. S. Rao and R. Shrock, Phys. Lett. 116B, 238 (1982); S.P. Misra and U. Sarkar, Bhubeneswer preprint (1982).
18. See, e.g. C. Dover, A. Gal and J. Richard, BNL preprint (1982) and references therein.
19. Georgi and Glashow, Ref. 1.
20. The 50 dimensional multiplet χ was introduced first by Arnellos and Marciano, Ref. 16, in order to generate $H-\bar{H}$ transitions.
21. F. Bucella, G.B. Gellmini, A. Masiero and M. Roncadelli, Max-Planck preprint (1982).
22. R.N. Mohapatra and G. Senjanović, "Spontaneous breaking of global B-L symmetry and matter-antimatter oscillations in grand unified theories", Phys. Rev. D (to appear).
23. J.C. Pati and A. Salam, Phys. Rev. D10, 275 (1974); R.N. Mohapatra and J.C. Pati, Phys. Rev. D11, 566; 2588 (1975); G. Senjanović and R.N. Mohapatra, Phys. Rev. D12, 1502 (1975). For a review of the basics of L-R symmetry, see G. Senjanović, Nucl. Phys. B153, 334 (1979).

24. R.E. Marshak and R.N. Mohapatra, Phys. Lett. 94B, 183 (1980).
25. R.N. Mohapatra and G. Senjanović, Phys. Rev. Lett. 44, 912 (1980) and Phys. Rev. D23, 165 (1981).
26. R.E. Marshak and R.N. Mohapatra, Phys. Rev. Lett. 44, 1316 (1980).
27. R.N. Mohapatra and G. Senjanović, "Higgs effects in grand unified theories", CCNY preprint (1982);
A. Raychaudhury and P. Roy, Univ. of Calcutta preprint (1982).
28. H. Georgi, Nucl. Phys. B156, 126 (1979).
29. F. del Aquila and L. Ibanez, Nucl. Phys. B177, 60 (1981). This paper was a first systematic study of Higgs effects on the problems of symmetry breaking in SO(10).
30. G. Senjanović, invited talk at the Neutrino Oscillations Workshop, Harvard Univ., 1982, to appear in the proceedings.
31. V. Elias, J.C. Pati and A. Salam, Phys. Rev. Lett. 40, 920 (1978);
V. Elias and S. Rajpoot, Phys. Rev. D20, 2445 (1979).
32. J.C. Pati, A. Salam and J. Strathdee, Nucl. Phys. B185, 1445 (1981);
J.C. Pati, unpublished
R.N. Mohapatra and M. Popović, CCNY preprint (1982);
A. Raychaudhury and U. Sarkar, Calcutta Univ. preprint (1982).
33. V. Elias, Phys. Rev. D19, 1896 (1976);
A. Buras, P.Q. Hung and J.D. Bjorken, Phys. Rev. D25, 805 (1982).
34. S. Rajpoot, Phys. Rev. D22, 2244 (1980);
N.G. Deshpande and R.J. Johnson, Univ. of Oregon preprint (1981);
R.W. Robinett and J.L. Rosner, Phys. Rev. D26, 2036 (1982);
S. Rajpoot, Phys. Lett. B (1982).
G. Fogleman and T.G. Rizzo, Phys. Lett. 113B, 240 (1982).
35. G. Senjanović and A. Šokorac, "Light lepto-quarks in SO(10)", BNL preprint (1982).
36. N.G. Deshpande and R.J. Johnson, Univ. of Oregon preprint (1982).
37. For a discussion of Higgs effects in the absence of the survival principle, see
D. Lüst, A. Masiero and M. Roncadelli, Phys. Rev. D25, 3096 (1982);
G. Costa, F. Feruglio and F. Zwirner, Univ. of Padua preprint (1982). For a general analysis, see
A. Šokorac, Boris Kidric Inst. (Belgrade) preprint (1982).

STATUS OF $SU(2)_L \times SU(2)_R \times U(1)_{B-L}$: PHENOMENOLOGY VERSUS THEORY

Ernest Ma
Department of Physics and Astronomy
University of Hawaii at Manoa
Honolulu, Hawaii 96822

ABSTRACT

The most general form of the effective low-energy neutral-current interactions for left-right gauge models is derived and compared with experimental data, resulting in the bounds $83 < M_{Z_1} < 116$ GeV and $M_{Z_2} > 205$ GeV. Constraints from SO(10) grand unification are briefly discussed.

INTRODUCTION

In this talk, I'll be addressing the following questions. (1) What are the neutral-current parameters of $SU(2)_L \times SU(2)_R \times U(1)_{B-L}$? (2) How are they constrained by present experimental data? (3) What are the model's most important predictions? (4) What does the model imply with regard to grand unification?

As a warm-up, let me go through these questions very quickly for the standard $SU(2)_L \times U(1)_Y$ electroweak gauge model.[1] The effective neutral-current interaction at low energies is well known to be given by

$$H_{NC}^{eff} = \frac{4G_F}{\sqrt{2}} J_{ZL}^2 , \tag{1}$$

where

$$J_{ZL} = J_{3L} - x_W J_{em} . \tag{2}$$

Hence there is only one free parameter x_W which is usually written as $\sin^2\theta_W$ and with a value of about 0.22, a good fit to all neutral-current data can be obtained.[2] This in turn gives us the predictions $M_W \simeq 83$ GeV and $M_Z \simeq 94$ GeV after radiative corrections,[3] and is also in good agreement with the prediction[4] of about 0.21 for x_W from SU(5) grand unification.[5]

NEUTRINO MASS IN THE LEFT-RIGHT MODEL

In $SU(2)_L \times SU(2)_R \times U(1)_{B-L}$, the left-handed (right-handed) fermions are doublets (singlets) under $SU(2)_L$ and singlets (doublets) under $SU(2)_R$. The left-handed partner of the electron is of course ν_e which is produced via the usual charged-current weak interaction, but the right-handed partner of the electron may or may not be ν_e. Suppose that ν_e is a Dirac neutrino with a nonzero mass, than ν_e is coupled to the electron via W_R as well as W_L, but if $M_{W_R} >> M_{W_L}$ and there is no mixing between W_R and W_L, then no present experimental result is sensitive to the presence of the right-handed component of ν_e, so a Dirac mass for ν_e cannot be excluded. On the other hand, if ν_e is a Majorana neutrino with a nonzero mass so that the right-handed partner of the electron is some other particle, then neutrinoless double beta decay will occur at some level in proportion to the neutrino mass.

It has been argued that if the neutrino has a nonzero mass, it will most likely have to be a Majorana mass so that one can explain why it is so small. Although such a claim does have theoretical merit, it should also be pointed out that there is no understanding of why, for example, m_e/m_b is as small as 10^{-4}, and here both the electron and the b-quark are indisputably Dirac particles. Therefore, one should not rule out a Dirac mass for the neutrino simply because $m_\nu/m_e \lesssim 10^{-4}$, i.e. $m_\nu \lesssim 50$ eV.

LEFT-RIGHT NEUTRAL-CURRENT ANALYSIS

In the $SU(2)_L \times SU(2)_R \times U(1)_{B-L}$ gauge model, the neutral-current interaction is given by

$$H_{NC} = g_L W_{3L} J_{3L} + g_R W_{3R} J_{3R} + \tfrac{1}{2} g_C C J_{B-L}, \tag{3}$$

where g_L, g_R, g_C are gauge couplings and W_{3L}, W_{3R}, C are the corresponding vector gauge bosons. The electromagnetic current is given by

$$J_{em} = J_{3L} + J_{3R} + \tfrac{1}{2} J_{B-L} , \tag{4}$$

so that the couplings must satisfy

$$\frac{1}{e^2} = \frac{1}{g_L^2} + \frac{1}{g_R^2} + \frac{1}{g_C^2} . \tag{5}$$

For convenience, let me define

$$x_L \equiv \frac{e^2}{g_L^2} , \; x_R \equiv \frac{e^2}{g_R^2} , \tag{6}$$

then the condition

$$x_L + x_R = 1 - \frac{e^2}{g_C^2} < 1 \qquad (7)$$

must hold. In terms of x_L and x_R, the photon is given by

$$A = x_L^{\frac{1}{2}} W_{3L} + x_R^{\frac{1}{2}} W_{3R} + (1 - x_L - x_R)^{\frac{1}{2}} C , \qquad (8)$$

and the weak bosons can be chosen to be

$$Z = (1 - x_L)^{\frac{1}{2}} W_{3L} - \left(\frac{x_L}{1 - x_L}\right)^{\frac{1}{2}} [x_R^{\frac{1}{2}} W_{3R} + (1 - x_L - x_R)^{\frac{1}{2}} C] , \qquad (9)$$

and

$$D = (1 - x_L)^{-\frac{1}{2}} [(1 - x_L - x_R)^{\frac{1}{2}} W_{3R} - x_R^{\frac{1}{2}} C] . \qquad (10)$$

With the notation

$$J_{ZL} \equiv J_{3L} - x_L J_{em} , \qquad J_{ZR} \equiv J_{3R} - x_R J_{em} , \qquad (11)$$

the weak-interaction part of Eq. (3) then becomes

$$H_{NC} = \frac{e}{[x_L(1 - x_L)]^{\frac{1}{2}}} \left\{ ZJ_{ZL} + D \left[\frac{x_L}{x_R(1 - x_L - x_R)} \right]^{\frac{1}{2}} \times [x_R J_{ZL} + (1 - x_L) J_{ZR}] \right\} . \qquad (12)$$

Consider now the most general mass-squared matrix for the Z and D vector bosons

$$M^2 = \frac{2e^2}{x_L(1 - x_L)} \begin{bmatrix} A & B \\ B & C \end{bmatrix} . \qquad (13)$$

At low energies, the effective neutral-current interaction is then given by[6]

$$H_{NC}^{eff} = \frac{4G_F}{\sqrt{2}} \left\{ (\rho_1 J_{ZL})^2 + (\rho_2 J_{ZL} + \eta J_{ZR})^2 \right\} , \qquad (14)$$

where

$$\frac{8G_F}{\sqrt{2}}\,\rho_1^2 = \frac{1}{2A}\,, \tag{15}$$

$$\frac{\rho_2}{\rho_1} = \left[\frac{C}{A} - \frac{B^2}{A^2}\right]^{-\frac{1}{2}} \left\{\left[\frac{x_L x_R}{1 - x_L - x_R}\right]^{\frac{1}{2}} - \frac{B}{A}\right\}\,, \tag{16}$$

and

$$\frac{\eta}{\rho_1} = \left[\frac{C}{A} - \frac{B^2}{A^2}\right]^{-\frac{1}{2}} \left[\frac{x_L(1 - x_L)^2}{x_R(1 - x_L - x_R)}\right]^{\frac{1}{2}} \,. \tag{17}$$

The most general form of the effective low-energy neutral-current interactions for left-right gauge models is thus Eq. (14) and it has five free parameters: x_L, x_R, ρ_1, ρ_2, and η.

RESTRICTIONS FROM HIGGS STRUCTURE

In the above, no assumption has been made with regard to the choice of Higgs-boson multiplets used to break $SU(2)_L \times SU(2)_R \times U(1)_{B-L}$ down to $U(1)_{em}$. If the only ones used are those with quantum numbers (I_L, I_R, B-L) equal to ($\frac{1}{2}$, 0, -1), (0, $\frac{1}{2}$, 1), ($\frac{1}{2}$, $\frac{1}{2}$, 0), (1, 0, 2), and (0, 1, -2), then

$$\eta > \rho_2,\ \tfrac{1}{2} < \rho_1^2 \leq 1\ . \tag{18}$$

If ($\frac{1}{2}$, $\frac{1}{2}$, 0) and (1, 0, 2) are left out, the special case $\rho_1 = 1$, $\rho_2 = 0$ is obtained which corresponds to having ν_L interactions the same as in the standard model. However, the ($\frac{1}{2}$, $\frac{1}{2}$, 0) is needed for fermion masses.

COMPARISON WITH DATA

Most previous analyses of left-right models assume a specific value for the ratio g_L/g_R as well as a definite set of Higgs bosons. Hence the five parameters are often severely constrained on purpose. The approach Vernon Barger, Kerry Whisnant and I took was different in that we simply used Eq. (14) in its most general form and compared it with all present available data on neutral-current interactions.[6,7] We performed our analysis first without and then with the constraints of Eq. (18).

We followed Kim, Langacker, Levine, and Williams[2] in treating the neutrino data as well as those on polarized electron scattering on deuterium. We used updated results on the weak charge from experiments on parity nonconservation in atomic transitions and on forward-backward lepton asymmetries in e^+e^- annihilation. We then obtained constraints on the five parameters x_L, x_R, ρ_1, ρ_2, and η to within one standard deviation of the best fit, which is given by[7]

$$\begin{aligned} x_L &= 0.232 \pm 0.007, \\ x_R &= 0.75 \pm 0.21, \\ \rho_1^2 &= 0.99 \pm 0.01, \\ \rho_2^2 &= 0.048 \pm 0.015, \\ \eta &= 0.17 \pm 0.05, \end{aligned} \tag{19}$$

where the errors quoted for each parameter only apply if other parameters are fixed at their best values, and the actual allowed region in parameter space is much bigger than what Eq. (19) indicates.

In Fig. 1, we show the allowed region in $x_L - x_R$ space, with and without the condition of Eq. (18). We also find in general

$$0.94 \le \rho_1^2 + \rho_2^2 \le 1.16, \tag{20}$$

as expected qualitatively by comparing Eq. (14) with Eq. (1). For other significant correlations between parameters, the reader is urged to consult Ref. 7.

NEUTRAL-BOSON MASS PREDICTIONS

In Fig. 2, we show the allowed region in $M_{Z_1} - M_{Z_2}$ space as predicted by the left-right model. We find the bounds

$$\begin{aligned} 83 < M_{Z_1} &< 116 \text{ GeV}, \\ M_{Z_2} &> 205 \text{ GeV}, \end{aligned} \tag{21}$$

in the general case, and

$$\begin{aligned} 89 < M_{Z_1} &< 100 \text{ GeV}, \\ M_{Z_2} &> 210 \text{ GeV}, \end{aligned} \tag{22}$$

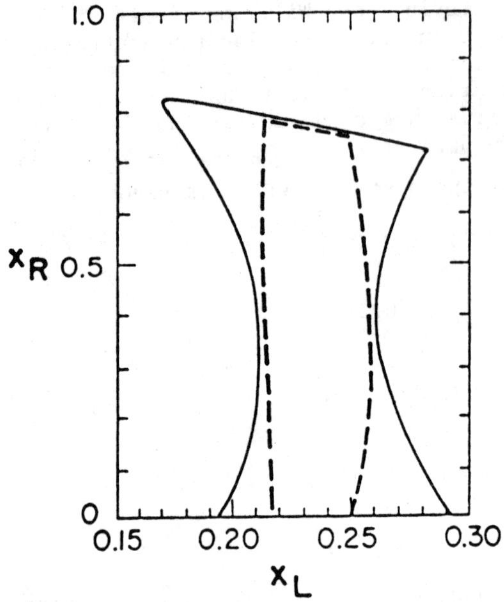

Fig. 1. Region in x_L - x_R space allowed by data. Solid boundary is for the general case. Dashed boundary is with the condition of Eq. (18).

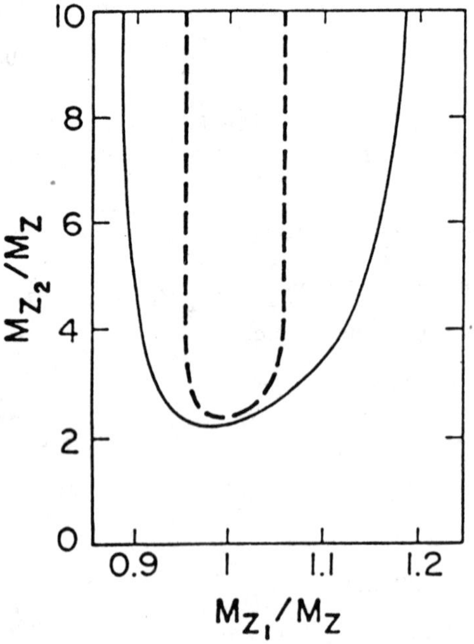

Fig. 2. Region in M_{Z_1} - M_{Z_2} space as predicted by left-right model. Solid and dashed boundaries as in Fig. 1.

with the condition of Eq. (18). The mass values are normalized to $M_Z = 93.8$ GeV, assuming that radiative corrections in the left-right model are about the same as those in the standard model. If there is no $W_L - W_R$ mixing, then we can use $M_{W_L} = (x_W/x_L)^{\frac{1}{2}} M_W$ to set the bounds

$$75 < M_{W_L} < 97 \text{ GeV} \tag{23}$$

as well.

SO(10) IMPLICATIONS

The gauge group $SU(2)_L \times SU(2)_R \times U(1)_{B-L}$ is naturally derived from SO(10) grand unification via the chain SO(10) breaking into $SO(6) \times SO(4)$ at M_U, then $SO(6) \sim SU(4)$ breaking into $SU(3)_{color} \times U(1)_{B-L}$ at M_C, and $SO(4) \sim SU(2)_L \times SU(2)_R$ breaking into $SU(2)_L \times U(1)_R$ at M_R. Let us leave out for now the contributions of the Higgs bosons, then the evolution equations[4] for the couplings are very simple and we get[8]

$$\frac{11\alpha}{3\pi} \ln \frac{M_U}{M_W} = \frac{1}{2} - \frac{4}{3}\frac{\alpha}{\alpha_s} - \frac{1}{2}(x_R - x_L), \tag{24}$$

$$\frac{11\alpha}{3\pi} \ln \frac{M_C}{M_W} = 1 - \frac{2}{3}\frac{\alpha}{\alpha_s} - x_L - x_R, \tag{25}$$

and

$$\frac{11\alpha}{3\pi} \ln \frac{M_R}{M_W} = x_R - x_L. \tag{26}$$

The exact magnitudes of the mass scales M_U, M_C, and M_R may be unknown, but we do know that $M_W \leq M_R$, $M_C \leq M_U$. From this, we get

$$x_R \leq 1 - x_L - \frac{2}{3}\frac{\alpha}{\alpha_s}, \quad (M_W \leq M_C) \tag{27}$$

$$x_R \geq 1 - 3x_L + \frac{4}{3}\frac{\alpha}{\alpha_s}, \quad (M_C \leq M_U) \tag{28}$$

$$x_R \geq x_L, \quad (M_W \leq M_R) \tag{29}$$

$$x_R \leq \frac{1}{3} + x_L - \frac{8}{9}\frac{\alpha}{\alpha_s}, \quad (M_R \leq M_U) \tag{30}$$

where α_S is the QCD analog of α, evaluated at $2M_W$. From the above, a trapezoidal region in x_L - x_R space is carved out as the allowed region from these theoretical considerations, as shown in Fig. 3.

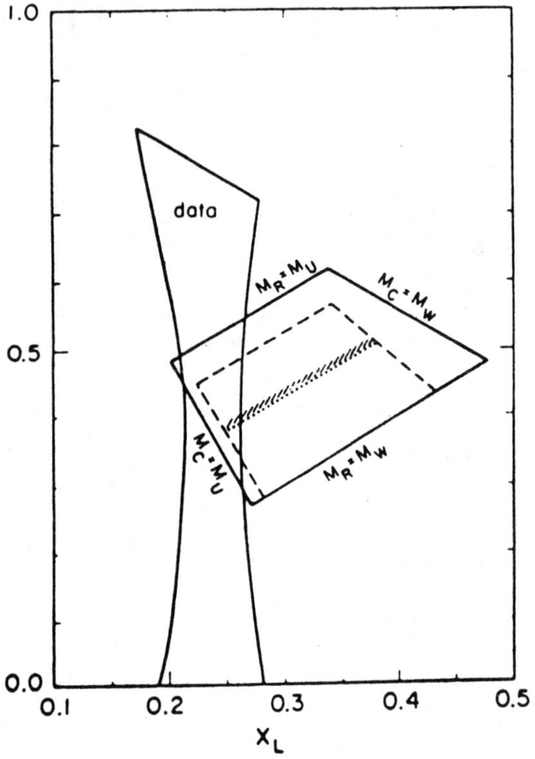

Fig. 3. Regions in x_L - x_R space allowed by data and by SO(10). Solid (dashed) quadrilateral is for the case without (with) Higgs contributions. Dotted line is for $M_U = 2 \times 10^{14}$ GeV.

The overlap with the region allowed by data tells us at once that $M_C/M_W > 10^7$ and $M_R/M_W > 25$. The bound on M_C so obtained is four orders of magnitude better than that from $K_L \to \mu e$.

The above-mentioned pattern of symmetry breaking in SO(10) can be realized with Higgs multiplets $\underline{54}$ at M_U, $\underline{210}$ at M_C, $\underline{45}$ at M_R, and $\underline{10}$, $\underline{16}$, and $\underline{126}$ at M_W. Their effect on Eqs. (27) to (30) is to reduce the allowed region in x_L - x_R space, as shown in Fig. 3, with the implications $M_C/M_W > 10^8$ and $M_R/M_W > 10^4$. From the present limit on the proton lifetime, i.e. 5×10^{30} years, it can be argued that $M_U > 2 \times 10^{14}$ GeV is a conservative estimate. The dotted line in Fig. 3 is for this value of M_U based on Eq. (24) with the Higgs contributions added, and the region above it is excluded. From the remaining very small region of overlap between experiment and theory, we then find $M_C/M_W > 10^{11}$, and $M_U/M_W < 10^{13}$ which corresponds to an upper limit of 10^{33} years for the proton lifetime. The bounds on the neutral vector bosons are tightened to

$$88 < M_{Z_1} < 93 \text{ GeV},$$
$$M_{Z_2} > 240 \text{ GeV}. \qquad (31)$$

This talk was based on work done in collaboration with V. Barger, K. Whisnant, N. G. Deshpande, and R. J. Johnson, and supported in part by the U.S. Department of Energy under Contract No. DE-AM03-76SF00235.

REFERENCES

1. S. Weinberg, Phys. Rev. Lett. 19, 1264 (1967); A. Salam and J. C. Ward, Phys. Lett. 13, 168 (1964); S. L. Glashow, Nucl. Phys. 22, 579 (1961); S. L. Glashow, J. Iliopoulos, and L. Maiani, Phys. Rev. D 2, 1285 (1970).
2. J. E. Kim, P. Langacker, M. Levine, and H. H. Williams, Rev. Mod. Phys. 53, 211 (1981).
3. W. J. Marciano and A. Sirlin, in Proc. of Second Workshop on Grand Unification, edited by J. P. Leveille et al. (Birkhäuser, Boston, 1981), p. 151.
4. H. Georgi, H. R. Quinn, and S. Weinberg, Phys. Rev. Lett. 33, 451 (1974).
5. H. Georgi and S. L. Glashow, Phys. Rev. Lett. 32, 438 (1974).
6. V. Barger, E. Ma, and K. Whisnant, Phys. Rev. Lett. 48, 1589 (1982).
7. V. Barger, E. Ma, and K. Whisnant, Phys. Rev. D 26 (Nov. 1, 1982).
8. V. Barger, E. Ma, K. Whisnant, N. G. Deshpande, and R. J. Johnson, Phys. Lett. B (in press).

COMPOSITE WEAK BOSONS AND STRUCTURE OF WEAK INTERACTIONS

R.N. MOHAPATRA†
Dept. of Physics, City College of the City University of New York
New York, New York 10031

ABSTRACT

A new approach to weak interactions is presented where possible similarities between weak and nuclear forces are exploited in a composite model of quarks and leptons. A new hypercolor dynamics, similar to quantum chromodynamics, but with a scale of order one TeV is postulated to act on the constituents of quarks and leptons and provide the binding mechanism. This leads to a composite picture of the W-bosons and provides an explanation of the origin of the Fermi constant. Understanding the "large" value of the weak mixing angle $\sin\theta_W$ seems to impose a constraint on the electric charge structure of the constituents.

†Work supported in part by National Science Foundation grant No. PHY78-24888 and a CUNY-FRAP award.

INTRODUCTION

There has been spectacular progress in our understanding of the various "elementary" particle interactions during the decade of the seventies. Quarks and leptons (eighteen quarks and six leptons) have been identified as a more fundamental layer of matter after the proton and the neutron with the basic forces of nature - the weak, electromagnetic and strong operating on them. The observed nuclear forces and β-decay forces etc. are understood as consequences of those basic interactions.

The nature of the so-called basic interactions are also now widely believed to be connected with a dynamical symmetry principle - the non-abelian gauge symmetries.[1] The strong interactions are supposed to arise from the $SU(3)_c$ gauge degree of freedom associated with color[2,3] (each quark coming in three colors and hence $SU(3)_c$) and electroweak interactions being consequences of flavor gauge degrees of freedom.[4] At low energies, it is sufficient to assume the electroweak gauge group to be $SU(2)_L \times U(1)$[5], (with parity being intrinsically broken in interactions), that uses only a subset of all the available gauge degrees of freedom. While the strong and electromagnetic forces are associated with unbroken local symmetries, the weak interactions in this framework must be associated with broken local symmetries, if it has to describe the real world.

Unbroken local symmetries have associated with them massless gauge bosons, which lead to long range forces, no long range component appears to be present in nuclear forces. Adding to this puzzle is of course the failure of the live quarks to show up in experiments to date. Both these difficulties are resolved by postulating the principle of confinement of color non-singlet degrees of freedom. Alternatively, in physical spectrum, only color singlet state can appear. This explains why quarks, which are colored ($SU(3)_c$ non-singlet) are not observed. The $SU(3)_c$ massless gauge bosons - the colored gluons, cannot be exchanged singly between the observed hadrons, which are color neutral. This explains the absence of Coulomb-like (one photon exchange) long range color forces.

The phenomenon of confinement is supposed to be a consequence of unbroken non-abelian nature of strong interactions and could therefore be considered as a new dynamical principle in constructing models.

On the other hand, in broken local symmetry theories that are supposed to describe the weak interactions, the spontaneous breaking of $SU(2)_L \times U(1)$, is introduced by hand - for example - by giving non-zero vacuum expectation value to scalar fields, $<\phi> \neq 0$.[1] This mass scale leads to the Fermi coupling constant G_F: $G_F \sim \frac{1}{<\phi>^2}$ and is adjusted to a value $\sim$ Tev to explain the weakness of weak interactions (i.e. $G_F \sim 10^{-5}(\text{Gev}^{-2})$ and predicts massive gauge bosons m_W and m_Z with masses: m_W = 83 Gev; m_Z = 93 Gev. The origin of this mass scale, however, remains a mystery.

There are also other unsatisfactory aspects of the standard $SU(2)_L \times U(1)$ model. For instance, origin of parity violation remains a mystery. Furthermore, the U(1)-quantum number has no physical meaning but is merely a parameter to be adjusted to fit the electric charge. A solution of both these mysteries is to postulate that weak interactions are intrinsically left-right symmetric with the associated broken local symmetry being $SU(2)_L \times SU(2)_R \times U(1)$[6]. It is then clear that the U(1)-generator is nothing but the quantum number B-L, with electric charge being given by:[7]

$$Q = I_{3L} + I_{3R} + \frac{B-L}{2} \tag{1}$$

This formula explains the small mass of neutrinoes as being a consequence of dominant V-A nature of weak interactions, and can be tested if neutrinoes are observed to be Majorana particles, say for example through observation of double β-decay processes.[8] Under certain extra assumptions, eqn.(1) also leads to baryon-non-conserving process of $\Delta B = 2$ type such as neutron-anti-neutron oscillations.[9] Experiments are currently under way to check both these predictions and will shed light on the nature of weak forces.

TECHNICOLOR AND ORIGIN OF G_F

It has often been speculated that the breaking of the local symmetries such as the ones presumably associated with weak interactions could have its origin in some unknown dynamical mechanism, that causes fermion bilinears $\psi\psi$ with non-vanishing weak quantum numbers to develop a vacuum expectation value. This would then bypass the need for scalar Higgs mesons, which, anyway have not been observed so far in experiments. The first concrete proposal about the nature of the dynamical mechanism responsible for symmetry breakdown was advanced by Susskind[10] and Weinberg[10]. They postulated the existence of a new superstrong force analogous to color forces (QCD) called Technicolor (QTD) operating on new kind of quarks, Q, called Techniquarks. The Q's are also assumed to be non-singlets under the electroweak gauge group G_W. Let us assume the scale parameter associated with QTD, Λ_{TC} to be of the order of a Tev i.e. α_{TC} $(\Lambda_{TC}) \approx 1$. In such a theory, one would expect from analogy with QCD that at $Q^2 \approx \Lambda_{TC}^2$, $\bar{Q}Q$ condensation would form:

$$< \bar{Q}Q > = \Lambda_{TC}^3 \tag{2}$$

If Q_L and Q_R are chosen to transform according to different representation of G_W, eqn.(2) would lead to a breakdown of G_W and one would obtain $m_W \approx g\ \Lambda_{TC}$, where g is the weak coupling constant.

While this is an interesting proposal, there are problems associated with this approach both of conceptual as well as practical nature.

In order to understand the fermion masses in this model, one needs to introduce extended technicolor,[11] which is broken by some unknown mechanism at a scale $\Lambda_{ETC} > \Lambda_{TC}$ (with roughly $\Lambda_{TC}/\Lambda_{ET} \approx \sqrt{\frac{m_f}{m_W}}$). The origin of Λ_{ETC} poses a conceptual problem for these kind of models. On the practical side, it has been argued that demanding the fermion mass spectrum be reproduced correctly constrains Λ_{ETC} so much that, one ends up with large flavor changin neutral current couplings.

Finally, of course, one is left with the uncomfortable feeling that to understand m_W, a whole plethora of new, heavy hadrons (not yet seen) has to be postulated and the new strong interaction has no other identifiable role in particle physics.

In this article, I describe a new approach to weak interactions recently advanced by Abbott and Farhi,[12] Fritzsch and Mandelbaum[13] and Barbieri, Masiero and myself,[14] which is based to a large extent on the lessons learned from the QCD approach to understanding nuclear forces and the hope that similar properties hold for a new strong force operating at the level of subquarks (or preons).[15] The immediate advantage of this approach over technicolor is that, the techniquarks are replaced by preons and the new strong interaction in addition to explaining the scale of weak interactions, also provides the binding force for subquarks forming quarks and leptons.

ANALOGY BETWEEN NUCLEAR FORCES AND WEAK INTERACTIONS

Forces observed in nature can be grouped into two categories: (i) long range forces such as gravitational and electromagnetic ones and (ii) short range forces such as nuclear forces ($R_S \sim$ Fermi) and weak interactions ($R_W \leq 10^{-3}$ Fermi). The long range forces are believed to be associated with unbroken local symmetries. Of the short range forces, nuclear forces are supposed not to be fundamental but residual effects of the basic strong force, QCD, the unbroken local $SU(3)_c$ interactions operating among the constituents of nucleons - the quarks. Since our idea is to develop a theory of the other short range force - the weak interactions in analogy with nuclear forces, let us review some basic facts about weak interactions and nuclear forces.

Our present knowledge of weak interactions is derived from experiments conducted for momentum transfers, $Q^2 << G_F^{-1} \approx (245\ \text{Gev})^2$ or at distances much longer compared to the range of weak interactions, i.e. $Q^{-2} >> R_W^2$). On the other hand, strong interactions have been explored starting from $Q^{-2} >> R_S$ (as in low energy nuclear physics) to $Q^{-2} << R_S$ (as in hadronic scattering at accelerator energies). We will derive our inspiration by studying the analogy between strong and weak forces for $Q^2 << R^{-2}$ for each case i.e. at low energies.

Two basic facts about nuclear forces operating between nucleons i.e.protons and neutrons is that, it almost exactly respects global SU(2) isospin symmetry i.e. scattering lengths after Coulomb effects are substracted are same for proton-proton and neutron proton scattering. The range of nuclear forces are determined by the exchange of vector, scalar, pseudoscalar strong interacting bosons - the ρ, ω, π, etc.

An understanding of all these are supposed to be provided by Quantum Chromodynamic theory of strong interaction. This theory is based on an unbroken non-abelian $SU(3)_c$ gauge theory operating on the 3-colors of each quark flavor. Thus, if for simplicity we choose two quark flavors (u_i, d_i) (i = 1,2,3, for color), the forces operate the same way on u and d and thus are symmetric under chiral rotations in the space of u and d. This leads to the chiral invariance group of strong interactions i.e. $SU(2)_L \times SU(2)_R \times U(1)_Y$. The only scale in such a theory is the parameter Λ_c, which is such that,

$$\alpha_c(\Lambda_c) \approx 1 \tag{3}$$

where α_c is the color analog of fine structure constant. The symmetry is spontaneously broken at $E \sim \Lambda_c$:

$$<\bar{q}q> = \Lambda_c^3 \tag{4}$$

Thus, all masses, not protected for symmetry reasons to be zero, are of order Λ_c:

$$\text{i.e. } m_\rho, m_p, m_\omega \approx \Lambda_c\, k \tag{5}$$

The pion which is the Goldstone boson associated with the breakdown of chiral symmetry acquires mass, once explicit chiral symmetry breaking terms - the quark masses are introduced: so

$$m_\pi^2 \approx m_q\, \Lambda_c \tag{6}$$

So, the range of nuclear forces is clearly $R_s \sim \frac{1}{\Lambda_c^{-1}}$ and strength of NN scattering amplitude $\approx 1; \approx 1/\Lambda_c^2 \approx 1(\text{Gev}^{-2})$, thus the forces are strong. So, QCD explains the global symmetry, the strength and the range of nuclear forces.

Turning now to weak interactions, if we look at the charged current processes such as β-decay or μ-decay:

$$n \to p + e^- + \bar{\nu}_e \tag{7}$$

and

$$\mu^- \to \nu_\mu + e^- + \bar{\nu}_e \tag{8}$$

they have a universal strength $\approx G_F = 1.027 \times 10^{-5}\, m_p^{-2}$. Similarly, if you look at neutral current scattering such as $\nu p \to \nu p$ or the parity violating piece of $ep \to ep$ scattering, they also have strength G_F. In fact, as has been observed by Bjorken,[16] the observed weak interaction phenomena at low energies can be understood by the following Lagrangian:

$$L_{wk} = \frac{4G_F}{\sqrt{2}} \left[\vec{J}_{\mu_L} \cdot \vec{J}_{\mu_L} - \frac{1}{2} \sin^2\theta_W\, \bar{\nu}\gamma_\mu (1 + \gamma_5)\nu\, J_\mu^{em} \right] \tag{9}$$

where

$$\vec{J}_{\mu_L} = \frac{1}{2} \left[\bar{Q}_L \gamma_\mu \vec{\tau}\, Q_L + \bar{\psi}_L \gamma_\mu \vec{\tau}\, \psi_L \right]$$

$$Q_L \equiv \begin{pmatrix} u_L \\ d_L \end{pmatrix} \quad ; \quad \psi_L \equiv \begin{pmatrix} \nu_L \\ e_L^- \end{pmatrix} \tag{10}$$

If we ignore the second term which could have a possible electromagnetic origin in substructure of neutrinoes, $L_{\omega k}$ is invariant under a global $SU(2)_L$ symmetry. If we add a possible right-handed current piece to $L_{\omega k}$ such as $4G_F/\sqrt{2}\,\epsilon\, (\vec{J}_{\mu R} \cdot \vec{J}_{\mu R})$ the approximate global symmetry becomes $SU(2)_L \times SU(2)_R$. This indicates that there exists a hidden symmetry of weak interactions that transforms $\nu \leftrightarrow e$ and $p \rightarrow n$ (or in quark language $u \rightarrow d$). If we denote this by a weak SU(2) symmetry, then, we can define multiplets: Q, ψ

$$Q = \begin{pmatrix} u \\ d \end{pmatrix} \qquad \psi = \begin{pmatrix} \nu \\ e^- \end{pmatrix} \tag{11}$$

and obtain an $SU(2)_W$ group. However, unlike strong interactions, weak interactions violate parity at low energies. This means, the symmetry generators involved at low energies are associated with the left-handed component of the weak isospin symmetry $SU(2)_W$ with possibly a small admixture of right-handed piece.

To proceed further, we recall that, in the case of strong SU(2) symmetry the Gell-Mann-Nishijima formula relating electric charge with Isospin is given by:

$$Q = I_3 + \frac{B}{2} \tag{12}$$

If we wish to take the nuclear force analogy for weak forces seriously, we would write a similar formula for electric charge:

$$Q = I_3^W + \frac{B-L}{2} \tag{13}$$

Since only the left-handed component of weak isospin is visible at low energy, a more physical expression for Q is:

$$Q = I_{3L}^W + I_{3R}^W + \frac{B-L}{2} \tag{14}$$

There are two ways to interpret this equation. Since no right-handed currents are visible at low energies, one possibility is to define a generator $2Y^W = I_{3R}^W + \frac{B-L}{2}$ and write

$$Q = I_{3L}^W + \frac{Y^W}{2} \tag{15}$$

Then the symmetry associated with weak interactions is $SU(2)_L \times U(1)_Y$. This is point of view advocated by the proponents of the standard $SU(2)_L \times U(1)$ gauge model of electroweak interactions. It is, of course, assumed further that $SU(2)_L \times U(1)_Y$ is a local symmetry.

Another point already alluded to before is to take the full equation seriously. In that case, a natural choice for the electroweak symmetry group is $SU(2)_L \times SU(2)_R \times U(1)_{B-L}$,[6,7] which is the left-right symmetry electroweak model advocated by us. Note that, the strong interaction symmetry is also approximate $SU(2)_L \times SU(2)_R \times U(1)$. This point of view elevates parity to the status of an exact symmetry of the Lagrangian, to be broken only by vacuum. The key to decide which symmetry is realized in nature, is of course whether neutrino is exactly massless or has a tiny mass m_ν. In fact, we have succeeded[17] in connecting the tiny m_ν to the scale at which right-handed interactions will manifest, i.e.

$$m_{W_R} \simeq \frac{m_q^2}{m_\nu} \tag{16}$$

Thus, we notice that, there is a certain degree of similarity between the properties of low energy nuclear forces and low energy weak interactions. Therefore, it is tempting to suggest that, like nuclear forces, weak interactions are also residual effects of a deeper underlying dynamics.

Adopting this point of view, one is still left with two apparent possibilities:

(i) The residual forces are associated with a broken local symmetry, so that one carry over all the known properties of $SU(2)_L \times SU(2)_R \times U(1)_{B-L}$ model discussed by us.[17] However, thre is no mathematical proof of this assumption that the low energy symmetry is a local symmetry and will not adopt this point of view.

(ii) The other possibility is to stay in close analogy with nuclear forces and write effective Lagrangian for composite fields, the quarks and leptons as one does in nuclear forces to write effective Lagrangian for the composite fields - the proton and neutron. The effective Lagrangian must however respect the global symmetry associated with the composite dynamics. Now, I give an explicit example of a mode of this type developed by Barbieri, Masiero and myself.[14] Before that, we present Table I where we summarize our stragegy: Our basic point of view is that, all fundamental interactions are associated with unbroken Yang-Mills theories and all observed short range forces are residual forces. This presupposes that quarks and leptons are composites made out of more fundamental particles, the preons. Before I proceed to discuss our model, I wish to summarize the limits on the compositeness sizes compatible with present low energy data.[18]

LIMITS ON COMPOSITENESS

The size of the subnuclear particles can be determined by looking at their electric form factors as is done for example by studying electron scattering of protons and neutrons. Similar studies for muons, electrons, and quarks[18] reveal that they point-like up to distances of order $\geq$ (1 Tev)$^{-1}$ probed up to now. The most sensitive limit on compositeness comes from the present data on (g-2) of muons, which is known to an accuracy of one part per hundred million and almost all of it accounted for by quantum electrodynamics, except for an amount $\lesssim 10^{-8}$. The contribution of lepton substructure could not therefore be larger than this.

To estimate this contribution, we note that (g-2) contribution vanishes in the limit of zero preon masses. This implies that

TABLE I

	Atomic Physics	Nuclear Physics	Weak Interactions
Basic Interactions	$U(1)_{em}$ Inv. Gauge Theory	Local $SU(3)_c$ Color Invariant Theory	Local Hyper-Color e.g $SU(4)_H$
Constituents on which the basic interactions act:	Protons and Electrons	Quarks (u,d,...)	Preons: e.g. F_u; ϕ_α, α = 1,2,3 lepton F_d
Composites	Electrically Neutral Atoms	Color Singlet Objects: Protons, Neutrons	Hyper Color Singlets: quarks, leptons
Global Symmetry	$U(1)_V$	$SU(2)_L x SU(2)_R$ x U(1) acting on protons and neutrons	$SU(2)_L x\ SU(2)_R$ x $SU(4)_c$ acting on quarks and leptons
Associated mass scale and size of Composites	$\sim m_e$	$\Lambda_{qcd} \simeq .1 - .4$ Gev $R_{p,n} \gtrsim 1/\Lambda_{qcd}$	$\Lambda_H \gtrsim 1$ Tev as inferred from (g-2) of muons. $R_{q,\ell} \gtrsim 1/\Lambda_H$
Relation between composite masses and scale of interactions:		$m_p, m_n \gtrsim \Lambda_{qcd}$	$m_{q,\ell} << \Lambda_H$ ('t Hooft anomaly constraint needed to explain this).

Table Caption: The possible analogy between atomic physics, nuclear physics and weak interactions are summarized.

$$\delta (g - 2) \approx \left(\frac{m_\mu}{\Lambda_H}\right)^2 \lesssim 10^{-8} \qquad (17)$$

which implies $\Lambda_H \gtrsim 1$ Tev. There are also constraints on compositeness coming from limits on decays such as $\mu^- \to e^- \gamma$. But they will be more model dependent and we do not discuss them here.

A COMPOSITE MODEL AND NATURE OF WEAK INTERACTIONS

In this Section, we will be a bit more technical and present a detailed composite model that realizes the programme outlined in previous sections. We will assume that the basic unbroken interactions of our theory are associated with the gauge group: $G_L \equiv SU(4)_H \times SU(3)_c \times U(1)_{em}$ operating on the following set of preons:[19] two sets of fermions:

$$F_u^\alpha \;:\; (4, 1, +\tfrac{1}{2})$$

$$F_d^\alpha \;:\; (4, 1, -\tfrac{1}{2}) \qquad (18)$$

and four scalar bosons

$$\phi_{\alpha,i} \;:\; (\bar{4}, 3, \tfrac{1}{6})$$

$$\phi_{\alpha,4} \;:\; (\bar{4}, 1, -\tfrac{1}{2}) \qquad (19)$$

The basic interaction Lagrangian involving these particles will be given by a local G-invariant renormalizable Lagrangian. As mentioned earlier, $SU(4)_H$ being an unbroken non-abelian group will be expected to have properties similar to QCD such as confinement of hypercolor non-singlet states, etc. So the observed spectrum will consist only of hypercolor singlets and color singlets. So, F_u F_d, ϕ are really not visible - just as quarks are not present in observed spectrum.

The quarks and leptons will be identified with the hypercolor singlet bound states:

$$u_i \equiv (F_u^\alpha \phi_{\alpha,i}) \;;\quad \nu \equiv (F_u^\alpha \phi_{\alpha,4})$$

$$d_i \equiv (F_d^\alpha \phi_{\alpha,i}) \;;\quad e^- \equiv (F_d^\alpha \phi_{\alpha,4}) \qquad (20)$$

We will further assume that the scale of hypercolor interactions will be given by Λ_H which will be the characteristic mass of hypercolor singlet preon composites F, $\phi \approx 1$ Tev unless they are protected by a symmetry. Thus the sizes of the quarks and leptons will be expected to be of order $\lesssim (1\ \text{Tev})^{-1} \approx 10^{-17}$ cm, since it arises from the binding force whose scale is $\approx$ 1 Tev.

Next problem we address is the smallness of quark and lepton masses in comparison with their inverse sizes. A satisfactory understanding of this feature requires the existence of an exact global chiral flavor symmetry of the hypercolor interactions, which will then be an exact symmetry also on the composites thus making the composite quarks and leptons exactly massless. As has been noted by 't Hooft and widely discussed in literature, this requires that certain triangle anomalies must be satisfied both by the preons and by the hypercolor singlet composite spin 1/2 fermions. This therefore not only explains the near masslessness of quarks and leptons, but also appears to restrict the spectrum of composite particles and could even provide an understanding of the generation structure.

It is therefore clear that we must study the global symmetry of the preon Lagrangian operating on the F's and ϕ's. We choose the most general gauge invariant, renormalizable Lagrangian involving F's and ϕ's. It is given by:

$$L = L_0 + L_m + L_\phi \tag{21}$$

$$L_0 = - \bar{F} \gamma_\mu D_\mu F - (\mathcal{D}_\mu \phi_c)^\dagger (\mathcal{D}_\mu \phi_c)$$

$$- (\mathcal{D}'_\mu \phi_\ell)^\dagger (\mathcal{D}'_\mu \phi_\ell) - \mu^2 \phi^\dagger \phi - \mu_\ell^2 \phi_\ell^\dagger \phi_\ell \tag{22}$$

$$L_m = m_{F_u} \bar{F}_u F_u + m_{F_d} \bar{F}_d F_d \tag{23}$$

$$L_\phi = \gamma \, \varepsilon^{\alpha\beta\gamma\delta} \varepsilon^{ijk} \phi_{c\alpha,i} \phi_{c\beta,j} \phi_{c\gamma,k} \phi_{\ell\delta} + \text{h.c.} \tag{24}$$

where

$$\phi_c \equiv \phi_{\alpha,i} \; : \; \phi_\ell \equiv \phi_{\alpha,4} \; ; \; \bar{F} \equiv (\bar{F}_u, \bar{F}_d)$$

$$D_\mu = \partial_\mu - \frac{i}{2} g_H \vec{\lambda}_H \cdot \vec{H}_\mu - \frac{i}{2} \tau_3 A_\mu \tag{25}$$

$$\mathcal{D}_\mu = \partial_\mu + \frac{i}{2} g_H \vec{\lambda}_H \cdot \vec{H}_\mu - \frac{i}{2} g_H \vec{\lambda}_H \cdot \vec{H}_\mu - \frac{i}{2} g_c \vec{\lambda}_c \cdot \vec{V}_\mu - \frac{ie}{6} A_\mu \tag{26}$$

$$\mathcal{D}' = \partial_\mu + \frac{i}{2} g_H \vec{\lambda}_H \cdot \vec{H}_\mu - \frac{ie}{2} A_\mu \qquad (27)$$

where $\vec{\lambda}_H$, $\vec{\lambda}c$ are $SU(4)_H$ and $SU(3)_c$ generators respectively; $\vec{H}_\mu$, $\vec{V}_\mu$ are the hypercolor and color gluons respectively; g_H and g_C are hypercolor and color gauge coupling constants. The crucial parameter of the model is the hypercolor scale Λ_H defined by the condition:

$$\alpha_H(\Lambda_H) \simeq 1 \qquad (28)$$

We assume Λ_H is of the order of a Tev.

Before proceeding further, we make a few parenthetical remarks about the scalar bosons. We remark that if $\mu_c, \mu_\ell >> \Lambda_H$, the scalar bosons decouple from the low energy Lagrangian and the spectrum has no quark or lepton composite states mentioned in (20). This theory is not of much use to us. We, therefore, restrict the mass scales $\mu_c, \mu_\ell \lesssim \Lambda_H$. As far as the m_{F_u} and m_{F_d} are concerned, we will assume them to give tiny mass to the electron, up and down quarks - i.e. in the limit of $m_F \to 0$, $m_e, m_u, m_d, m_\nu \to 0$. We will consider the chiral symmetries of this theory in the limit of $m_F \to 0$. Including L_ϕ the global symmetry of the preon Lagrangian is (in the limit of $e, g_c \to 0$)

$$G \equiv SU(2)_L \times SU(2)_R \times U(1)_F \times SU(4)_c \qquad (29)$$

Of course, once we switch on e and g_c, the global symmetry becomes:

$$G' = U(1)_L \times U(1)_R \times U(1)_F \times U(1)_{\phi_c - 3\phi_\ell} \qquad (30)$$

In order to study the bound state spectrum of hyper-color-bound preon states, we switch off electromagnetism and color forces, and consider the constraints resulting from the global chiral symmetry group G. We see that if we assume $m_{F_u}, m_{F_d} = 0$, one generation of quarks and leptons:

$$\psi^a_{L,R,i} = \begin{pmatrix} u_1 & u_2 & u_3 & \nu \\ d_1 & d_2 & d_3 & e^- \end{pmatrix}$$

$$i = u, d$$
$$a = 1 \ldots 4 \qquad (31)$$

give the same $U(1)_F\,[SU(2)_{L,R}]^2$ anomaly as the preons (F_u, F_d). This explains the near masslessness of quarks and leptons.

STRUCTURE OF WEAK INTERACTIONS IN THE COMPOSITE MODEL

As mentioned earlier, the only constraint on the structure of weak interaction in this model is the global symmetry, G of the hypercolor Lagrangian in equations (21) to (27). If we ignore the effect of $U(1)_{em}$ and $SU(3)_c$ interactions, $G \equiv SU(2)_L \times SU(2)_R \times U(1)_F \times SU(4)_c$ is an exact global symmetry of the theory; the weak interactions in this model must be left-right symmetric.[6,17] To explain the V-A theory of low energy weak interactions, we must introduce condensation of right-handed preonic operators such as:

$$< O_R > \equiv < \bar{F}_R \gamma_\mu F_R \bar{F}_R \gamma_\mu F_R > = V_R^6 \qquad (32)$$

while keeping

$$< O_L > \equiv < \bar{F}_L \gamma_\mu F_L \bar{F}_L \gamma_\mu F_L > = 0 \qquad (33)$$

where $F = \binom{F_u}{F_d}$

We wish to remark here that this pattern of dynamical symmetry breaking represents a departure of our theory from QCD in that, the QCD condensates conserve parity whereas our condensate will break parity. This may be plausible, since our theory unlike QCD has scalar bosons as basic constituents. To demonstrate how this will suppress the V + A weak interactions; we note that, at low energies (i.e. for $E \ll \Lambda_H$), the effective Lagrangian for the theory will be expressed as G-singlet of the hypercolor composite fermions $\psi_{L,R}$ and will consist of 4, 6, 8, ... point functions of $\bar{\psi}$ with dimensions provided by appropriate powers of Λ_H^{-1}. For example, one typical class of terms involving the composite fermions that respects the global symmetry G is

$$L_{eff} = \frac{1}{\Lambda_H^2} [\bar{\psi}^a_{i,L} \gamma_\mu \psi^a_{j,L} \bar{\psi}^b_{j,L} \gamma_\mu \psi^b_{i,L} G(\frac{O_L}{\Lambda_H^6})$$

$$+ \bar{\psi}^a_{i,R} \gamma_\mu \psi^a_{j,R} \bar{\psi}^b_{j,R} \gamma_\mu \psi^b_{i,R} G(\frac{O_R}{\Lambda_H^6})]$$

$$+ \ldots\ldots \qquad (34)$$

where

$$G(x) = \sum_n A_n x^n \qquad (35)$$

On substituting eqn.(32) and (33), this leads to the effective weak interaction Hamiltonian of the form:

$$L_{wk}^{eff} = \frac{a_o}{\Lambda_H^2} \bar{\psi}^a_{i,L} \gamma_\mu \psi^a_{j,L} \bar{\psi}^b_{j,L} \gamma_\mu \psi^b_{i,L}$$

$$+ \frac{G\left(\frac{v_R^6}{\Lambda_H^6}\right)}{\Lambda_H^2} \bar{\psi}^a_{i,R} \gamma_\mu \psi^a_{j,R} \bar{\psi}^b_{j,R} \gamma_\mu \psi^b_{i,R} \qquad (36)$$

One can imagine $G\left(\frac{v_R^6}{\Lambda_H^6}\right) << a_0$ in which case the right-handed currents effects in weak interactions are suppressed. For example, if $\phi_n = (-1)^n a_0$ we would get, $G(x) = \frac{a_0}{1+x} << a_0$.

There are of course other possible four Fermi terms compatible with the global symmetry. Those will be the new arbitrary parameters of the model for reasonable values of which observed weak interaction data will not lead to any trouble. We will not discuss them any further here and refer the reader to reference[14] for detailed analysis.

We now discuss the problem of fermion masses in this model. In order to generate quark and lepton masses, we add the L_{mass} to the Lagrangian. This gives, $m_u \approx m_d \approx m_e \approx m_{F_{u,d}}$. This will also give Dirac mass to the neutrino, m_ν^D of the order of m_{F_u}. While this provides an understanding of u,d and e masses, neutrino masses are much too small compared with m_ν^D. To understand the neutrino mass, we must introduce a further condensate of the following type:

$$< N_R^T C^{-1} N_R > = M_N^3 \qquad (37)$$

This generates the following mass matrix for neutrinoes:[17]

$$\begin{pmatrix} 0 & m_e \\ m_e & m_N^3/\Lambda_H^2 \end{pmatrix} \qquad (38)$$

This predicts the existence of a heavy neutral Majorana lepton with mass $m_{N_R} \simeq m_N^3/\Lambda_H^2$ as in the $SU(2)_L \times SU(2)_R \times U(1)_{B-L}$ models and leads to light Majorana neutrino mass, m_ν:

$$m_\nu \approx \frac{m_e^2 \Lambda_H^2}{M_N^3} \quad (39)$$

If we assume for definiteness $m_N \lesssim \frac{1}{2}\Lambda_H$, $m_{N_R} \simeq \frac{1}{8}\Lambda_H$

$$m_\nu \simeq 8\, m_e^2/\Lambda_H \approx 2\text{ev for } \Lambda_H \simeq 200 \text{ Gev}$$

The condensate (37) breaks the exact global symmetry associated with global lepton number, thus leading the existence of a zero mass particle, the Majoron. As it had been shown earlier by Chikashige, Mohapatra and Peccei[20], such a particle is practically invisible even if it exists. This enables us to build a consistent composite model that accommodates observed fermion mass spectra without any conflict with observations.

COMPOSITE WEAK BOSONS AND WEAK INTERACTIONS

In the composite dynamical picture of weak interactions, the W-bosons that mediate weak interactions are composite objects like ρ- mesons. So, as explained earlier, $m_W \approx \Lambda_H \approx G_F^{-1/2}$ thus providing dynamical understanding of the weakness of weak interactions. We now discuss how one can study weak interactions using an effective Lagrangian involving the weak bosons, by a vector dominance type approach as suggested by Hung and Sakurai[21]. The generalization of the vector-dominance approach to our model has been discussed by Barbieri and this author.[22]

The basic idea of this approach is to write a G-invariant effective Lagrangian involving quark and lepton weak currents, the $W_{L,R}$-bosons which transform as the weak currents under $SU(2)_L \times SU(2)_R$ weak interactions and a "photon" field $\tilde{A}_\mu$ for simplicity, we will consider the parity breaking coming only from eqn.(32) and (33) and ignore the effects of (37) on $W_L - W_R$ mixing. Under this set of hypothesis, we can write the $L_{\omega k}$ as:

$$L_{wk} = L_0 + L_{mix} + L' \quad (40)$$

where

$$L_0 = -\frac{1}{4}\tilde{F}_{\mu\nu}\tilde{F}_{\mu\nu} - \frac{1}{4}\vec{W}_{\mu\nu,L}\cdot\vec{W}_{\mu\nu,L} - \frac{1}{4}\vec{W}_{\mu\nu,R}\cdot\vec{W}_{\mu\nu,R}$$

$$-\frac{1}{2}m^2_{W_L}\vec{W}_L\cdot\vec{W}_L - \frac{1}{2}m_{W_R}\vec{W}_R\cdot\vec{W}_R$$

$$+ g(\vec{J}_{\mu_L}\cdot\vec{W}_{\mu_L} + \vec{J}_{\mu,R}\cdot\vec{W}_{\mu,R}) + \tilde{e}J^{em}_{\mu}\tilde{A} \tag{42}$$

$$L_{mix} = -\frac{1}{2}\lambda\tilde{F}_{\mu\nu}(W^3_{\ \ ,L} + W^3_{\mu\nu,R}) \tag{43}$$

L' stand for other G-invariant terms involving $W_{L,R}$, A_μ that are not very relevant for description of weak interaction at low energies. They may be relevant for example in W W-scattering etc. The γ-W mixing parameter λ is responsible for the electromagnetic piece of the neutral weak current in eqn.(8). This Lagrangian has to be diagonalized to obtain the physical vector boson fields and the physical photon A_μ. This will also give rise to the neutral current weak Lagrangian. We show that for $m_{W_R} \gtrsim m_{W_L}$, the neutral current Langrangian is indistinguishable from the $SU(2)_L \times U(1)$ model to zeroth order in $(m_{W_L}/m_{W_R})^2$, provided, we identify

$$\text{"}\sin^2\theta_W\text{"} = \frac{e\lambda}{g} \tag{44}$$

The positivity of the mass spectrum requires that $|\lambda| < \sqrt{2}$. We also find the following constraints on $W_L^{\pm}$ and Z-boson masses i.e.

$$m_{W_L^{\pm}} \lesssim \sqrt{2}\ m_{W_L} \tag{45}$$

compelling evidence for the composite W-boson picture will, of course, be the observation of a strong decay mode of the Z-boson such as $Z \to W + \pi_H$ where π_H represent some pseudo-Goldstone boson of the composite model.

WEAK MIXING ANGLE AND COMPOSITE W-BOSONS

In the preceding pages, we have developed a picture of weak interactions, where W-bosons have properties similar to the ρ-meson in QCD. This similarity, however, cannot be stretched too far. For instance, we note that $m_\rho >> \Lambda_c$, whereas we would expect (and would require in γ-W^3 mixing models) that $m_W \lesssim \Lambda_H$. Furthermore, the ρ-photon mixing parameter, $\lambda_{\gamma-\rho}$(the analog of λ) can be estimated using ρ-dominance of isovector piece of electromagnetic current and experimental data to be:

$$\lambda^2_{\gamma-\rho} \simeq \frac{e^2}{g^2_\rho} \simeq \frac{1}{300} \tag{46}$$

Since $\sin\theta_W \simeq e\lambda/g$, if λ were to be of the same order as $\lambda_{\gamma\rho}$, $\sin\theta_W$ would be predicted to be too small and would pose difficulties for the composite W-boson picture presented here.

In this section we address this question. To start with, we note that there is an important difference between QCD and QHCD in the composite spectra they predict. In the case of QCD, the associated chiral $SU(2)_L \times SU(2)_R$ symmetry is spontaneously broken leading to composite fermion masses (the nucleons) $m_N \gg \Lambda_c$. On the other hand, for QHCD, the chiral symmetry remains exact, thereby leads to massless composite fermions - the quarks and leptons. We argue[23] below that, it is this asymmetry in composite spectra, that may hold the key to understanding the magnitude of $\sin\theta_W$.

From eqn.(41), we note that, the electromagnetic current J^{em}_μ can be written as:

$$J^{em}_\mu \simeq \lambda\, m^2_W\, W^3_\mu + J^{em}_\mu\,(q,\ell,W) \tag{47}$$

where J^{em} contains pieces bilinear in the fields and we have ignored the contribution of W_R since for all our considerations, we stay much below m_{W_R}. In order to study λ, we have to study the two point function involving J^{em}_μ:

$$(q^2\,\delta_{\mu\nu} - q_\mu q_\nu)\,\Pi(q^2) = i\int e^{iq\cdot x}\, d^4x\, \langle 0|\,T\,(J^{em}_\mu(x)\, J^{em}_\nu(0))|0\rangle \tag{48}$$

The asymptotic behaviour of $\Pi(q^2)$ is dictated by asymptotic freedom of underlying QHCD exactly as in the QCD. Using this property, Shifman et al[24] have written down sum rules, known as Asymptotic Freedom Sum Rules (AFSR), which is extremely successful in describing low energy hadronic physics. An outstanding preciction[24] of the AFSR in QCD is the value of $g^2_\rho/4\pi \simeq 2.3$ (to be compared with the experimental value of $2.36 \pm .18$).

The analog of this sum rule for QHCD is:

$$\int ds\, e^{-\frac{s}{M^2}} R(s) = M^2 \sum_F Q_F^2 \left[1 + O\left(\alpha_H(M^2), \frac{\Lambda_H^2}{M^2}\right)\right] \tag{49}$$

where $R(s) = \sigma(e^+e^- \to \text{hadrons})/\sigma(e^+e^- \to \mu^+\mu^-)$ and Q_F denotes the sum over charges of the preons, the constituents of quarks and leptons. In order to study λ, we note that, the left-hand side receives contributions from the continuum of quarks and leptons and the W-boson. We choose $M >> \Lambda_H$ and assume $m_W \simeq \Lambda_H$. If we further assume that, the preon continuum starts above M, eqn.(49) leads to

$$\sum_{i=q,\ell} Q_i^2 + 12\pi^2 \frac{\lambda^2}{e^2} \left(\frac{m_W^2}{M^2}\right) \simeq \sum_F Q_F^2 \tag{50}$$

It is clear from (50) that, if we define

$$A \equiv \sum_{\substack{i=\text{quarks,}\\ \text{leptons}}} Q_i^2 - \sum_{F=\text{preons}} Q_F^2 \tag{51}$$

(A factor of 1/4 is understood for a scalar preon.) To obtain reasonable λ, A must be large and negative. Otherwise, i.e. if $A > 0$, the only way λ can be large is if $m_W << \Lambda_H$, so that the W-boson contribution is within the uncertainties in our approximation. In the case that, $A < 0$, we find,

$$\lambda^2 \simeq -\frac{\alpha}{3\pi} A \left(\frac{M}{m_W}\right)^2 \tag{52}$$

This provides a new constraint on the composite models. It turns out that, most of the existing models do not satisfy this constraint. However, pure fermion type composite models[25] or supersymmetric extensions[26] of ref.14 do satisfy this constraint. A possible (but presumably unlikely) exception to these constraints may arise if, the preonic continuum starts at $s \simeq \Lambda_H^2$. Again in this case $m_W << \Lambda_H$ and we lose control on the parameter λ.

CONCLUSION

In this paper, we have tried to develop the point of view, that the only fundamental interactions of nature are those associated with exact local symmetries of nature. We derive our inspiration from the apparent success of QED as the basis of atomic physics and QCD as the basis of nuclear physics. We propose that weak interactions are similar to nuclear forces and owe their origin to the existence of an unbroken hypercolor local symmetry which is responsible for the binding of preons to quarks and leptons exactly as QCD binds quarks to form nucleons. We have used scalar bosons and fermions as the preons and work for values of scalar mass $\mu_\phi \lesssim \Lambda_H$ and end up with a left-right symmetric theory of electro-weak interactions without broken gauge interactions. We find that consistency of the model requires the neutrinoes to be Majorana particles and consequently, we predict a zero mass pseudoscalar boson - the Majoron, whose couplings to ordinary matter are so weak that it remains invisible. Weak interactions in this model can be described by means of a vector-dominance type picture with W-bosons being composite objects. First of all, in analogy with the ρ-meson, W-bosons must have mass of order of the scale of Hypercolor forces which is $\approx R_{g,\ell}^{-1}$, the inverse radius of quarks and leptons. This provides a natural explanation of the Fermi coupling constant. Finally, we show that for these models to explain the large value of $\sin\theta_W$, the sum over (charge)^2 for the preons must be less than that over quarks and leptons. This, we believe, may provide additional guide to model building.

I wish to thank R. Barbieri for collaboration and for many useful discussions.

REFERENCES

1. C.N. Yang and R.L. Mills, Phys. Rev. 96, 191 (1954).
2. O.W. Greenberg, Phys. Rev. Lett. 13, 598 (1964);
   M.Y. Han and Y. Nambu, Phys. Rev. 139, B 1006 (1965).
3. D. Gross and F. Wilczek, Phys. Rev. D8, 3633 (1973);
   H.D. Politzer, Phys. Rev. Lett. 30, 1346 (1973).
4. For extensive references and review, see,
   P. Langacker, Phys. Reports, 72, 185 (1981);
   R.N. Mohapatra, Fortschritt der Physik (to appear) (1983);
   H. Fritzsch and P. Minkowski, Phys. Reports, (1982);
   J.C. Taylor, "Gauge Theories of Weak Interactions", Cambridge University Press (1976).
5. S.L. Glashow, Nucl. Phys. 22, 579 (1961)
   S. Weinberg, Phys. Rev. Lett. 19, 1264 (1967);
   A. Salam in Elementary Particle Theory, ed. N. Sartholm (Almquist and Wikskells, 1968), p.367.
6. J.C. Pati and A. Salam, Phys. Rev. D10, 275 (1974);
   R.N. Mohapatra and J.C. Pati, Phys. Rev. D11, 566, 2558 (1975);
   G. Senjanovic and R.N. Mohapatra, Phys. Rev. D12, 1502 (1975).
7. R.E. Marshak and R.N. Mohapatra, Phys. Lett. 94B, 222 (1980).
8. For a review, see S.P. Rosen and H. Primakoff, Purdue Preprint (1980); K. Nishiura, Osaka University Preprint (1981).
9. R.N. Mohapatra and R.E. Marshak, Phys. Rev. Lett. 44, 1316 (1980).
10. L. Susskind, Phys. Rev. D20, 2619 (1979);
    S. Weinberg, Phys. Rev. D19, 1277 (1979).
11. S. Dimopoulos and L. Susskind, Nucl. Phys. B155, 237 (1979);
    E. Eichten and K. Lane, Phys. Lett. 90B, 125 (1980).
12. L. Abbott and E. Farhi, Phys. Lett. 101B, 69, (1981).

13. H. Fritzsch and G. Mendelbaum, Phys. Lett. 102B, 319 (1981).
14. R. Barbieri, R.N. Mohapatra and A. Masiero, Phys. Lett. 102B, 319 (1981)
15. For a recent review, see H. Harari, Weizmann Preprint (1982).
16. J.D. Bjorken, Phys. Rev. D19, 335 (1979).
17. R.N. Mohapatra and G. Senjanovic, Phys. Rev. Lett. 44, 912 (1980); Phys. Rev. D23, 165 (1981).
18. S.J. Brodsky and S.D. Drell, Phys. Rev. D22, 2236 (1980); G.L. Shaw, D. Silverman and R. Slausky, Phys. Lett. 94B, 57 (1980).
19. J.C. Pati and A. Salam, Phys. Rev. D10, 275 (1974); O.W. Greenberg and J. Sucher, Phys. Lett. 99B, 339 (1981); K. Matsumoto and K. Kakazu, Prog. Theor. Phys. 65, 390 (1981); 64, 1490 (1980).
20. Y. Chikashige, R.N. Mohapatra and R. Peccei, Phys. Lett. 98B, 265 (1981).
21. P.Q. Hung and J.J. Sakurai, Nucl. Phys. B143, 81 (1978).
22. R. Barbieri and R.N. Mohapatra, Phys. Rev. D25, 2419 (1982).
23. R. Barbieri and R.N. Mohapatra, CCNY-Preprint CCNY-HEP-82/14 (1982).
24. M. Shifman, A. Vainstein, L. Okun and V. Zakharov, Nucl. Phys. B147, 385 (1978).
25. L. Abbott, E. Farhi and A. Schwimmer, MIT Preprint (1982); G. Bordi, R. Casalbouni, D. Dominici and R. Gatto, Univ. of Geneva Preprint (1982).
26. R. Barbieri, Scuola Normale Superiore Preprint (1982).

AMPLITUDE ZEROES - TESTS OF THE MAGNETIC MOMENT OF THE W BOSON AND THE QUARKS

Mark A. Samuel
Oklahoma State University, Stillwater, Oklahoma 74078

ABSTRACT

We consider the general phenomenon of amplitude zeroes, first discovered in the process $d\bar{u} \to W^-\gamma$. The effect of the quarks having an anomalous magnetic moment on the observed zeros (dips) in the angular distribution for the process $d\bar{u} \to W^-\gamma$ ($p\bar{p} \to W^-\gamma X$) is studied. It is found that for small values ($\leqslant 10^{-3}$) of the quark anomaly [$a \equiv (g-2)/2$] the distributions are practically unaffected and as a increases, the dips gradually disappear. This study might provide us with a way of obtaining an upper bound for the anomalous magnetic moment of the W bosons. Using spin-0 particles, we investigate the circumstances for amplitude zeroes to occur in the physical region and, as well, their location in phase space. Futhermore, we ask whether or not these zeroes persist in higher order. The zeroes persist for internal line radiation and for radiation from internal bubbles. However, when we consider the 1-loop correction to the scalar three-point function (with a photon attached in all possible ways), it is shown, by an explicit calculation, that amplitude zeroes do not persist in general.

INTRODUCTION

A few years ago, it was discovered[1] that the angular distribution for the process $d\bar{u} \to W^-\gamma$ ($u\bar{d} \to W^+\gamma$) vanishes at a certain angle, provided the anomalous magnetic moment parameter $\kappa = g - 1$ for the W has the value assigned by gauge theories, namely $\kappa = 1$. The angle at which this zero occurs is independent of the photon energy and is given by (for massless quarks)

$$\cos\theta = 1 - \frac{2Q}{Q+Q'} = \begin{cases} 1/3 & d\bar{u} \to W^-\gamma \\ -1/3 & u\bar{d} \to W^+\gamma \end{cases}$$

where θ is the angle between the quark and the photon directions in the $W\gamma$ center-of-mass frame and Q (Q') are the quark (anti-quark) charges. (See Fig. 1)

This peculiar behavior was proposed as a means of measuring the magnetic moment of the W in $p\bar{p}$ and pp collisions. As can be seen in Fig. 2, $p\bar{p}$ is very sensitive to κ but in pp the sensitivity is washed out. One can, however, restore the sensitivity to κ in pp collisions by binning the events according to the longitudinal momentum of the $W\gamma$ system[2], as is shown in Fig. 3.

More recently, it was pointed out that the zero also occurs (for scalar quarks as well as the standard spin-1/2 quarks) in radiative decays of the W where, in this case, the energy distribution vanishes

0094-243X/83/990165-22 $3.00

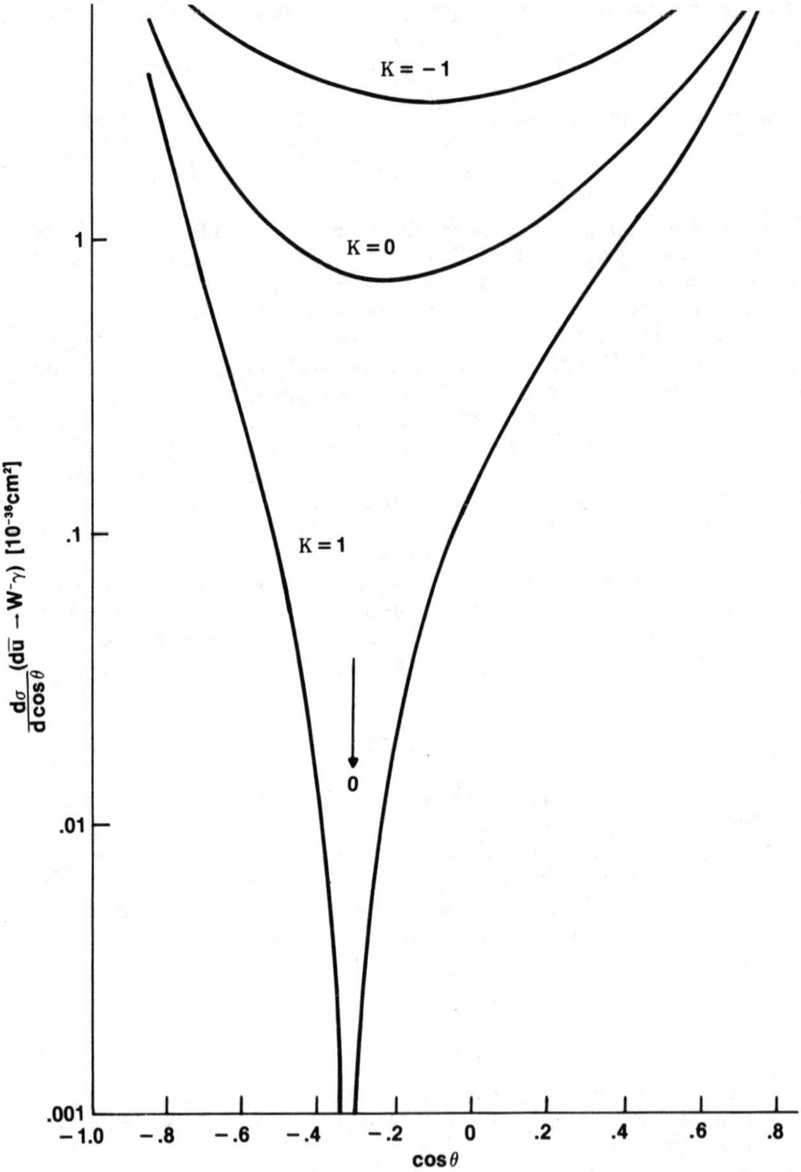

FIG. 1. The differential cross section for $d\bar{u} \to W^-\gamma$. θ is the angle between W^- and d, or between γ and $\bar{u}$, in the c.m. frame. $\sqrt{s}$ = 200 GeV and M_W = 85 GeV/c^2.

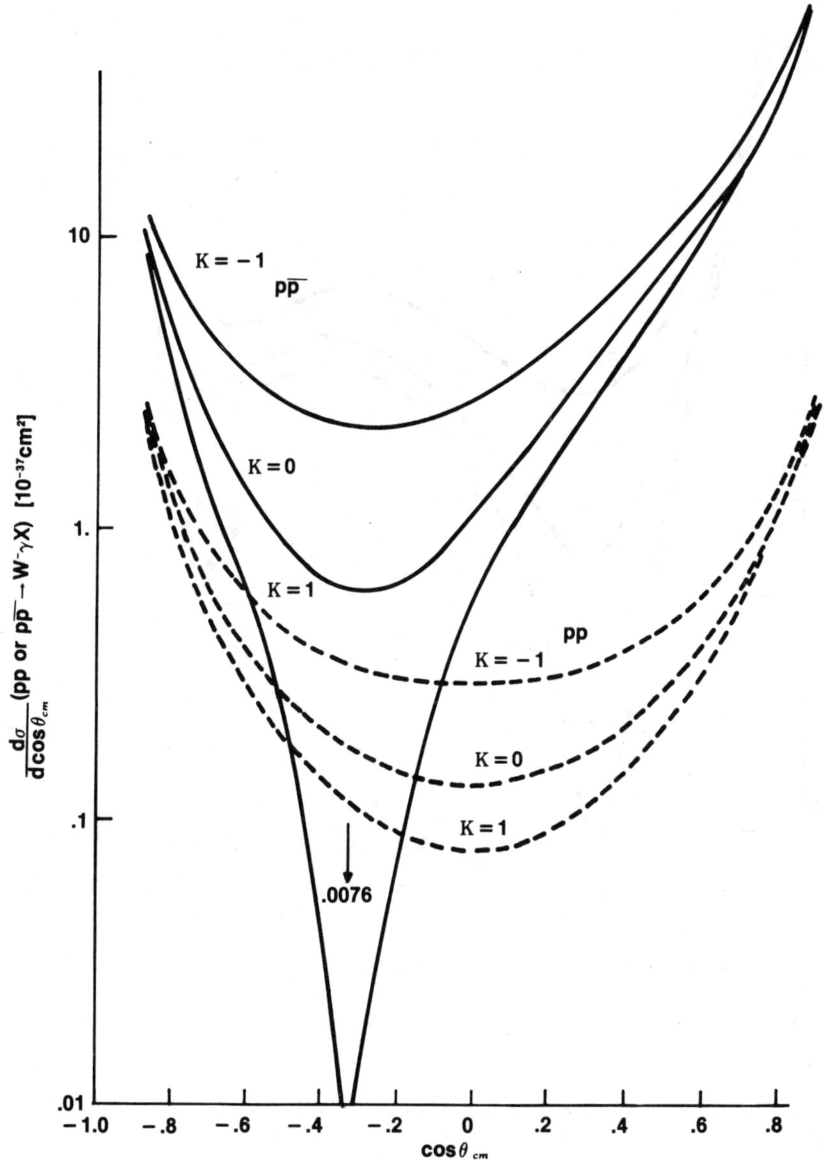

FIG. 2. The differential cross section for $pp \to W^- \gamma X$ and $p\bar{p} \to W^- \gamma X$, with a photon energy cut $E_\gamma > 30$ GeV. $\theta_{c.m.}$ is the angle between the W^- and the proton direction in the $W^- \gamma$ c.m. frame, $\sqrt{s}$ = 540 GeV and M_W = 85 GeV/c^2.

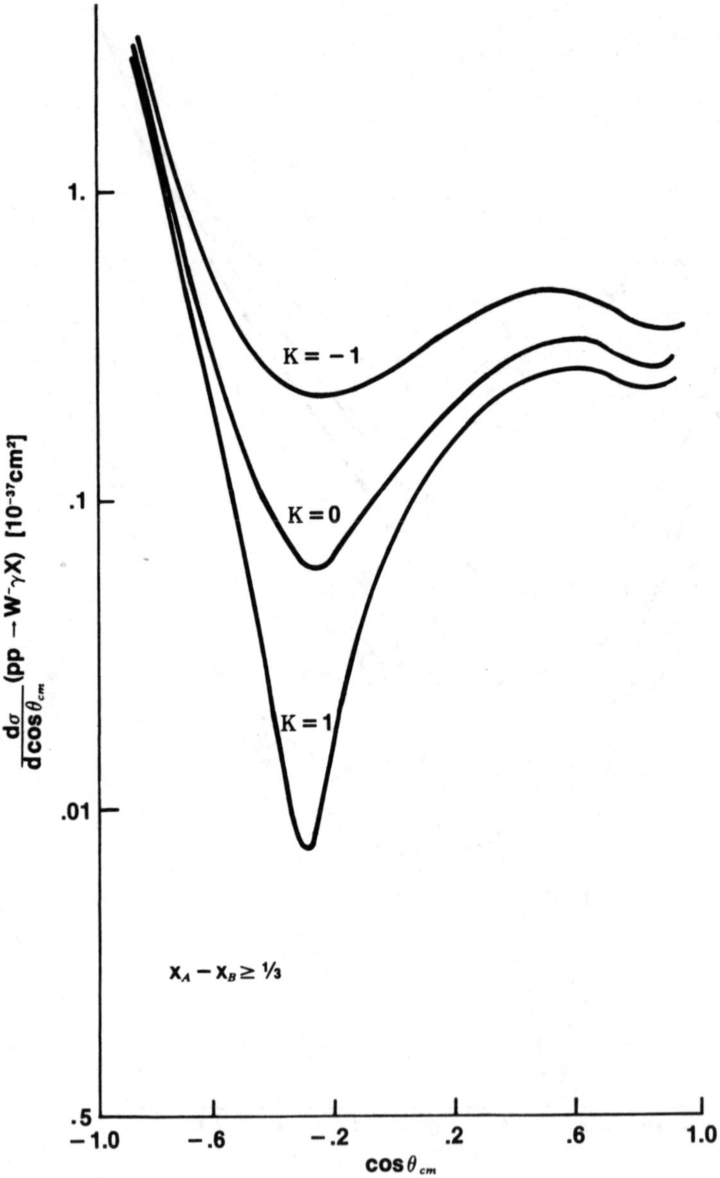

FIG. 3. The differential cross section for $pp \to W^-\gamma X$ with binning of events. The events included have $X_A - X_B \geq 1/3$ and $E_\gamma > 30$ GeV. $\theta_{c.m.}$ is the angle between the W^- and one of the incident proton beams in the $W^-\gamma$ c.m. frame. $\sqrt{s} = 800$ GeV and $M_W = 77.5$ GeV/c^2.

along a certain line in the Dalitz plot.[3] These zeroes are quite remarkable- the lowest-order amplitude vanishes for each spin state and the position of the zero is independent of the photon energy. (For massless quarks, it depends only on the quark charges). The amplitude zero, and the related amplitude factorization,[4] provides a check on the magnetic moments of both the W and the quarks[5] and the position of the zero enables a direct measure of fractional quark charges by real photons.[6]

It has been shown that the tree diagrams for these reactions have a factorization property which is quite general.[7] In this paper we first study the effect of a nonzero quark anomalous magnetic moment on the amplitude zero and, hence, on the angular distributions for both $d\bar{u} \to W^-\gamma$ and $p\bar{p} \to W^-\gamma X$. Then we investigate the general conditions for a zero to occur in the physical region. Using spin-0 for the incoming and outgoing charged particles, and standard coupling to the photon, we will see that the zeroes are essentially due to the complete destructive interference of the radiation patterns.[8] Finally, we present some results which address the following question: Do amplitude zeroes persist in higher order?

AMPLITUDE ZEROES AND THE QUARK MAGNETIC MOMENT

As discussed above, the effect of a non-zero anomalous magnetic moment of the W on the zero has been investigated and it offers a way of measuring the W magnetic moment. Here we will study the effect of a quark anomalous magnetic moment on the angular distributions for both $d\bar{u} \to W^-\gamma$ and $p\bar{p} \to W^-\gamma X$, assuming that the magnetic moment of the W boson has its gauge value $\mu_W = e/M_W$. We shall study the effect as a function of the quark anomaly defined as $a = (g - 2)/2$. For simplicity, we have chosen $a_{\bar{u}} = a_d$. We have used the standard coupling of the W boson with the other particles. The point is that if, in an experiment a deviation from the results predicted by the standard model (with no quark or W-boson anomaly) is observed, it would be of interest to know how much of this deviation could be attributed to the quark anomaly and thus place an upper bound on the W-boson anomalous magnetic moment. Of related interest, anomalous magnetic moments provide very tight constraints on possible composite models of quarks and leptons.[9]

The details of this calculation have been presented elsewhere.[5] We will confine ourselves here to a brief discussion of the results. Figure 4 shows the differential cross section for $d\bar{u} \to W^-\gamma$ as a function of $\cos\theta$. We have taken the values $\sqrt{s} = 200$ GeV, $M_W = 77.4$ GeV, and $m_u = m_d = 0.3$ GeV. As expected, one finds the zero at $\cos\theta = -1/3$ for $a = 0$. For values $a \sim 10^{-3}$ a significant change in the distribution appears and this increases as a increases. Beyond $a \sim 10^{-2}$ even a dip is hardly noticeable and the whole curve moves up higher.

Figure 5 shows the differential cross section for $p\bar{p} \to W^-\gamma X$ as a function of $\cos\theta_{c.m.}$ where $\theta_{c.m.}$ is the angle between the $W^-$and the proton beam in the $W^-\gamma$ c.m. frame. We have chosen the c.m. energy to be $\sqrt{s} = 540$ GeV, $M_W = 77.4$ GeV, and $m_d = m_u = m_s = 0.3$ GeV. We also have introduced a cut in the photon energy $E_\gamma > 30$ GeV. The effect

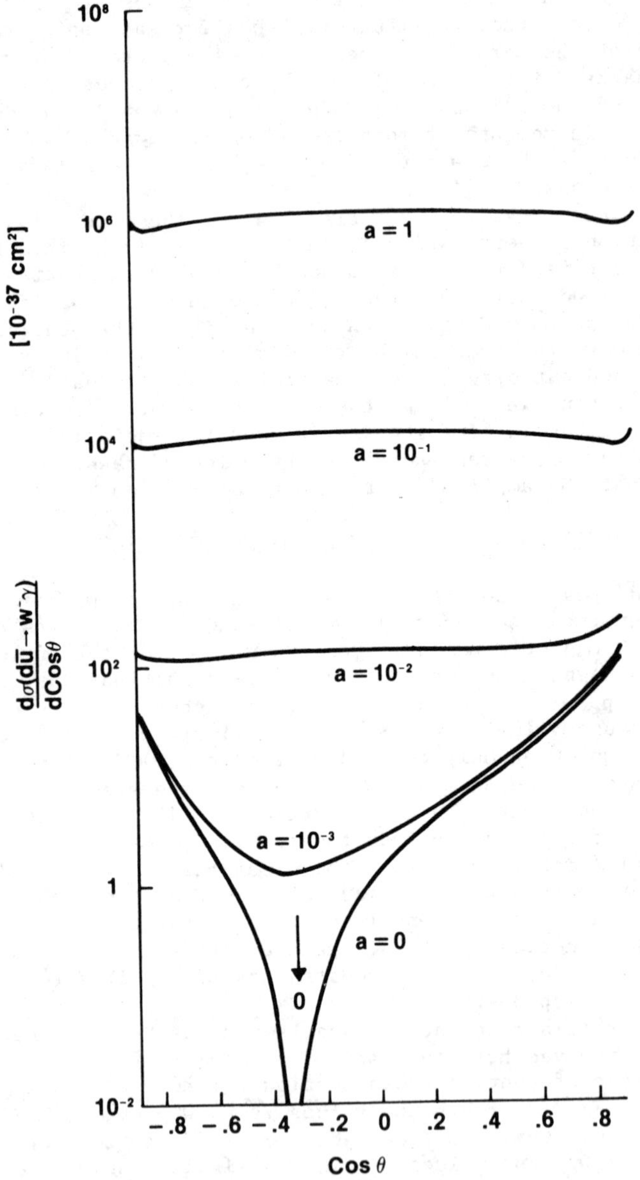

FIG. 4. The differential cross section for $d\bar{u}\to W^-\gamma$. θ is the angle between the W^- and the d in the c.m. frame. $\sqrt{s}$ = 200 GeV and M_W = 77.4 GeV/c^2.

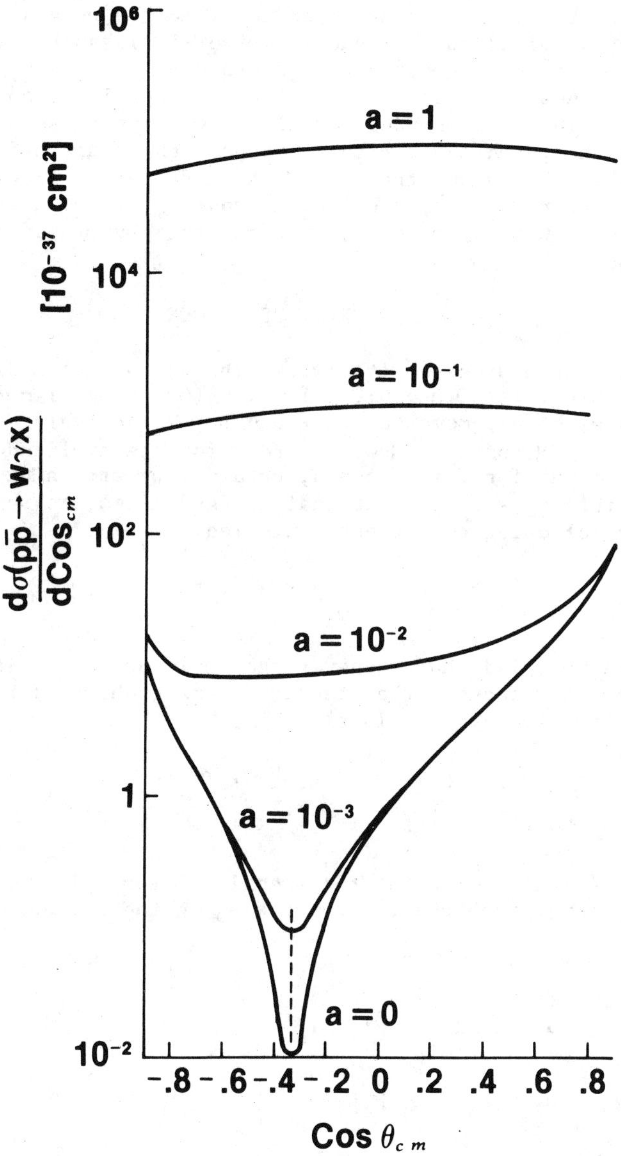

FIG. 5. The differential cross section for $p\bar{p}\to W^-\gamma X$ with a photon energy cut $E_\gamma > 30$ GeV. $\theta_{c.m.}$ is the angle between the W^- and the proton beam in the $W^-\gamma$ c.m. frame. $\sqrt{s}$ = 540 GeV and M_W = 77.4 GeV/c^2.

of putting the strange-quark mass equal to the u-(and d-) quark mass has been estimated to introduce an error of less than 10%. Looking at Fig. 5, we again observe very sharp dips (at $\cos\theta_{c.m.} = -1/3$) for small a values. For larger values of a, the dip disappears, as expected. The parton-model formula for the $p\bar{p}$ cross section in terms of the p and $\bar{p}$ distribution functions is given in Ref 1.

We have not shown the $pp \to W^-\gamma X$ differential cross section-but here, even for small a, the dips at $\cos\theta_{c.m.} = \pm 1/3$ tend to be washed out. One can however, overcome this by proper binning of events as in Fig. 3.

AMPLITUDE ZEROES

Consider a process with 1 real photon and n+1 additional external particles Q and Q_i, i=1, ...n (Q and Q_i also denote their charges) with four-momentum k, P and p_i respectively. The masses of Q and Q_i are M and m_i. Now, in order for the amplitude for the set of tree graphs for this process, obtained by attaching the photon in all possible ways to the external charged lines, to vanish, we must have the following conditions satisfied:

$$\frac{k.P}{Q} = \frac{k.p_1}{Q_1} = \frac{k.p_2}{Q_2} = \ldots. = \frac{k.p_n}{Q_n} \qquad (1)$$

It can be verified that whenever the amplitude for this process contains, as a factor, the standard bremsstrahlung form (The photon polarization four-vector is ε)

$$A_o = \left(\sum_{i=1}^{n} \frac{Q_i p_i}{k.p_i} - \frac{QP}{k.P} \right). \ \varepsilon \qquad (2a)$$

(This is certainly true at the tree level for spin-0 charges) the zeroes occur, since one can rewrite A_o in the following symmetrized form.

$$A_o = \frac{1}{2k.P} \sum_{i,j}^{n} f_{ij}\, g_{ij} \qquad (2b)$$

where $f_{ij} \equiv Q_i\, k.p_j - Q_j\, k.p_i$

$$\text{and } g_{ij} \equiv \left(\frac{p_i}{k.p_i} - \frac{p_j}{k.p_j}\right) . \ \varepsilon \qquad (2c)$$

Now it can easily be seen that if one imposes the zero conditions (eq (1)), $A_o = 0$ and the amplitude vanishes. (The above forms for A_o describe the decay $Q \to Q_1 + Q_2 + \ldots + Q_n + \gamma$, but it is trivial to write A_o for other related processes).

In addition, of course, we must also ensure energy-momentum conservation. Actually four-momentum and electric charge conservation always ensure that one of the n equations in eq (1) above is trivially satisfied. One can see from eq (1) that a necessary condition for a zero to exist is that all of the charges must be of the same sign or neutral (For $m_i = 0$, this is also sufficient). For this reason, in everything which follows, we will assume that we have no opposite sign charges. We will now consider in some detail several cases of interest.[10] Since the essence of amplitude zeroes can be seen at the scalar level, in the following, we will confine ourselves mainly to scalar charges.

$Q \rightarrow Q_1 + Q_2 + \gamma$:

Using the zero conditions (eq. (1) we obtain a line of zeroes given by

$$Q_1(1-X_1) = Q_2(1-X_2) - Q\Delta \qquad (3)$$

where the X_i are the scaled-energy variables

$$X_i = 2E_i/M$$

and

$$\Delta = \frac{m_1{}^2 - m_2{}^2}{M^2} \qquad (4)$$

Whether or not these zeroes are in the physical region in a given situation can be determined by constructing the phase space boundaries which, of course, depend on the masses. Figure 6 shows 2 representative cases with $Q_1 < Q_2$, the solid line for the massless case and the dotted line for the massive case, $m_1 > m_2 > 0$. For the massless case we have the line given by

$$Q_1(1-X_1) = Q_2(1-X_2) \qquad (5)$$

which is in the physical region, for any $Q_1/Q_2 \geqslant 0$ and which agrees with the result found by Grose and Mikaelian[3] for the decay $W^- \rightarrow du\gamma$. Notice that for equal masses, $m_1 = m_2 \neq 0$ the line of zeroes is identical to the one for the massless case (eq. (5)). Furthermore, for the general massive case, $m_1 \neq m_2$ and $m_i \neq 0$, $i=1,2$, the slope of the line is identical to that in the massless (or equal mass) case, i.e. Q_1/Q_2, but the line is shifted down by $(Q/Q_2)\Delta$, for $\Delta > 0$. (In figure 6, to be specific, we have chosen $Q_1 > Q_2$ and $m_1 > m_2$, however, it is, of course, very easy to discuss the other cases as well.) Finally, we would like to point out that, in the case of a massless neutral charge $m_1 = Q_1 = 0$, the zero line is $X_2 = 1$ and for $m_2 = Q_2 = 0$, it is $X_1 = 1$.

$Q + Q' \rightarrow Q_1 + \gamma$:

In this case, the amplitude zero occurs if the photon direction relative to the incident beam (Q) direction is given by

$$\cos\theta = \left(\frac{Q'}{\beta} - \frac{Q}{\beta'}\right)/Q_T \qquad (6)$$

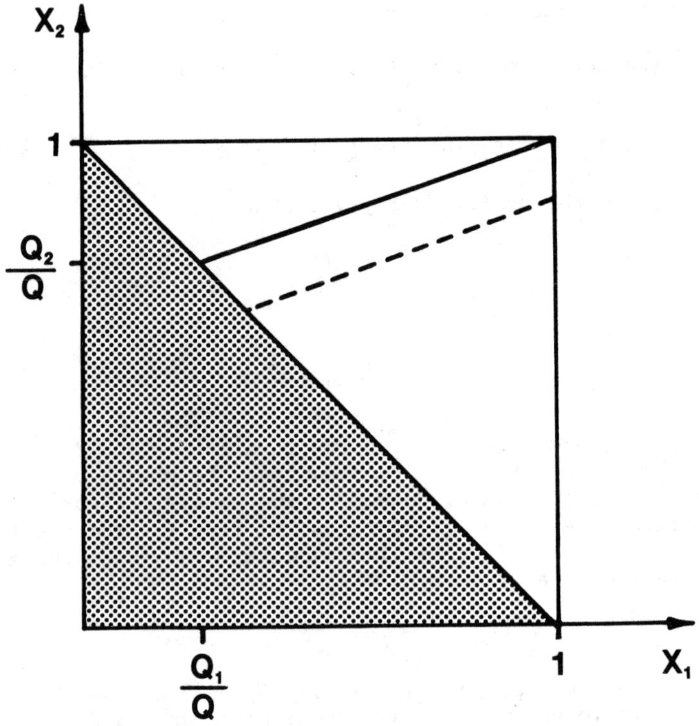

FIG. 6. The line of zeroes is shown for the radiative decay $Q \to Q_1 + Q_2 + \gamma$, with $Q_1 \leqslant Q_2$. The solid line represents the case $m_1 = m_2$, while the dashed line is for $m_1 > m_2$.

where

$$Q_T = Q + Q' \tag{7}$$

and

$$\beta' = \sqrt{1-\frac{M'^2}{E'^2}}$$
$$\beta = \sqrt{1-\frac{M^2}{E^2}} \tag{8}$$

are the velocities of Q' and Q respectively.

In the case of equal-mass incident particles, M=M', eq. (5) becomes

$$\cos\theta = \frac{1}{\beta}\left(1-\frac{2Q}{Q_T}\right) \tag{9}$$

In the massless limit, M=M'=0, this becomes

$$\cos\theta = \left(1-\frac{2Q}{Q_T}\right) \tag{10}$$

This is the result found in the original discovery[1] of the zero, where, for $d\bar{u}\rightarrow W^-\gamma$, $Q = Q_d = -1/3$, and

$$\cos\theta'=-\cos\theta=-1/3. \tag{11}$$

We have also directly verified for $d\bar{u}\rightarrow W^-\gamma$ that eqs. (6) and (9) give the correct zero direction. It can be seen from eqs. (6) and (9) that contrary to the massless case, in the case of massive incident particles, the zero direction does depend on the incident energies (as does the question of whether or not the zeroes are in the physical region). For example, in $d\bar{u}\rightarrow W^-\gamma$, for $M=M'\neq 0$, eq. (11) becomes

$$\cos\theta' = -\frac{1}{3\beta} \tag{12}$$

and the zero is in the physical region if and only if $\beta\geq 1/3$.

<u>$Q + Q' \rightarrow Q + Q' + \gamma$:</u>

It is interesting to note that <u>for equal charge to energy ratios</u>, $\frac{Q'}{X'} = \frac{Q}{X}$, the amplitude zero, <u>if it occurs at all</u>, must occur at $\cos\theta=0$. i.e. The photon direction must be perpendicular to the incident beam directions, <u>independent of charges, energies and masses</u>.

If we now take the non-relativistic limit for the incident particles, β and $\beta'\rightarrow 0$, the conditions for an amplitude zero include $Q/M = Q'/M'$ as well as the conditions on the final particles. It should be emphasized that the zero here, if it does occur in the physical region, is independent of photon direction (and, of course, photon energy). i.e. The amplitude identically vanishes!

In particular, if we now specialize to the non-relativistic collision $Q + Q' \to Q + Q' + \gamma$ we find that the zero conditions are satisfied if and only if $Q/M = Q'/M'$. i.e. The amplitude identically vanishes in non-relativistic collisions of particles with equal charge to mass ratios. Thus the zero conditions (eq. (1)) are a generalization of the non-relativistic result from classical electromagnetism that electric dipole radiation vanishes in collisions of particles with the same charge-to-mass ratio.[11,12]

A very interesting paper by S. J. Brodsky and R. W. Brown has recently been published[11] in which the following theorem is given: "Consider single photon emission by an external particle system (spin$\leqslant$1) with the internal particles unspecified. The photon couplings to the particles are to be those of gauge theory. The interactions of the particles among themselves can involve any number of fields with constant or single derivative couplings, and the derivative couplings must be of gauge theory form. Then any set of tree graphs, defined by photon emission in all possible ways, vanishes if the quantities $k \cdot p_i/Q_i$ are the same for all particles." i.e. eq. (1) is satisfied. They further state that neutral particles can be included, provided they are massless, they can propagate along the photon direction and their spin terms vanish in that configuration. Thus we have the spin-independence of the amplitude zeroes.

DO AMPLITUDE ZEROES PERSIST IN HIGHER ORDER?

Amplitude zeroes have been studied at the tree level for scalar charges[10], as well as spin-1/2 and spin-1.[11] We feel that it is now time to go to higher order and try to answer the question posed above in the title of this chapter. Since the essence of amplitude zeroes can be seen at the scalar level, in the following, we will confine ourselves (with one exception) to scalar charges, leaving to a later time the question of extending our results to higher spin. We will return to this point in the conclusions. The results reported in this chapter have been presented elsewhere.[13]

There are several special cases in which general requirements such as gauge invariance, angular momentum conservation and parity conservation are sufficient to demonstrate that, under the zero conditions, the amplitude vanishes to all orders:

(a) $Q \to Q + \gamma$: Here it is easy to show that the amplitude identically vanishes.

(b) $Q \to Q_1 + Q_2 + \gamma$ with $Q_1 = Q_2$: The amplitude, in general contains, as a factor,

$$A = E\left(\frac{p_2}{k.p_2} - \frac{p_1}{k.p_1}\right) . \varepsilon \qquad (13)$$

In lowest order, one can see from eqs. (2b) and (2c) that

$$E = \frac{f_{12}}{2k.P} \tag{14}$$

and hence, E is anti-symmetric under the combined interchange $(Q_1, p_1) \leftrightarrow (Q_2, p_2)$. Thus, A is symmetric under this interchange in lowest order and must remain symmetric to all orders. E must remain anti-symmetric, and hence, for $Q_1 = Q_2$ and $k.p_1 = k.p_2$, $E = 0$ and the zeroes persist, independent of m_i.

(c) $Q \to Q_1 + Q_2 + \gamma$ with $Q_1 = 0$: Here, the zero conditions require the neutral particle to be massless and to propagate along the photon direction.[11] This is a generalization of case (a), in fact, for $Q \to Q$ + any number of neutrals + γ the zeroes must persist to all orders.

(d) The soft photon limit: Yennie et al[14] have shown that, in general, in the soft photon limit, the amplitude contains, as a factor, the standard bremsstrahlung form (eq (2a)) and, hence, the zeroes persist. (See eqs (2b) and (2c)).

These special cases will be used as a check of our calculations, which we now describe.

It has been shown[11] that the zeroes persist for internal line radiation (for particles with spin$\leqslant$1). This goes through, because the product of the two propagators, arising from the internal line to which the photon is attached, can be expressed as a difference of propagators. Then by exploiting charge conservation at each vertex, the total amplitude for all the diagrams, including external as well as internal line radiation, can be cast into the standard form given in eq. (2a).

For example,[10] consider radiative decays

$$Q \to Q_1 + (Q_{int} \to Q_2 + Q_3 + \ldots + Q_n) + \gamma$$

where Q_{int} represents a virtual particle of mass m and, of course,

$$Q_{int} = Q - Q_1 \tag{15}$$

Otherwise, the notation is the same as before. The total amplitude for this process, obtained by attaching the photon in all possible ways to the external lines and the internal line is

$$A_{TOT} \propto -ie. \quad \left[\frac{QP}{k.p} - \frac{Q_1 P_1}{k.p_1} - \frac{Q_{int}\ (P-p_1)}{k.\ (P-p_1)}\right]$$

$$/((P-p_1-k)^2-m^2) - \left[\ \sum_{i=2}^{n} \frac{Q_i p_i}{k.p_i}\right.$$

$$- \frac{Q_{int}\,(P-p_1)}{k.\,(P-p_1)} \,]/((P-p_1)^2-m^2) \tag{16}$$

It can easily be verified that under the previous zero conditions (eq. (1)), the quantity in each square bracket in eq. (16) vanishes and, hence, $A_{TOT} = 0$. Thus, amazingly, the zeroes persist at the same location in phase space, independent of the mass of the internal particle.

We shall now discuss some examples in ϕ^3 and ϕ^4 theories which do have amplitude zeroes. We begin with a ϕ^3 theory and we will consider (i) the decay $Q \to Q_1 + Q_2$ with one scalar bubble attached to, say particle 2 (Fig. 7b). The two-point function will be denoted $\Sigma(p)$ and is given by

$$\Sigma(p) = \int d^N \ell [\ell^2-m^2]^{-1}[(\ell+p)^2-m^2]^{-1} \tag{17}$$

For convenience, we have chosen the same mass m for both scalar particles in the bubble. We then attach an external photon in all possible ways, which, in this case, leaves us with a gauge-invariant set of five diagrams.

As an extension of this, we consider (ii) the decay $Q \to Q_1 + Q_2 + Q_3$ with one scalar bubble attached to the internal line. We can think of this as a two-step process. Namely first consider the decay $Q \to Q_{12} + Q_3$ with the system Q_{12} being virtual and then the, decay $Q_{12} \to Q_1 + Q_2$.

Finally, we consider (iii) the topologically equivalent process in a ϕ^4 theory. (Fig. 7c). We have not shown the tadpole diagrams. They are trivial. Recall that a tadpole diagram does not depend on any external momentum and that photon bremsstrahlung from one automatically vanishes.

The reason the amplitude zeroes persist is due to the fact that the vertex function

$$\Lambda_\mu\,(p+k,p,k)\big|_{k^2=0} = \frac{p_\mu}{2k.p}\,[\Sigma(p+k) - \Sigma(p)]. \tag{18}$$

Using this fact, we easily obtain the amplitude $A^{(i)}$ for the first process:

$$A^{(i)} \propto \frac{f_{12}\,g_{12}}{k.P}\,\frac{\Sigma(p_2)}{p_2^2 - m_2^2} - Q_2\,\frac{p_2.\varepsilon}{k.p_2}\,\frac{p_2^2 - m_2^2}{[\,(p_2+k)^2-m_2^2]^2} \tag{19}$$

If particle Q_2 is real, $p_2^2 = m_2^2$, and the last term in eq. (19) then vanishes. The amplitude, therefore, contains, as a factor, the lowest-order amplitude A_0 and so the zeroes persist. Using the same equation for a virtual particle, we obtain for the second process ($p_{12} \equiv p_1 + p_2$):

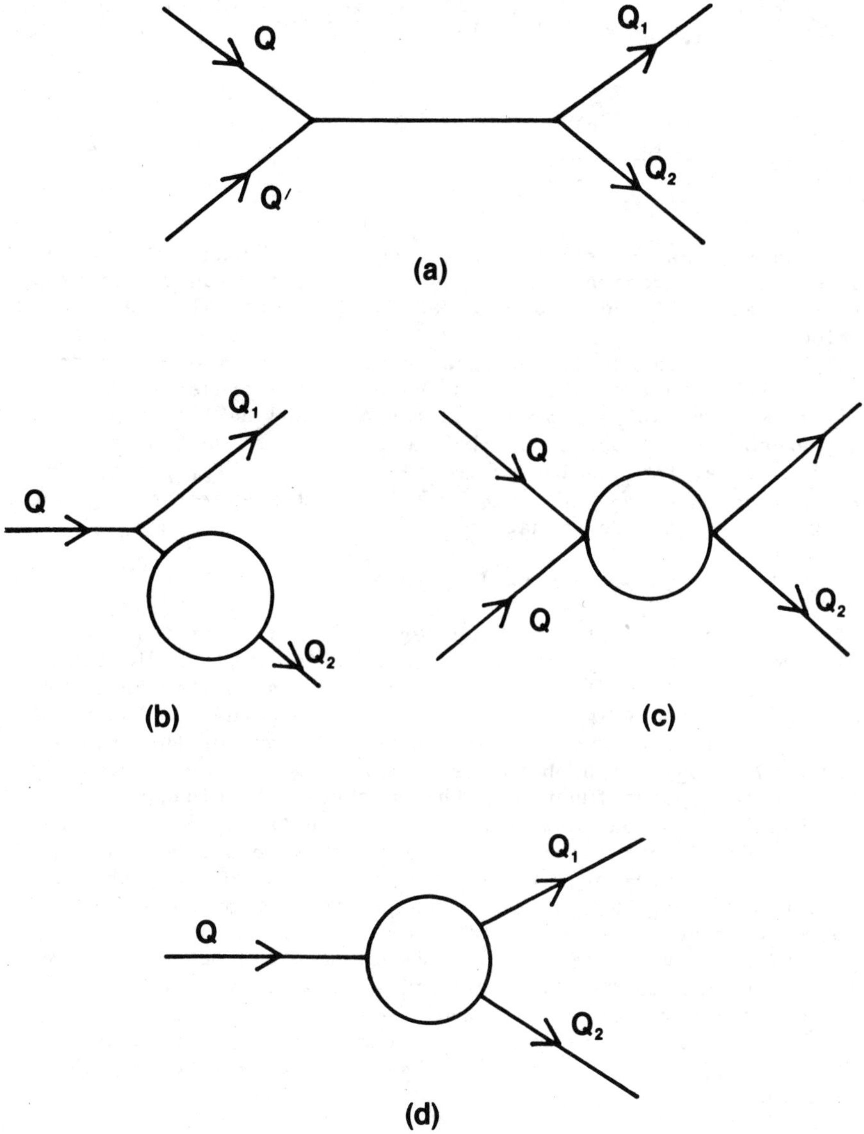

FIG. 7. Feynman diagrams for various processes involving charged scalar particles. To obtain the amplitude of interest, in each case we attach an external photon in all possible ways.

$$A^{(ii)} \propto \frac{1}{k.P}\left[-\frac{k.p_3}{k.P_{12}} f_{12}\, g_{12} + f_{13}\, g_{13} + f_{23}\, g_{23}\right]\frac{\sum(p_{12})}{[p_{12}^2 - m_o^2]^2}$$

$$+ \frac{f_{12}\, g_{12}}{k.p_{12}} \frac{\sum(P_{12} + k)}{[(P_{12}+k)^2 - m_o^2]^2} \qquad (20)$$

where m_o is the mass of the internal line. That the amplitude $A^{(ii)}$ vanishes under the zero conditions is in agreement with the previous result that an amplitude zero must persist for internal line radiation.

We have also carefully verified that if we replace one of the scalar particles in the bubble by a photon (and include diagrams with seagull terms) the above expressions for $A^{(i)}$ and $A^{(ii)}$ do not change, except that $\sum(p)$ is now the electromagnetic self-mass function. Hence, the conclusion is the same. Finally, for the ϕ^4 theory we find that the amplitude $A^{(iii)}$ has the same form as in eq. (18), except that the denominators

$(p_{12}{}^2 - m_o{}^2)^{-2}$ and $((p_{12}+k)^2 - m_o{}^2)^{-2}$ are absent.

In order to see that amplitude zeroes do not persist in general, when we include loops beyond bubbles (eg. triangles, box diagrams, etc), we will consider (iv) the 1-loop correction to the scalar three-point function (Fig. 1d). Inserting an external photon in all possible ways gives us two groups of diagrams: (a) the three-point function $\Lambda(P,p_1,p_2)$, with photon bremsstrahlung from an external leg and (b) the four-point function with inner bremsstrahlung, $\phi_\mu(P,p_1,p_2,k)$. The reason we don't get an amplitude zero for general charges Q_1 and Q_2 is related to the fact that ϕ_μ can't be reduced to a difference of three-point functions. ϕ_μ depends on two external momenta, $p_{1\mu}$ and $p_{2\mu}$ say, while in the previous examples Λ_μ depended only on one momentum.

In order to avoid complicating the computation needlessly, we restrict ourselves to the case where $m_1 = m_2 = 0$. We introduce the variables $r = 1 - x_1$, $s = 1 - x_2$ and $t = 1 - r - s$, where $x_i = 2p_i.P/M^2$ are scaling variables and $\rho \equiv (2m/M)^2$. It is relatively easy to check the gauge invariance of the amplitude. We can write the amplitude as:

$$A^{(iv)} \propto \frac{M(Q_i,r,s)}{2t}\, g_{12}$$

with

$$M(Q_i,r,s) = \frac{Q_1}{Q} M(r,s) - \frac{Q_2}{Q} M(s,r)$$

and

$$M(r,s) = \frac{r-s}{r+s} T(\frac{t}{\rho}) + T(\frac{r}{\rho}) + \frac{t-r}{t+r} [T(\frac{1}{\rho}) - T(\frac{s}{\rho})]$$

$$+ I_o (\frac{s}{\rho},\frac{t}{\rho},\frac{1}{\rho}) - I_o(\frac{r}{\rho},\frac{t}{\rho},\frac{1}{\rho}) - I_o(\frac{r}{\rho},\frac{s}{\rho},\frac{1}{\rho}). \quad (21)$$

The T and I_o functions are one-parameter integrals representing essentially the scalar three- and four-point functions, respectively. The explicit expressions exist in the literature.[15] Since $I_o(\frac{r}{\rho},\frac{s}{\rho},\frac{1}{\rho})$ is symmetric under the interchange $r \leftrightarrow s$, the scalar function $M(Q_i,r,s)$ is antisymmetric under the combined interchange $(Q_1,r) \leftrightarrow (Q_2,s)$. The amplitude, therefore, has a zero in the equal charge case (r=s). If we consider the case of one neutral particle ($Q_1 = 0$) in the final state and let $s \to 0$, we find that M tends to zero linearly with s. This means that in this case as well, the zeroes persist. (This is also true, of course, for $Q_2 = 0$ with $r \to 0$). These results are in agreement with the previous statements made for the special cases (b) and (c).

To see whether or not the zeroes persist for $Q_1 \neq Q_2$ and $Q_i \neq 0$, we have evaluated $|M(Q_i, r,s)|^2$ numerically under the zero condition ($Q_1 r = Q_2 s$) and displayed it as a function of Q_1/Q for different values of r in Fig. 8. For $Q_1/Q = 1/2$ (the equal charge case) and $Q_1/Q = 0$ (the neutral charge case) it indeed goes to zero for all values of r. For r=0.05 the function is already very small. This reflects the soft photon limit ($r,s \to o$), for which we find $|M|^2 \sim (rs)^2$. This can be seen clearly in Fig. 9, where we have plotted $\log|M|^2$ vs. log r. That we obtain a zero in the soft photon limit is expected from a previous discussion (See (d) above). It is, however, a useful check on our computations. However, the main point to be seen from Fig. 8 is that the result is not identically zero and, hence, amplitude zeroes do not persist in general.

SUMMARY AND CONCLUSIONS

We have presented some recent results, which clarify the phenomenon of amplitude zeroes, first discovered in the process $d\bar{u} \to W^- \gamma$. The effect of the quarks having an anomalous magnetic moment on the zeros (dips) in the angular distribution for the process $d\bar{u} \to W^- \gamma (p\bar{p} \to W^- \gamma X)$ has been studied. For small values ($\leq 10^{-3}$) of the quark anomaly the distributions are practically unaffected and as a increases, the dips gradually disappear. This study might provide us with a way of obtaining an upper bound for the anomalous magnetic moment of the W bosons.

Using spin-0 for the incoming and outgoing charged particles, we have seen that the amplitude zeroes are essentially due to the complete destructive interference of the radiation patterns. A necessary condition for amplitude zeroes to occur in the physical region is that the quantities $k.pi/Q_i$ must be the same for all the external charged particles. Hence, a necessary condition is that

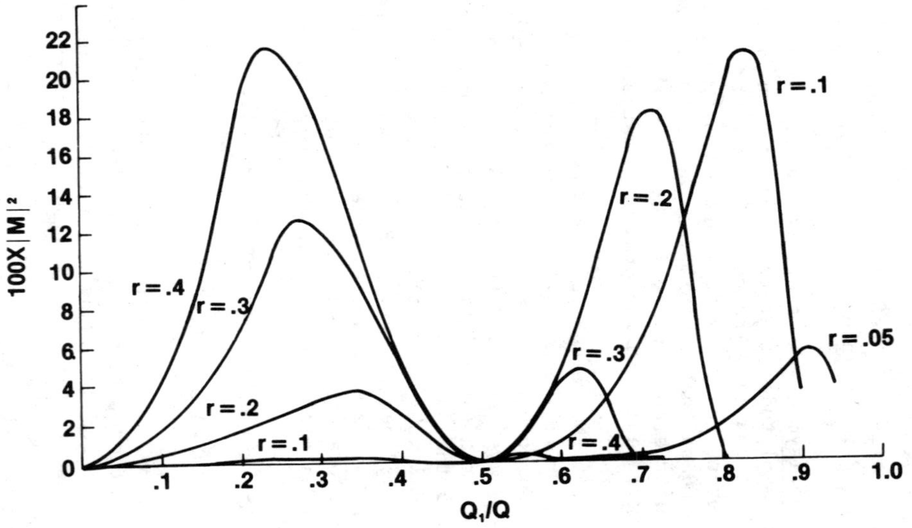

FIG. 8. 100 x $|M(Q_i,r,s)|^2$ vs. Q_1/Q for various values of r. (s is fixed by the zero condition $Q_1r = Q_2s$). We have chosen the parameter $\rho = .1$.

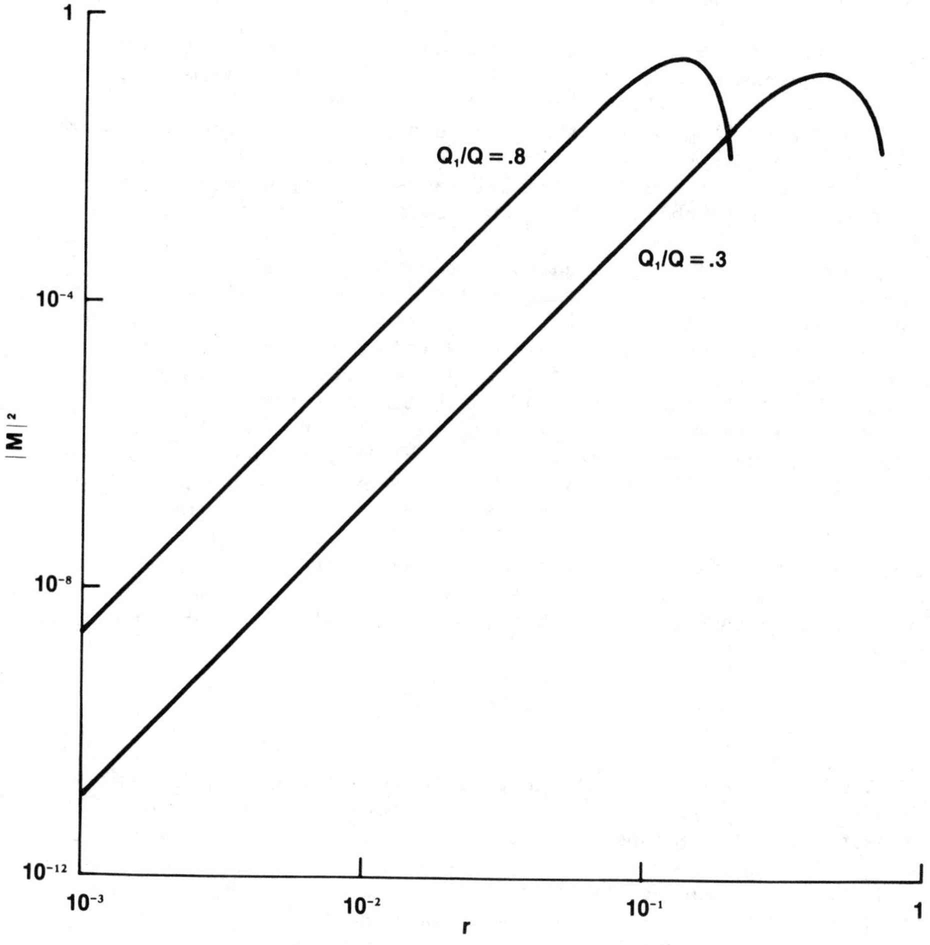

FIG. 9. Log - log plot of $|M|^2$ vs. r for two values of Q_1/Q. (s is fixed by the zero condition $Q_1 r = Q_2 s$). This shows clearly that in the soft photon limit $|M|^2 \sim r^4$. We have chosen the parameter $\rho = .1$.

there be no opposite-sign charges. At the tree level, if all but 1 of the particles is massless ($m_i = 0$), this condition is also sufficient. These results remain valid for particles with spin$\leqslant 1$, even if one includes radiation from internal lines.

We have studied the decay process $Q \rightarrow Q_1 + Q_2 + \gamma$ for scalar charges and have found that the amplitude zeroes persist in higher order for graphs involving internal bubbles, provided γ emission from all the charged lines is included. However, we find that even at the one-loop level, in general, the zero is spoiled if we consider graphs involving internal triangles (not counting the photon). Nevertheless, there are some special cases where the zero persists to all orders, and in these cases, our results are consistent with this.

Though we have not yet analyzed theories with spin-1/2 and spin-1 particles, we expect that here the situation will be more complicated. The analysis of the process $d\bar{u} \rightarrow W^- \gamma$ with an assumed W or quark anomalous magnetic moment [1,5] indicates that the zero will probably be spoiled if the set of gauge invariant diagrams involves a contribution to an anomalous W or quark magnetic moment.

The point we want to stress (and this is borne out by the examples that have been worked out so far) is that the study of the prototype scalar theory is useful in connection with the possible persistence of amplitude zeroes- if the scalar theory shows that the zero persists, perhaps the zero will be there when realistic (with spin, etc.) theories are considered, though there is no guarantee. However, if the scalar theory yields a spoiling of the zero, it is almost certain that in a realistic theory as well, the zero would not persist.

ACKNOWLEDGEMENTS

I would like to acknowledge my collaborators Morten Laursen, Achin Sen and Gary Tupper. Part of this work was done during a pleasant stay at the Aspen Center for Physics. It is a pleasure to thank Sheldon Glashow for very interesting discussions and encouragement to pursue the quark anomaly effect. I am also happy to acknowledge valuable discussions with Stan Brodsky, Bob Brown, Jacques Leveille, Karnig Mikaelian, Kim Milton and Lincoln Wolfenstein. This work was supported by the U.S. Department of Energy under Contract No. EY-76-S-05-5074.

FOOTNOTES AND REFERENCES

1. K. O. Mikaelian, M. A. Samuel and D. Sahdev, Phys. Rev. Lett. 43 746 (1979).

2. F. Paige, private communication.

3. T. R. Grose and K. O. Mikaelian, Phys. Rev. D23, 123 (1981).

4. Z. Dongpei, Phys. Rev. D22 2266 (1980); C. J. Goebel, F. Halzen and J. P. Leveille, Phys. Rev. D23, 2682 (1981).

5. M. L. Laursen, M. A. Samuel and A. Sen, Phys. Rev. D26, 2535 (1982).

6. For a recent calculation which shows that, if the photon is virtual, the zeroes are spoiled, see K. O. Mikaelian, Phys. Rev. D25, 66 (1982).

7. Independent of whether or not amplitude zeroes occur in the physical region, the amplitude can be written in the generalized-factored form shown in eq. (2b). For n=2 (i.e.-the case of 3 external charged particles and 1 real photon) eq. (2b) can be rewritten with just 1 term (In general there are n-1 terms):

$$A_0 = k.p_i \left(\frac{Q}{k.p} - \frac{Q_1}{k.p_1}\right)\left(\frac{P_2}{k.p_2} - \frac{p_1}{k.p_1}\right) \cdot \varepsilon$$

   This is referred to as factorization in refs. 4.

8. The possibility that the zeroes are essentially due to the complete destructive interference of the radiation patterns and could be seen at the scalar level was first suggested to the author by Stan Brodsky more than two years ago. Subsequent calculations performed by the author soon afterwards verified the correctness of this idea.

9. G. L. Shaw, D. Silverman, and R. Slansky, Phys. Lett. 94B, 57 (1980); S. J. Brodsky and S. D. Drell, Phys. Rev. D22, 2236 (1980); M. Bander, T. W. Chiu, G. L. Shaw, and D. Silverman, Phys. Rev. Lett. 47, 549 (1981). For a recent calculation of the effect of quark or τ form factors at large q^2 in e^+e^- annihilation (as distinct from the static moments considered in this paper), see D. Silverman and G. L. Shaw, University of California, Irvine, Report No. 82-11, 1982 (unpublished).

10. For more details, see M. A. Samuel, "Amplitude Zeroes", Oklahoma State University Research Note 133 (1982), unpublished.

11. S. J. Brodsky and R. W. Brown, Phys. Rev. Lett. 49, 966 (1982).

12. "Classical Electrodynamics", J. D. Jackson, pg. 735 (2nd edition), John Wiley & Sons, N.Y. (1975); "The Classical Theory of Fields," L. D. Landau and E. M. Lifshitz, pgs. 175 and 189 (4th edition), Pergamon Press (1975).

13. M. L. Laursen, M. A. Samuel, A. Sen and G. Tupper, "Do Amplitude Zeroes Persist in Higher Order?", Oklahoma State University Research Note 137 (1982), unpublished.

14. D. R. Yennie, S. C. Frautschi and H. Suura. Annals of Physics 13 379 (1961).

15. M. L. Laursen, M. A. Samuel, G. B. Tupper and A. Sen, "Z° Decay into Two Gluons and a Photon for Massive Quarks," Oklahoma State University Research Note 135 (1982), to be published in Phys. Rev. D; V. Constantini, B. De Tollis and G. Pistoni, Nuovo Cimento 2A 722 (1971).

NEUTRINO OSCILLATIONS IN GRAND UNIFIED THEORIES

K.A. Milton
Oklahoma State University, Stillwater, OK 74078

ABSTRACT

The natural scale of ν_e mixing angles in SO(10) is small, of order $\theta_c/3$. It is possible to achieve large ν_e mixing angles θ, but such solutions are unstable, in the sense that $\alpha\partial\tan\theta/\partial\alpha \sim 100$, where α is a Majorana mass parameter. This difficulty reflects the quark mass hierarchy. On the other hand, ν_μ–ν_τ mixing could naturally be substantial. These qualitative features would be expected to hold any grand unified model where the Gell-Mann-Ramond-Slansky mechanism generates naturally small neutrino masses.

INTRODUCTION

The last year has seen a significant improvement in the experimental bounds on neutrino masses and mixing angles as inferred from neutrino oscillation experiments[1]. Already ruled out are the earlier indications of terrestrial scale neutrino oscillations[2]. Of course, there remains the solar neutrino problem[3] as well as the Russian tritium β-decay measurement of the mass of the anti-electron-neutrino[4], both of which suggest that the neutrino system possesses an interesting mass matrix. The next few years will see advances on this problem from several fronts: β-decay measurements[5], neutrinoless double β-decay experiments[6], and improved limits on neutrino oscillations[7].

It is therefore appropriate to ask what guide theory can offer on this question. The standard theory of neutrino masses is provided by a grand unified theory (GUT), such as SO(10)[8], where the smallness of the observed neutrino masses arises from the Gell-Mann-Ramond-Slansky (GRS) mechanism[9]. We will concentrate here on SO(10) because it is the "minimal" model in which non-zero neutrino masses emerge[10], simply from the appearance, automatically, of a right-handed neutrino and hence a Dirac mass. This is not to say that neutrino masses cannot occur in SU(5)[11], or of course in higher groups[12]. However, our essential qualitative conclusions should be independent of the group structure (provided the GRS idea applies), because of the general expectation that the Dirac mass matrix for the neutrinos should reflect the hierarchical structure of the quark and charged lepton mass matrices.

In examples we have shown[13] that the typical values of the Kobayashi-Maskawa[14] (KM) mixing angles for the left-hand neutrinos are

$$|\theta_1| \sim \theta_c/3, \quad |\theta_2| \lesssim \theta_c, \quad |\theta_3| \lesssim \pi/4 \tag{1}$$

where θ_c is the Cabibbo angle. This means that the KM matrix relating flavor and mass eigenstates is simply ($c_3 = \cos\theta_3$, $s_3 = \sin\theta_3$)

$$U \simeq \begin{pmatrix} 1 & \theta_1 c_3 & \theta_1 s_3 \\ -\theta_1 & c_3 & s_3 \\ 0 & -s_3 & c_3 \end{pmatrix} \quad (2)$$

apart from CP violating phases. U is in turn related to the probability amplitude for flavor α at time 0 to turn into flavor β at time t by[15]

$$\langle \nu_\beta(t) | \nu_\alpha(0) \rangle = \delta_{\alpha\beta} + \sum_{i \neq j} (e^{i\Delta_{ij}} - 1)\, U_{\beta i}^* \, U_{\alpha i}, \quad (3)$$

where

$$\Delta_{ij} = (m_i^2 - m_j^2)\, t/2p, \quad (4)$$

t being the oscillation time, p the neutrino momentum, and m_i the neutrino mass eigenvalues. If we further make the most likely assignment[13]

$$m_1 \ll m_2 \ll m_3, \quad (5)$$

we obtain from (2) ν_e oscillation formulas only slightly more complicated than in two generation mixing ($\Delta_3 = \Delta_{13}$ with $m_1 = 0$)

$$\begin{aligned} P(\nu_e \to \nu_\mu) &= 4\theta_1^2\, s_3^4 \sin^2(\Delta_3/2), \\ P(\nu_e \to \nu_\tau) &= 4\theta_1^2\, s_3^2\, c_3^2 \sin^2(\Delta_3/2), \\ P(\nu_e \to \nu_e) &= 1 - 4\theta_1^2\, s_3^2 \sin^2(\Delta_3/2). \end{aligned} \quad (6)$$

These may well be at an observable level in the next round of experiments[7], which may be sensitive to "$\sin^2 2\theta$" $= 4\theta_1^2 = 0.02$, although it should be recognized that it may be hard in GUTs to even reach m_3=1eV[13].

GELL-MANN-RAMOND-SLANSKY MECHANISM

Having thus set the stage, we ask if the small ν_e mixing implied by (1) is general. Affirmative answers have been given[16,17]: large ν_e mixing can only be achieved in SO(10) with an unreasonable adjustment of parameters. The purpose of this report is to sharpen the degree of unreasonableness required.

The GRS mechanism[9] for naturally small neutrino masses is the use of the following form for the 6×6 neutrino mass matrix acting on $\psi_\nu = (\nu_e, \nu_\mu, \nu_\tau, \nu_e^c, \nu_\mu^c, \nu_\tau^c)$:

$$M^\nu = \begin{pmatrix} 0 & \nu \\ \nu & M \end{pmatrix}. \quad (7)$$

The Dirac mass matrix ν, in SO(10), is related to that of the up quarks; for definiteness we take a phenomenologically acceptable form[18]

$$\nu = \begin{pmatrix} 0 & a & 0 \\ a & 0 & m \\ 0 & m & b \end{pmatrix}, \tag{8}$$

where $|a| = (m_u\, m_c)^{1/2}$, $|b| = m_t$, $|m| = 3(m_c\, m_t)^{1/2}$. On the other hand, there are no constraints on the Majorana mass matrix M,

$$M = \begin{pmatrix} E & A & F \\ A & C & D \\ F & D & B \end{pmatrix} \tag{9}$$

(we ignore CP-violating phases). All we know is that det $M \neq 0$ (otherwise we will get Dirac neutrinos with masses $\sim m_q$), and that the scale of M is of order 10^9–10^{15} GeV. (The former scale might occur if the Witten mechanism[19] were operative.) Small ν_e mixing is suggested by (8), but one would think that that situation could be overturned by large ratios occuring between the elements of M. Essentially, as we will show, that is not the case.

The essential feature of the GRS mechanism is insertion of a zero left-handed Majorana mass initially [the zero in (7)]. This is done to avoid the introduction of an additional very small neutrino mass scale. However, terms will appear in that portion of the mass matrix via radiative corrections. But it appears[20] that typically the radiatively induced left-handed Majorana mass matrix is negligible relative to that obtained by block diagonalizing (7).

The only accessible neutrinos are the left-handed ones, which are described by the 3 × 3 mass matrix

$$\mu = -\nu\, M^{-1}\, \nu, \tag{10}$$

whose elements are ($-\det M^{-1}$ is suppressed on the RHS)

$$\begin{aligned} \mu_{11} &= a^2\alpha \quad , \\ \mu_{12} = \mu_{21} &= -a(m\beta + a\delta), \\ \mu_{13} = \mu_{31} &= a(m\alpha - b\beta), \\ \mu_{22} &= m^2\gamma + 2ma\xi + a^2\epsilon, \\ \mu_{23} = \mu_{32} &= m(b\gamma - m\beta) + a(b\xi - m\delta), \\ \mu_{33} &= m^2\alpha - 2mb\beta + b^2\gamma, \end{aligned} \tag{11}$$

in terms of the abbreviations

$$\begin{aligned} \alpha &= EB - F^2, \quad & \beta &= ED - AF, \quad & \gamma &= EC - A^2, \\ \delta &= AB - DF, \quad & \xi &= AD - CF, \quad & \epsilon &= CB - D^2. \end{aligned} \tag{12}$$

CRITERIA FOR LARGE ν_e MIXING

The diagonalizing transformation is given in terms of the eigenvectors of μ, which we write as

$$\psi = (1,\ x,\ y). \tag{13}$$

Large ν_e mixing is described by one of the following situations:

$$\begin{aligned} &y \sim 1,\ z \ll 1, \\ &y \ll 1,\ z \sim 1, \\ &y \sim 1,\ z \sim 1. \end{aligned} \tag{14}$$

Our strategy for determining the circumstances under which (14) is possible is to first eliminate λ from the eigenvalue equation $M\psi=\lambda\psi$:

$$\begin{aligned} 0 = {} & \alpha[a^2y + mayz] + \beta[-am(y^2-1)-abyz+m^2z] \\ & + \gamma[-m^2y-mbz] + \delta[-a^2(y^2-1) + amz] \\ & + \xi[-2amy-abz] + \varepsilon[-a^2y], \end{aligned} \tag{15a}$$

$$\begin{aligned} 0 = {} & \alpha[-am-m^2z] + \beta[m^2y + 2mbz] \\ & + \gamma[-b^2z-mby] + \delta[-a^2yz + amy] \\ & + \xi[-aby], \end{aligned} \tag{15b}$$

where we have dropped terms which can never be significant when (14) holds, in view of

$$a/m \simeq 1/200. \tag{16}$$

For any given $y,z \lesssim 1$ eqns. (15) are two constrants on the five unknown parameter ratios β/α, γ/α, δ/α, ξ/α, ε/α, which can always be solved. So large mixing is always possible either by achieving large ratios or cancellations.

It is the stability of these solutions which is at issue. So we compute the derivatives $\partial y/\partial\alpha_i$, $\partial z/\partial\alpha_i$, $\{\alpha_i\} = \{\alpha,\beta,\gamma,\delta,\xi,\varepsilon\}$. The coefficient matrix for $(\partial y/\partial\alpha_i,\ \partial z/\partial\alpha_i)$ is A where

$$\begin{aligned} A_{11} &= (a^2 + maz)\alpha - (2amy + abz)\beta - m^2\gamma \\ &\quad -2a^2y\delta - 2am\xi - a^2\varepsilon, \\ A_{12} &= amy\alpha + A_{21},\ A_{21} = m^2\beta-mb\gamma+am\delta-ab\xi, \\ A_{22} &= -m^2\alpha + 2mb\beta - b^2\gamma - a^2y\delta. \end{aligned} \tag{17}$$

A stable solution (one in which a small change in a parameter leads to a small change in mixing angles) is one for which

$$\alpha_i\partial(y,z)/\partial\alpha_i \sim 1; \tag{18}$$

these being eight more constraints, they cannot be solved in general. We therefore will have for some parameters

$$\alpha_i\partial(y,z)/\partial\alpha_i \sim m/a\ , \tag{19}$$

that is, a 1% change in the parameters destroys the mixing. It is therefore very unnatural to expect large ν_e mixing to occur in these theories. We expect that there are two natural scales for y (or z):

$$y \sim a/m \sim 0.01,$$

$$y \sim m/a \sim 100; \qquad (20)$$

significant derivation from these natural scales requires an unnatural balancing act.

Let's make this argument more explicit for a case when $z \sim 1$ (if $z \ll 1$, $y \sim 1$ the analysis is similar). Then the system (15) reduces to

$$0 = yA_{11} + zA_{21},$$

$$0 = yA_{21} + zA_{22}, \qquad (21)$$

implying $A_{11}\, A_{22} - A_{21}{}^2 = 0$ and hence

$$\det A = A_{21}\; amy\alpha \qquad (22)$$

by (17). Equations (15) and (21) now yield

$$\alpha(\partial z/\partial\alpha) = (m/a)(z/y)(A_{11}/A_{21}) + z$$

$$= -(m/a)(z/y)^2 + z \quad , \qquad (23)$$

which is large $[O(m/a)]$ unless

$$y \sim (m/a)^{1/2}, \qquad (24)$$

smaller than the scale suggested in (20) but still consistent with small mixing. But in fact this is not the natural mixing scale (20) because it is now easy to see that

$$\alpha(\partial y/\partial\alpha) \sim (m/a)^{1/2}. \qquad (25)$$

So the result (19)-(20) follows.

CONCLUSION

Large ν_e mixing can of course be accomodated in SO(10). It is however, a priori very unlikely, because the adjustment of parameters required to achieve it means that a small (~1%) change in the large Majorana masses will destroy the mixing. Experiments therefore should be designed to seek ν_e mixing angles at the level of 1/3 θ_c.[21] Muon-tau neutrino mixing is essentially unconstrained by SO(10). These qualitative features should transcend SO(10), and hold in any grand unified model when the GRS mechanism applies.

This work was supported in part by the U.S. Department of Energy, and by a Starter Grant from the Dean, College of Arts and Sciences, OSU.

REFERENCES

1. J.L. Vuilleumier, talk at this conference.
2. F. Reines, H.W. Sobel, and E. Pasierb, Phys. Rev. Lett. 45, 1307 (1980); D. Silverman and A. Soni, ibid. 46, 467 (1981).
3. J.N. Bahcall et al., Phys. Rev. Lett. 45, 945 (1980).
4. V.A. Lyubimov, E.G. Novikov, V.Z. Nozik, E.F. Tretyakov, and V.S. Kosik, Phys. Lett. 94B, 268 (1980).
5. See talks at this conference by H. Ravn, T. Bowles, B. Robinson, P. Seiler, R. Graham, O. Fackler, and R. Raghavan.
6. See talks at this conference by T. Kirsten, C. Liguori, and D. Caldwell.
7. See talks at this conference by A. Mann, D. Jovanovic, A. Pevsner and M. Baldo-Ceolin.
8. H. Fritzch and P. Minkowski, Ann. Phys. (NY) 93, 193 (1975); H. Georgi, in Particles and Fields - 1974 (APS/DPF Williamsburg) ed. C.E. Carlson (AIP, New York, 1975).
9. M. Gell-Mann, P. Ramond, and R. Slansky, in Supergravity, eds. P. van Nieuwenhuizen and D.Z. Freedman (North Holland Publishing Company, 1979), P. 315; T. Yanagida in Proceedings of Workshop on the Unified Theory and the Baryon Number of the Universe, KEK, 1979, eds. O. Swada and A. Sugamota (unpublished).
10. R. Barbieri, D.V. Nanopoulos, G. Morchio, and F. Strocchi, Phys. Lett. 90B, 91 (1980).
11. R. Barbieri, J. Ellis, and M.K. Gaillard, Phys. Lett. 90B, 249 (1980); A. Zee, ibid. 93B, 389 (1980); L. Wolfenstein, Nucl. Phys. B175, 93 (1980); L. Wolfenstein, Proceedings of the Neutrino Mass Miniconference and Workshop, Telemark, Wisconsin, October 2-4, 1980; G. Lazarides and Q. Shafi; Phys. Lett. 99B, 113 (1981).
12. G.L. Shaw and R. Slansky, Phys. Rev. D 22, 1760 (1980); P. Ramond, Proceedings of Workshop on Weak Interactions as Probes of Unification, Blacksburg, Virginia, December 4-6, 1980.
13. S. Hama, K. Milton, S. Nandi, and K. Tanaka, Phys. Lett. 97B, 221 (1980); K. Milton, S. Nandi, and K. Tanaka, Phys. Rev. D 25, 800 (1982); J.A. Harvey, D.B. Reiss and P. Ramond, Nucl. Phys. B199, 223 (1982); T. Yanagida and M. Yoshimura, Phys. Lett. 97B, 99 (1980); K. Kanaya, Prog. Theor. Phys. 64, 2278 (1980).
14. M. Kobayashi and T. Maskawa, Prog. Theor. Phys. 49, 652 (1973).
15. A. DeRujula, M. Lusignoli, L. Maiani, S.T. Petcov, and R. Petronzio, Nucl. Phys. B168, 54 (1980).
16. K. Milton and K. Tanaka, Phys. Rev. D 23, 2087 (1981); Gauge Theories, Massive Neutrinos, and Proton Decay (Orbis Scientiae 1981) ed. A. Perlmutter (Plenum, N.Y., 1981), p. 207.
17. T. Goldman and G.J. Stephenson, Phys. Rev. D 24, 236 (1981).
18. J.A. Harvey, P. Ramond, and D.B. Reiss, Phys. Lett. 92B, 309 (1980).
19. E. Witten, Phys. Lett. 91B, 81 (1980).
20. G. Branco and A. Masiero, Phys. Lett. 97B, 95 (1980); Y. Tomozawa, UM-HE 80-17.
21. This residual ν_e mixing comes from the charged lepton sector. See Ref. 16.

MINIMALLY EXTENDED ELECTROWEAK GAUGE THEORIES IN SO(10) AND E_6

R. W. Robinett*
Physics Department, University of Wisconsin, Madison, WI 53706

J. L. Rosner
Enrico Fermi Institute, University of Chicago, Chicago, IL 60637

ABSTRACT

The possibility of minimally extending the standard $SU(2)_L \times U(1)$ electroweak theory within the context of SO(10) and E_6 grand unification by adding U(1) factors is explored. The neutrino neutral current interactions in these schemes are arranged to coincide with the standard model predictions. Limits on the masses of the extra Z's generated by these U(1) factors are obtained by considering other parity violating effects. Additional Z's as light as 2.5-3.0 times the standard model Z° mass are allowed.

INTRODUCTION

In this work we first consider a particular breakdown of the grand unified group SO(10) which contains the standard $SU(3)_c \times SU(2)_L \times U(1)$ model plus an additional unbroken U(1) at energies above the weak scale $G_F^{-1/2}$ which gives a 2-Z electroweak model. The group structure of SO(10) is briefly discussed in order to better understand the charge to which the new U(1) is coupled. We investigate a particular breakdown of this electroweak group leading to the standard model neutrino interactions and also set limits on the breakdown scale of the second Z by examining parity violation in heavy atoms and $e^+e^- \to \mu^+\mu^-$. We also discuss the possible production of the second Z in $p\bar{p}$ and e^+e^- collisions. A breakdown of E_6 is also considered which yields two additional unbroken U(1)'s above $G_F^{-1/2}$ (giving a 3-Z model) and a similar phenomenological analysis is performed.

ONE EXTRA Z IN SO(10)

Many authors have considered minimally extending the standard $SU(2)_L \times U(1)$ electroweak gauge group by adding U(1)'s[1] or SU(2)'s.[2] In the context of grand unified theories (GUTs) the standard electroweak model is contained in the minimal grand unification group based on SU(5);[3] any extension requires a larger unification group such as SO(10)[4,5] which is the next smallest rank group compatible with grand unification.

There are two breakdowns of SO(10); one by the left-right symmetric path,

$$SO(10) \xrightarrow{M_u} SO(6) \times SO(4) \approx SU(4) \times SU(2)_L \times SU(2)_R \,, \qquad (1)$$

allows for the possibility of fractionally charged lepto-quark bosons $B^{\pm 2/3}$ (which mediate $K_L^\circ \to \mu^+ e^-$) in the desert in addition to right-handed W's and a Z. We examine here the breakdown

$$SO(10) \xrightarrow{M_u} SU(5) \times U(1)_\chi \tag{2}$$

(via a $\underset{\sim}{45}$ of Higgs fields) where the SU(5) is broken further as usual but the $U(1)_\chi$ is left unbroken. This leads to the effective low energy electroweak gauge group

$$G = SU(2)_L \times U(1)_Y \times U(1)_\chi \ , \tag{3}$$

with no new physics in the desert except one additional $\underset{\sim}{Z}$. (This group has also been considered by Deshpande and Iskandar[6] and Masiero.[7]) A plot of the running couplings ($\alpha^{-1}(Q^2)$) versus energy ($t = \ell n(Q^2/4M_W^2)$) in Fig. 1 illustrates that in this scheme the SO(10) breaks down at a scale somewhere below the Planck mass, $M_p \simeq 10^{19}$ GeV (Fig. 1(b)), but above the standard SU(5) scale M_x (Fig. 1(a)). The coupling constant renormalization prediction for

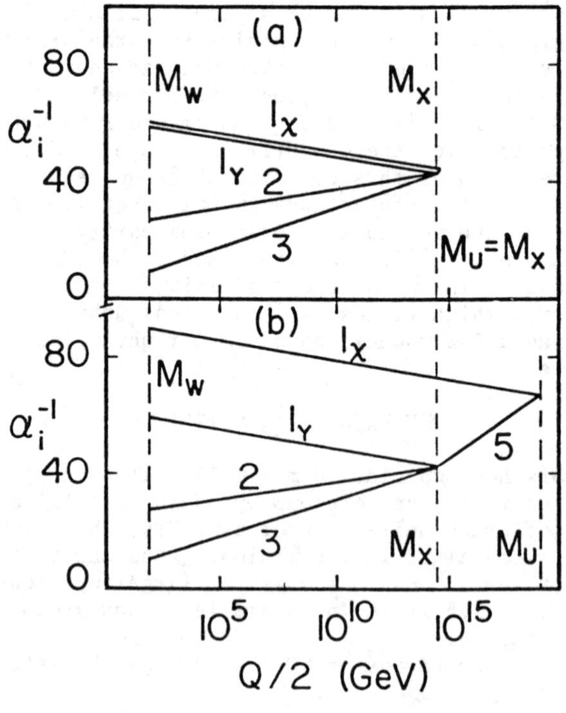

Fig. 1. Running couplings for $SO(10) \xrightarrow{M_u} SU(5) \times U(1)_\chi$ when (a) $M_u = M_x$ and (b) $M_u = M_p$, the Planck mass.

M_X is changed by the presence of the $U(1)_\chi$, even if unbroken down to low energies, only by the effects of Higgs fields in two loop order; the extra Z hides itself well in this respect. There is thus a range in the low energy value of the χ coupling, g_χ, due to the uncertainty in the initial breakdown scale. If the second Z were found and its g_χ measured, one could work backwards to gain information on M_u.

The generators of SO(2n) "fill out" those of SU(n)

$$\begin{aligned} SU(n): &\quad \pm(\underset{\sim}{e}_i - \underset{\sim}{e}_j) \\ SO(2n): &\quad \pm(\underset{\sim}{e}_i \pm \underset{\sim}{e}_j) \end{aligned} \tag{4}$$

where the $\underset{\sim}{e}_i$ are n-dimensional unit vectors. The representations of SO(10) can then be thought of as five-dimensional vectors; for example, the $\underset{\sim}{16}$ containing the fermions is

$$\underset{\sim}{16}_f = (\pm\frac{1}{2}, \pm\frac{1}{2}, \pm\frac{1}{2}, \pm\frac{1}{2}, \pm\frac{1}{2}) \quad \text{(odd \# of minus signs)} \tag{5}$$

$$= \underset{\sim}{1}(-5) + \underset{\sim}{10}(-1) + \underset{\sim}{5}^*(3) \ . \tag{6}$$

(Equation (6) gives the (unnormalized) Q_χ charges of the SU(5) subgroups of SO(10) for the $\underset{\sim}{16}_f$.) In this language the operation of CP is equivalent to changing all signs in a vector while P consists of changing just the last sign. The diagonal generators can also be expressed as 5D vectors whose dot product with the representation members gives the value of the corresponding charge. For the SU(3) color subgroup of SU(5) we have

$$I_{3c} = (\frac{1}{2}, -\frac{1}{2}, 0, 0, 0) \tag{7}$$

$$Y_c = \frac{1}{2\sqrt{3}}(1, 1, -2, 0, 0) \tag{8}$$

while for the usual $SU(2)_L \times U(1)$ subgroup we have

$$I_{3L} = (0, 0, 0, \frac{1}{2}, -\frac{1}{2}) \tag{9}$$

$$Y = \frac{1}{\sqrt{60}}(-2, -2, -2, 3, 3) \ . \tag{10}$$

The additional $U(1)_\chi$ from SO(10) is generated by the vector

$$\chi = \frac{1}{\sqrt{10}}(1, 1, 1, 1, 1) \tag{11}$$

so that the χ charge is (up to normalization) just the sum of the entries of each representation vector. Thus there is one member of the $\underset{\sim}{16}_f$ with 5 minus signs ($Q_\chi = -5$), 10 next nearest neighbors with 3 signs ($Q_\chi = -3 + 2 = -1$), etc.

In order to break the symmetry G to the unbroken $U(1)_{EM}$ we allow, for the moment, the general Higgs structure in Table I and examine how these Higgs fields can be embedded in SO(10). We allow

Table I. Higgs fields quantum numbers

Fields	I_{3L}	Y	χ	VEV
ϕ_i	1/2	-1	$Q_\chi^{(i)}$	$v_i/\sqrt{2}$
ψ_j	0	0	$Q'^{(j)}_\chi$	$V_j/\sqrt{2}$

any number of $SU(2)_L$ doublets (ϕ_i) with arbitrary χ charges and vacuum expectation values (VEVs) to ensure that the condition $\rho = 1$ is naturally satisfied. We also choose SU(5) singlets (ψ_j) in order to break the combined $U(1)_Y \times U(1)_\chi$ symmetry. If we ask that the resulting neutral current interactions be identical to the standard model predictions <u>for neutrinos</u>, then the only further condition imposed is on the "average" value of the doublets' χ charge, i.e.

$$\overline{Q}_\chi \equiv (\sum Q_\chi^{(i)} v_i^2)/(\sum v_i^2) = 3 \ . \tag{12}$$

This is most easily satisfied by a single SO(10) 16 which is the only representation up to at least dimension 210 which allows for $Q_\chi = 3$. Similarly, the $\underset{\sim}{16}$ is also the smallest representation to contain an SU(5) singlet; thus we use two $\underset{\sim}{16}$'s with VEVs $\langle\phi\rangle = v/\sqrt{2}$ and $\langle\psi\rangle = V/\sqrt{2}$. (16's are not sufficient to give fermions a mass at tree level. For a novel way to use $\underset{\sim}{16}$'s in this respect see the talk by Leung, Robinett and Rosner in these proceedings.)

With these Higgs fields we obtain the usual charged current interactions and the neutral current interaction

$$H^{NC}_{SO(10)} = \frac{4G_F}{\sqrt{2}} [(J_{3L} - \sin^2\theta_W J_{em})^2 + \frac{1}{R} (J_{3R} - \frac{3}{5} \cos^2\theta_W J_{em})^2] \tag{13}$$

where $G_F = (\sqrt{2}\, v^2)^{-1}$ and $R = V^2/v^2$.[8] It gives the usual result for neutrinos since $J_{3R}(\nu) = J_{em}(\nu) = 0$. The interaction (13) would look more left-right symmetric if we identify $\sin^2\theta_W$ with e^2/g_L^2 and $3\cos^2\theta_W/5$ with e^2/g_R^2 and this is in fact what we obtain when we consider an $SU(2)_L \times U(1)_R \times U(1)_{B-L}$ gauge group (also in SO(10)) broken by the same set of Higgs fields. After rotating the diagonal generators into this basis, the Georgi-Weinberg theorem[9] applies making the form of Eq. (13) obvious. (See Ref. 10 and the appendix of Ref. 5 for a discussion of the neutral current interactions in different breakdowns of the same grand unified group.)

We can then use ν data to fix $\sin^2\theta_W = .23 \pm .01$ (before radiative corrections) and use other parity-violating effects to limit R. Polarized electron-deuteron scattering[11] happens to give almost no limit on R because the coefficient of the R^{-1} term in (13) for this process is small. Parity violation in bismuth[12] gives a limit $R \gtrsim 10$ while recent results from groups at DESY[13] (CELLO, JADE, MkJ, and TASSO) on the forward-backward asymmetry, A_{FB}, in $e^+e^- \to \ell^+\ell^-$ give $R \gtrsim 7$. This translates into limits on the masses of the two Z's as follows; the lightest Z_1 (now nearly the Z°) must have $1.0 \geq M_1/M_{Z°} \gtrsim 0.98$ while the heavier Z_2 (now mostly the Z_χ boson) must satisfy $M_2/M_{Z°} \gtrsim 2.5$-3.0 where the range comes from the range in coupling constant g_χ. This is another example of 2-Z theories where if M_1 varies as $(1-\varepsilon)M_{Z°}$ then M_2 goes as $\varepsilon^{-1/2} M_{Z°}$.[14]

The production of the Z_χ in $f\bar{f} \to Z_\chi$ is proportional to the Q_χ^2 of the fermions. Using (6) we obtain the relative production strengths (or equivalently the $Z_\chi \to f\bar{f}$ branching ratios)

$$\begin{array}{ccccccc} u\bar{u} & : & d\bar{d} & : & e^+e^- & : & \nu\bar{\nu} \\ 3(1^2+1^2) & : & 3(1^2+3^2) & : & 1^2+3^2 & : & 3^2 \\ 6 & : & 30 & : & 10 & : & 9 \end{array} \tag{14}$$

so the Z_χ couples to d's five times more strongly than to u's. This leads to a ratio of production cross sections for the heavier Z_2 ($\simeq Z_\chi$) to the lighter Z_1 ($\simeq Z°$) in $p\bar{p}$ collisions

$$\frac{\sigma(p\bar{p} \to Z_2)}{\sigma(p\bar{p} \to Z_1)} \simeq \frac{1}{40} \tag{15}$$

where the cross sections are evaluated at the <u>same</u> <u>s/M^2</u> and $M_2 = 3M_{Z°}$ is assumed. Had the couplings of the Z_χ to quarks been the same as for the Z° this would have been expected to be approximately $3^{-2} = 9^{-1}$.

The Z_2 also hides itself well in A_{FB} even at energies above the Z_1 pole (see Fig. 2). The effect of the Z_2 is much more dramatic in the average μ^+ longitudinal polarization (or equivalently the asymmetry with polarized electrons) in $e^+e^- \to \mu^+\mu^-$ as seen in Fig. 3.

We see then that an additional U(1) can be added to the standard electroweak gauge group within the context of grand unification and that the resulting extra Z can be as light as 2.5-3.0 times the standard model Z° mass without affecting any existing low energy experiment. The allowed deviation of the light Z_1 mass from the Z° mass is of the same order of magnitude as the expected radiative

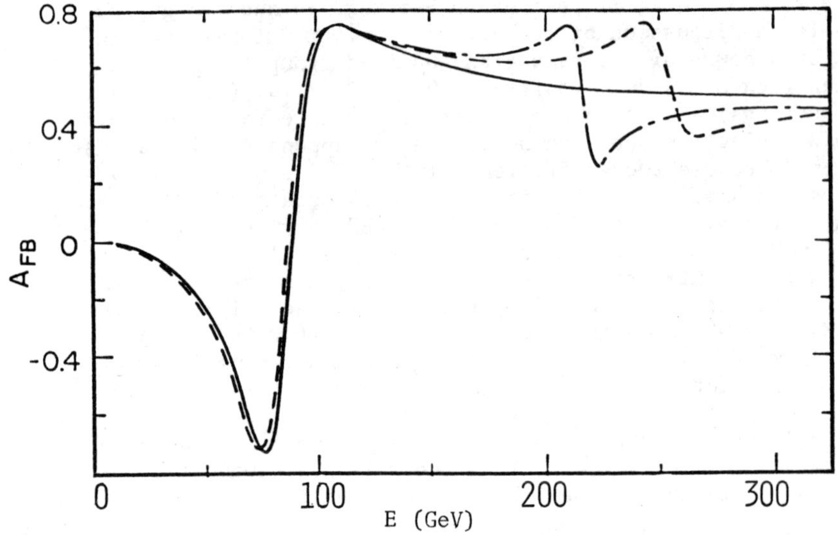

Fig. 2. Forward-backward asymmetry in $e^+e^- \rightarrow \ell^+\ell^-$ versus energy. Solid line is the standard model, dash-dot line and dashed line are for $M_2 \simeq 2.5\ M_{Z^\circ}$ and $M_2 \simeq 3.0\ M_{Z^\circ}$ respectively.

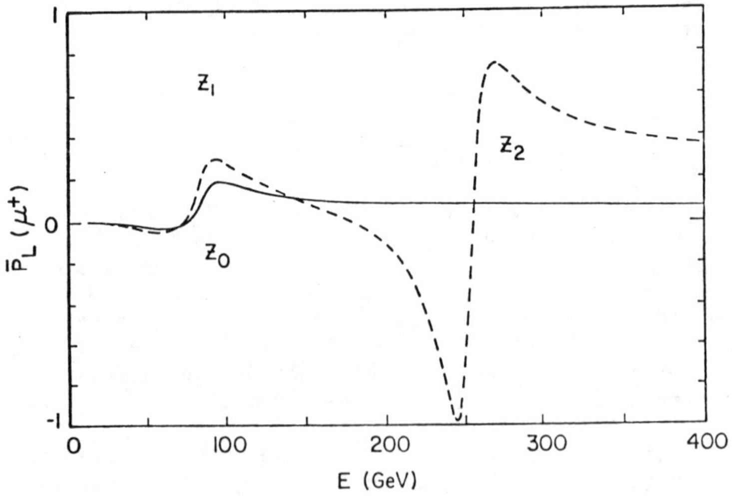

Fig. 3. Average longitudinal μ^+ polarization in $e^+e^- \rightarrow \mu^+\mu^-$ versus energy. Solid line is standard model, dashed line is for $M_2 \simeq 2.5\ M_{Z^\circ}$.

corrections to the Z° mass[15] and its width, so careful measurements will be necessary to rule out a second Z.

TWO EXTRA Z'S IN E_6

If one thinks (as suggested by Dynkin diagram language) of SU(5) as "E_4" and SO(10) as "E_5", then it is natural to consider E_6 as a candidate for grand unification.[16] Much of the discussion of minimally extended electroweak theories in SO(10) given already can then be extended to E_6.[17] There is, for example, a left-right symmetric breakdown of E_6

$$E_6 \xrightarrow{M_u} SU(3)_c \times SU(3)_L \times SU(3)_R \longrightarrow \\ SU(3)_c \times SU(2)_L \times U(1)_L \times SU(2)_R \times U(1)_R \tag{16}$$

but we consider the "minimal" breakdown

$$E_6 \xrightarrow{M_u'} SO(10) \times U(1)_\psi \xrightarrow{M_u} SU(5) \times U(1)_\chi \times U(1)_\psi \tag{17}$$

and the resulting 3-Z electroweak theory based on

$$\tilde{G} = SU(2)_L \times U(1)_Y \times U(1)_\chi \times U(1)_\psi \ . \tag{18}$$

The E_6 group representations and diagonal generators can now be expressed as six-dimensional vectors (see Ref. 17) but we simply note here that the fermions are placed in the E_6 $\underset{\sim}{27}$ with

$$\underset{\sim}{27}_f = \underset{\sim}{16}(1) + \underset{\sim}{10}(-2) + \underset{\sim}{1}(4) \tag{19}$$

where the numbers in parentheses are the (unnormalized) ψ charges. We break $\underset{\sim}{G}$ with the same Higgs fields as in SO(10) adding an SO(10) $\underset{\sim}{1}$ to complete the symmetry breakdown. These Higgs are most easily embedded in three E_6 $\underset{\sim}{27}$s which again give neutral current interactions for neutrinos as in the standard model. We find that

$$H^{NC}_{E_6} = H^{NC}_{SO(10)} + \frac{G_F}{4\sqrt{2}\,\tilde{R}} (J_{3L} - J_{3R} - \sqrt{6}\, J_\psi + \delta\, J_{em})^2 \tag{20}$$

where $\delta = (3 - 8 \sin^2\theta_W)/5$ and $\tilde{R} = \tilde{V}^2/v^2$ with $\tilde{V}$ the VEV of the SO(10) singlet Higgs. As before, the left-right symmetric formulation gives a similar neutral current interaction up to the identification of various couplings. Experiments discussed previously limit the two heavy Z's to be above (2.5-3.0) $M_{Z°}$ and the lightest Z_1 to be within 2% of the Z° mass.

The relative strengths for producing the Z_ψ in $f\bar{f} \to Z_\psi$ (or the $Z_\psi \to f\bar{f}$ branching ratios) are given by the ψ charges of the fermions. All the members of the SO(10) $\underset{\sim}{16}$ of ordinary fermions in the $\underset{\sim}{27}$ of Eq. (19) couple with equal ψ charge (1) to give

$$\begin{array}{ccccccc} u\bar{u} & : & d\bar{d} & : & e^+e^- & : & \nu\bar{\nu} \\ 3(1^2+1^2) & : & 3(1^2+1^2) & : & 1^2+1^2 & : & 1^2 . \\ 6 & : & 6 & : & 2 & : & 1 \end{array} \quad (21)$$

The Z_ψ couples to u's and d's with equal strength and so for the same mass it is slightly easier to produce in $p\bar{p}$ collisions than the Z_χ boson.

CONCLUSIONS AND OUTLOOK

The standard $SU(2)_L \times U(1)$ model is well-known to be in agreement with all low energy experimental data but we have seen that electroweak physics beyond the standard model but still in the context of grand unification at energies above 200 GeV is hard to rule out. This reinforces the important notion that accelerator experiments testing electroweak physics also probe grand unification.

We have also noticed that the neutral current interactions in SO(10) with a single additional Z can look very much like those in the left-right symmetric formulation with additional W's so it is important to distinguish between extra U(1)'s and SU(2)'s. (There is a recent proposal to improve the limit on the ratio M_R/M_L by using positron polarimetry.[19] Such precision weak-decay experiments are still important and will continue to complement accelerator searches.)

Proton decay experiments also touch on low energy electroweak physics. The rate is governed by the overall unification scale and in some models[20] a low mass left-right symmetry restoration implies a unification scale larger than the SU(5) M_X and an unobservably long proton lifetime.

There are, of course, other non-GUT alternatives for extended electroweak interactions. We have already mentioned various models with extra U(1)'s[1] and SU(2)'s[2] not embedded in GUTs. Many authors have discussed the possibility of the compositeness of the weak bosons either because the weak interaction $SU(2)_L$ is really strong and confining[21] or because the weak bosons are bound states of preons in a new hypercolor interaction[22] (the ρ's of some super-QCD). In either case one expects excited states giving higher mass W's and Z's. There is also the possibility that in theories in more than four dimensions that the dimensional reduction may naturally take place at the weak scale, $G_F^{-1/2}$.[23] In this case, new dimensions of physics may quite literally open up at accelerator energies and heavier W's and Z's may appear as excited modes on the higher dimensional manifold. Experiments at the CERN $p\bar{p}$ collider will soon begin to tell the story.

ACKNOWLEDGMENTS

This research was supported in part by the University of Wisconsin Research Committee with funds granted by the Wisconsin Alumni Research Foundation, and in part by the Department of Energy under contract DE-AC02-76ER00881.

REFERENCES

1. See, E. H. de Groot, G. J. Gounaris and D. Schildknecht, Phys. Lett. 85B, 399 (1979).
2. See, e.g., V. Barger, W. Y. Keung and E. Ma, Phys. Lett. 94B, 377 (1980).
3. H. Georgi and S. L. Glashow, Phys. Rev. Lett. 32, 438 (1974).
4. H. Georgi, in Particles and Fields - 1974, Proc. of the meeting of the APS Division of Particles and Fields, Williamsburg, Virginia, ed. by C. E. Carlson (AIP, New York, 1975), p. 575; H. Fritzsch and P. Minkowski, Ann. Phys. 93, 193 (1975).
5. Much of this section is based on R.W. Robinett and J.L. Rosner, Phys. Rev. D25, 3036 (1982).
6. N. G. Deshpande and D. Iskandar, Phys. Rev. Lett. 42, 20 (1979); Phys. Lett. 87B, 383 (1979); Nucl. Phys. B167, 223 (1980).
7. A. Masiero, Phys. Lett. 93B, 295 (1980).
8. This has also been obtained by the authors of Ref. 6.
9. H. Georgi and S. Weinberg, Phys. Rev. D17, 275 (1978).
10. R. W. Robinett, Wisconsin preprint MAD/TH/45 (to appear in Phys. Rev. D).
11. C. Y. Prescott et al., Phys. Lett. 77B, 347 (1978); 84B, 524 (1979).
12. L. M. Barkov and M. S. Zolotorev, Phys. Lett. 85B, 308 (1979).
13. M. Davier, Proc. of the XXI Int. Conf. on High Energy Physics, Paris, 1982 (to be published).
14. F. Del Aguila and A. Mendez, Nucl. Phys. B189, 212 (1981).
15. W. J. Marciano and A. Sirlin, Nucl. Phys. B189, 442 (1981); C. H. Llewellyn Smith and J. F. Wheater, Phys. Lett. 105B, 486 (1981).
16. See, e.g., F. Gürsey and M. Serdaroğlu, Nuovo Cim. 65, 337 (1981) and references therein.
17. For a complete discussion see R. W. Robinett, Wisconsin preprint MAD/TH/44 (to appear in Phys. Rev. D).
18. R. Slansky, Phys. Rep. 79C, 1 (1981).
19. M. Skalsey et al., Phys. Rev. Lett. 49, 708 (1982).
20. T. G. Rizzo and G. Senjanović, Phys. Rev. D24, 704 (1981); 25, 235 (1982).
21. L. F. Abbot and E. Fahri, Phys. Lett. 101B, 69 (1981); Nucl. Phys. B189, 547 (1981).
22. H. Fritzsch and G. Mandelbaum, Phys. Lett. 102B, 319 (1981).
23. G. Chapline and R. Slansky, Los Alamos preprint LA-UR-82-1076 (to appear in Nucl. Phys. B).

SECOND-ORDER FERMION MASSES

C. N. Leung
School of Physics and Astronomy
University of Minnesota, Minneapolis, MN 55455

R. W. Robinett
Department of Physics, University of Wisconsin, Madison, WI 53706

J. L. Rosner
Enrico Fermi Institute, University of Chicago, Chicago IL 60637

ABSTRACT

An SO(10) model with the lowest-energy symmetry breaking arising from two members of a 16-plet of Higgs bosons can lead to fermion masses if, in addition to fermion 16-plets, there are heavy fermions belonging to SO(10) singlets. Some consequences of this model for neutrino masses are explored briefly.

I. INTRODUCTION

The simplest grand unified model in which all observed fermions belong to a single type of representation of the unifying group is SO(10). In this group all left-handed fermions belong to 16-dimensional spinor representations. The SU(5) content (5^*+10+1) of these representations entails a charge-conjugate partner $N_L^c \in 1$ of the ordinary left-handed neutrino $\nu_L \in 5^*$ and, as a result, neutrinos can (and almost always do) acquire masses in such models. The question then arises as to why the neutrino masses are so much smaller than other observed fermion masses, at least within corresponding 16-plets.

One possibility for small neutrino masses[1] which arises naturally in SO(10) is that N acquires a large Majorana $\Delta I_{weak}=0$ mass M, while the $\Delta I_{weak}=1/2$ mass ($\bar{N}_R \nu_L$ + h.c) is of the order of a typical Dirac mass m. Both first-order and second-order mechanisms have been explored for generating M.[2] The small neutrino mass then arises from the diagonalization of a mass matrix

$$M = \begin{bmatrix} 0 & m \\ m & M \end{bmatrix} \qquad (1)$$

in the basis space of $\begin{bmatrix} \nu \\ N \end{bmatrix}$.

Another possibility which we would like to mention briefly here (it is still under investigation) is that all fermion masses (for 16-plet members, at least) arise to second order in the expectation values of Higgs fields. This is a natural consequence of a particular SO(10) symmetry breaking chain described elsewhere in this conference.[3] However, the mechanism could be more general. In such a model, properties of products of group representations prevent first-order Higgs contributions to fermion masses. We describe these properties in Sec. II, compare models in Sec. III, and summarize in Sec. IV.

0094-243X/83/990202-05 $3.00

II. SO(10) EQUIVALENCE CLASSES

Every group has representations belonging to distinct equivalence classes.[4] The product of two representations then leads to representations all of which belong to a specific class. A well-known example is triality for SU(3). In SO(10) there are four classes, which may be labeled by integers;

n	Example of representation
0	1, 45 (adjoint)
1	16 (spinor)
2	10 (vector), 126
3	16* (spinor)

In the product of representations

$$R_1 \times R_2 = \bigoplus_j R_j \qquad (2)$$

every representation R_j is characterized by an n_j such that

$$n_j = n_1 + n_2 \quad (\text{mod } 4) \qquad (3)$$

Fermion mass terms for 16-plets in SO(10) must transform as the product of 16 x 16 = 10 + 120 + 126. Since n(16) = 1, all the representations on the right-hand side must have n=2. One may generate fermion masses either via Higgs mesons belonging to the 10, 120, or 126,[5] or via higher-order contributions of Higgs mesons belonging to n=1 representations (say, the 16).

III. COMPARISON OF MODELS

The "standard" picture of fermion masses in SO(10)[2,6] makes use of a 10-plet to generate the dominant $\Delta I_W=1/2$ contributions to Dirac masses, and (say) a 126-plet to generate a large $\Delta I_W=0$ Majorana mass. The mass matrix is then of the form (1). The eigenvalues are

$$m_- \simeq m^2/M \qquad (4)$$

$$m_+ \simeq M \qquad (5)$$

We identify the usual neutrino with the mass eigenstate m_-. In many such models,[1,6] M is typical of grand unified scales, though it could be smaller by powers of the coupling constant.[2] The neutrino mass then tends to be tiny and the particle corresponding to m_+ is unobservable.

An alternative model is suggested by the SO(10) symmetry-breaking chain described in Refs. 3 and 7. Let us suppose that we have succeeded in breaking down SO(10) to SU(3) x SU(2) x U(1) x U(1).[8] This can be accomplished entirely via n=0 representations,[7] so that no fermion in the 16-plet has acquired a mass. Now we introduce two neutral Higgs bosons belonging to a 16-plet, one (call it ϕ) transforming as a weak isodoublet (like ν), and the other (call it ψ) transforming as a weak isosinglet (like N). The corresponding vacuum expectation values are:

$$\langle\phi\rangle = v \text{ (weak isodoublet)} \quad (6)$$

$$\langle\psi\rangle = V \text{ (weak isosinglet)} \quad (7)$$

Both these Higgs bosons are needed to break SU(2) x U(1) x U(1) all the way down to $U(1)_{e.m.}$; a single one will not do. The standard Weinberg-Salam Higgs field is ϕ, so v takes on its value in the standard model. The constraints of low-energy neutral-current phenomenology require V to be at least a few times v.[3,7]

In contrast to the usual interpretation of the Weinberg-Salam Higgs field, ϕ in the present model cannot by itself generate fermion masses, according to the considerations of Sec. II. (No trilinear coupling of 16-plet spinors is allowed.) Fermion masses must be at least second order in 16-plet Higgs vacuum expectation values here.

A simple way in which the desired contribution of 16-plet Higgs fields to fermion masses may be achieved is to introduce an SO(10) singlet fermion F with mass M_F. In the basis space of (ν, N) the tree graph of Fig. 1 with an intermediate

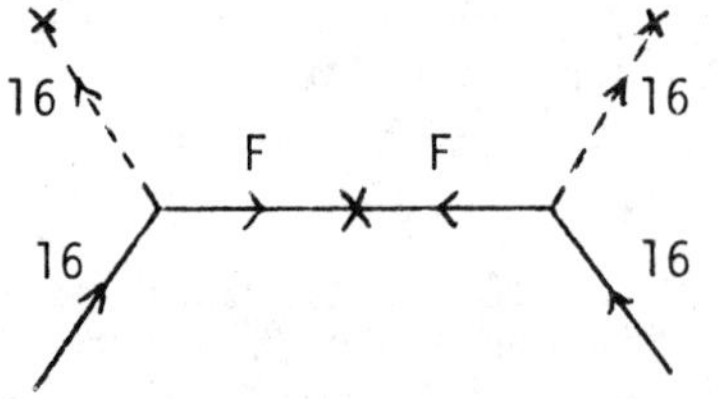

Fig. 1. Tree graph leading to neutral fermion masses. solid lines: fermions; dashed lines: Higgs.

fermion F then leads to a mass matrix of the form

$$M_{(2 \times 2)} = \frac{1}{M_F}\begin{bmatrix}(gv)^2 & gvGV \\ gvGV & (GV)^2\end{bmatrix}, \quad (8)$$

where g and G are the couplings of ϕ and ψ to $\bar{\nu}F$ and $\bar{N}F$, respectively. This mass matrix has a zero eigenvalue, leading to the conclusion that the neutrino has zero mass in the tree approximation. Loop diagrams[9] of Fig. 2 will not in general leave

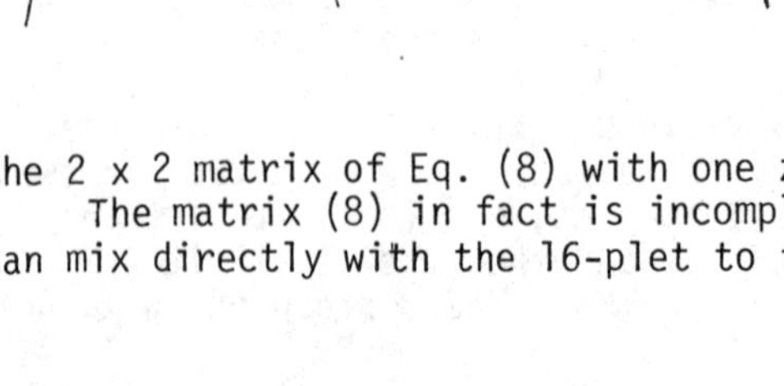

Fig. 2. Loop diagrams affecting fermion masses. Wavy line: gauge boson.

the 2 x 2 matrix of Eq. (8) with one zero eigenvalue.[10]

The matrix (8) in fact is incomplete, since the fermions F also can mix directly with the 16-plet to first order in Higgs vacuum expectation values. Taking account also of such effects, and including the effects of the radiative corrections depicted in Fig. 2,[11] we find in the (ν, N, F) basis the mass matrix

$$M = \begin{bmatrix} \frac{V_2^2(1+\varepsilon_1)}{M_F} & \frac{V_1V_2(1+\varepsilon_2)}{M_F} & V_2 \\ \frac{V_1V_2(1+\varepsilon_2)}{M_F} & \frac{V_1^2(1+\varepsilon_3)}{M_F} & V_1 \\ V_2 & V_1 & M_F \end{bmatrix} , \qquad (9)$$

where $V_1 \equiv GV$, $V_2 \equiv gv$. This matrix has two zero eigenvalues in the limit $\varepsilon_i \to 0$. When $\varepsilon_i \neq 0$, the eigenvalues may be identified for $V_2 << V_1 \lesssim M_F$ as

$$m_{small} \simeq \frac{(\varepsilon_1\varepsilon_3-\varepsilon_2^2)V_2^2}{\varepsilon_3 M_F} ,$$

$$m_{medium} \simeq \frac{\varepsilon_3 V_1^2}{M_F} , \qquad (10)$$

$$m_{large} \simeq M_F .$$

Loops contribute also to Dirac masses of charged leptons and quarks, which we expect also to be $\mathcal{O}(\varepsilon_2V_1V_2/M_F)$ in models where the dominant contributions of Fig. 2 come from loop momenta of order M_F. Thus one still obtains the relation $m(\nu)\ m(N) = \mathcal{O}(m^2_{Dirac})$, but the neutral lepton N need not be so heavy as in some schemes.[1,2,6]

Mixing among generations will occur whenever the singlet fermions F (one for each "family") can mix with one another. One then expects the patterns of Kobayashi-Maskawa mixing angles[12] to be correlated with those of leptons.

IV. CONCLUSIONS

We are examining a model in which two Higgs field vacuum expectation values are needed to give rise to all observed fermion masses. This type of model arises naturally in SO(10) when the lowest-mass-scale symmetry breaking occurs via 16-plet Higgs fields. An early conclusion of this model is that the neutral lepton N (the partner of ν in SO(10)models) need not be so heavy as conventionally thought, and might be detectable at accelerator energies.[13]

ACKNOWLEDGEMENTS

We thank E. Ma and L. Wolfenstein for helpful discussions. This work was supported in part by the U. S. Department of Energy under Contracts EY-76-C-02-1764, DE-AC02-76ER 00811, and DE-AC02 -82ER 40073. C. N. Leung thanks the Graduate School, University of Minnesota, for partial support thorugh a Doctoral Dissertation Fellowship.

REFERENCES

1. M. Gell-Mann, P. Ramond, and R. Slansky, in Supergravity, edited by D. Z. Freedman and P. van Nieuwenhuizen (North-Holland, Amsterdam, 1979), p. 315; T. Yanagida, in Proc. of the Worksop on the Unified Theory and the Baryon Number in the Universe, eds. G. Sawada and A. Sugamoto (KEK, Japan, 1979), R. N. Mohapatra and G. Senjanovic, Phys. Rev. Lett. 44, 912 (1980).
2. E. Witten, in First Workshop on Grand Unification (Proc. of Conf. in Durham, N. H., April, 1980), edited by Paul H. Frampton, Sheldon L. Glashow, and Asim Yildiz (Math. Sci. Press, Brookline, Mass., 1980), p. 275, and references therein.
3. R. Robinett, these proceedings.
4. R. Slansky, Phys. Reports 79, 1 (1981).
5. For the forms of couplings to these multiplets see P. Langacker, Phys. Reports 72, 185 (1981).
6. J. A. Harvey, D. B. Reiss, and P. Ramond, Nucl. Phys. B199, 223 (1982).
7. R. Robinett and J. Rosner, Phys. Rev. D25, 3036 (1982).
8. Models particularly close to ours have been described by N. G. Deshpande and David Iskandar, Phys. Rev. Lett. 42, 20 (1979); Phys. Lett. 87B, 383 (1979); Nucl. Phys. B167, 223 (1980.
9. Similar diagrams have been considered by S. M. Barr, Phys. Rev. D24, 1895 (1981), Ibid , D25, 1904 (1982).
10. This was realized clearly only after discussions with L. Wolfenstein, whom we thank.
11. Diagrams involving external SO(10) singlet fermions are not affected by the corrections illustrated in Fig. 2.
12. M. Kobayashi and T. Maskawa, Prog. Theor. Phys. 49, 652 (1973).
13. Light neutral leptons have been discussed recently by M. Gronau, SLAC-PUB-2965, 2967 (1982), submitted to Phys. Rev. Lett.

RECENT THEORETICAL PROGRESS ON THE ββ DECAY

E. Takasugi
Inst. of Physics, College of General Education, Osaka University
Toyonaka, Osaka, Japan

ABSTRACT

The general feature of the ββ decay is reviewed and the recent theoretical progress is discussed.

INTRODUCTION

On the direct measurement of the ββ decay, more than 15 projects are proposed[1] and some of them have already been reporting the preliminary data.[2] In a couple of years, the life-time as large as 10^{22} - 10^{23} years is expected to be reached. In consonant with these experimental development, it is requested to establish the theoretical predictions with further precision.

In this talk, I will give a review of the general feature of the ββ decay and discuss the recent theoretical development. There are three main directions in the recent works;(i) the investigation of the type of interactions to cause the ββ decay[3], (ii) the improvement of the decay formula by the use of the relativistic Coulomb distorted electron wave function including the nuclear finite size effect[4,5], and (iii) the elaborate evaluation of the nuclear matrix element[6]. About (iii), Prof. Stephenson will discuss later so that I will mainly concentrate on (ii) and brefely discuss on (i).

Experimentally, Heidelberg group has reported a new geochemical observation[7] of the the ratio of ^{128}Te to ^{130}Te half-lives. I will only give our result from this data on the neutrino mass and the right-handed parameters.[5] Prof. Kirsten will give the detail discussion of this data in a later talk.

My talk essentially follows the recent work of our group[5]. The earlier works will be found in the references of the review article by Prof. Primakoff and Rosen,[8] and our previous articles.[9]

THE TYPE OF INTERACTION

Recently, there has been the discussion of the Higgs particle contribution to the ββ decay.[3] Contrary to the general expectation, the triplet Higgs particles, which cause the lepton number violation, in the SU(2) **x** U(1) scheme turns out to contribute negligibly because parameters are intimately related to the small neutrino mass.[3] The singlet Higgs particle such as in the Zee model does not contribute either.[10,11] Therefore it may be natural to suppose the interaction consisting of the vector V and the axial-vector A currents only. The most general form of the V and A interaction will be written by assuming that neutrinos are lighte as follows:

$$H_W = G_F \cos\theta_c/\sqrt{2}\,\{ j_{L\mu} J_L^{+\mu} + \kappa\, j_{L\mu} J_R^{+\mu} + \eta\, j_{R\mu} J_L^{+\mu} + \lambda\, j_{R\mu} J_R^{+\mu} \}, \qquad (1)$$

where

$$j^{\mu}_{L,R} = \bar{e}\gamma^{\mu}(1 \mp \gamma_5)\nu_{eL,R}\,, \quad \nu_{eL} = \sum U_{ej} N_{jL}\,, \quad \nu_{eR} = \sum V_{ej} N_{jR}\,. \qquad (2)$$

Here J_L (J_R) is the left-(right-)handed hadronic current, N_j is the mass eigenstate neutrino with m_j, and U and V are the mixing matrices . The parameter λ is due to the existence of $SU(2)_R$ gauge particle W_R, and η and κ are from the mixings between W_L-W_R and between ordinary and mirror particles (leptons and qurks). If there is no mirror particle, then $\eta = \kappa$ and they vanish in the case of no W_L-W_R mixing. Even though there is no W_R, η and κ may not be zero, but $\lambda = 0$.

STRUCTURE AND SELECTION RULE OF ββ DECAY

For the two neutrino mode $(\beta\beta)_{2\nu}$, i.e., $(A,Z-2) \rightarrow (A,Z)+2e^-+2\bar{\nu}_e$, it is sufficient to keep the V-A current part only.

Because of the parity conservation, all leptons are taken to be in S-wave state. After replacing the energy of the intermediate nuclear state with the average value, using the closure approximation, and performing the Fierz transformation to the lepton part, the amplitude of the $0^+ \rightarrow J^+$ transition is expressed by[*,5]

$$R \propto \int d\vec{x}\,d\vec{y}\,\langle J^+ | J_L^{+\mu}(\vec{x})\, J_L^{+\nu}(\vec{y}) | 0 \rangle \{ 2(K+L)\, \bar{\Psi}(E_1)(1+\gamma_5)\psi^c(E_2)\, \bar{\Phi}(\omega_1)\gamma_\nu \gamma_\mu (1-\gamma_5)\Phi^c(\omega_2)$$
$$+ \tfrac{1}{2}(K-L)\, \bar{\Psi}(E_1)\sigma_{\alpha\beta}(1+\gamma_5)\psi^c(E_2)\, \bar{\Phi}(\omega_1)\gamma_\nu \sigma^{\alpha\beta}\gamma_\mu (1-\gamma_5)\Phi^c(\omega_2) \}, \qquad (3)$$

where

$$K = [\langle E_n \rangle - M_i + E_1 + \omega_1]^{-1} + [\langle E_n \rangle - M_i + E_2 + \omega_2]^{-1}, \qquad (4)$$

and $L = K(E_1 \leftrightarrow E_2)$. Here ψ and ϕ are the S-wave functions of electron and neutrino. Since the lepton part is the rank 0 operator, the the relevant operator to the nuclear transition is the hadronic part. It should be noted that K+L part only contributes to the $0^+ \rightarrow 0^+$ transition. Thus we obtain

$$\frac{\Gamma(0^+ \rightarrow 2^+)_{2\nu}}{\Gamma(0^+ \rightarrow 0^+)_{2\nu}} \sim \left(\frac{K-L}{K+L}\right)^2 \cdot (PR)^4 \ll 1\,. \qquad (5)$$

For the $0^+ \rightarrow 0^+$ transition, the nuclear tensor operator is $J_L^{+\mu} J_{L\mu}^{+}$ which does not change the spin and thus Δ - N transition in the N^*-mechanism is forbidden as discussed in Refs.5,6,9.

* We used that the S-wave function is independent of the position r_n. See Eq.(17).

The result is summarized as follows:

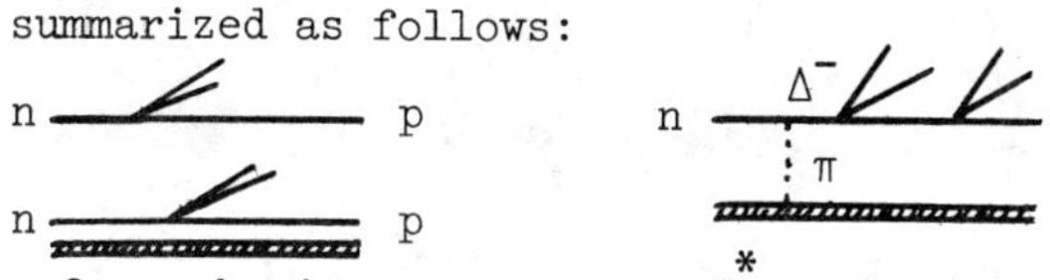

$(\beta\beta)_{2\nu}$	2n-mechanism	N^*-mechanism
$0^+ \to 0^+$	O	X
$0^+ \to 2^+$	o	o

The small circle indicates that the transition is suppressed considerably. The cross means that the transition is forbidden. The fact that the $0^+ \to 0^+$ transition in the 2n-mechanism dominates over other ones is extremely important when analyzing the data obtained in the geochemical method, which includes all modes and transitions. This also is important for extracting the information of the nuclear matrix element from the direct measurement of the $(\beta\beta)_{2\nu}$ mode.

For the neutrinoless mode $(\beta\beta)_{0\nu}$, i.e., $(A,Z-2) \to (A,Z)+2e^-$, all V and A terms should be taken into account.

In the second order perturbation, the amplitude is characterised by the structure of the lepton vertices as shown in Fig.1.

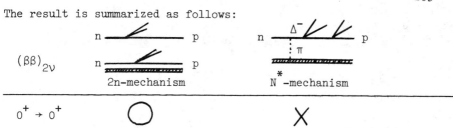

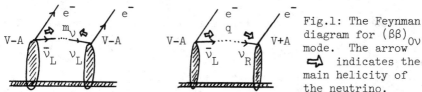

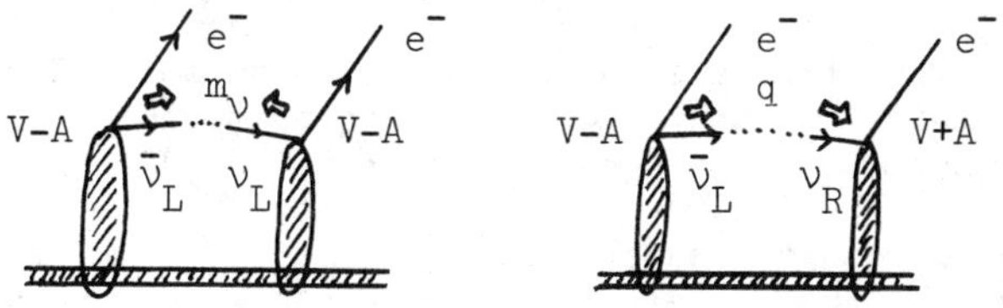

Fig.1: The Feynman diagram for $(\beta\beta)_{0\nu}$ mode. The arrow ⇨ indicates the main helicity of the neutrino.

If both lepton vertices are V-A or V+A, the mass of the propagating neutrino will contribute because at one vertex the anti-neutrino is emitted and at the other the neutrino is absorbed. Because the helicity of two neutrinos should match, non-zero mass is required in addition to that the neutrino is Majorana fermion. If the neutrino is massless, the emitted anti-neutrino has the helicity 1 and the absobed neutrino has -1 so that the transition is forbidden. On the other hand, if one of the vertex is V-A and the other is V+A, the momentum of the neutrino will contribute. Thus it is convenient to reorganize the interaction in term of the lepton vertices, i.e.,

$$H_W = G_F \cos\theta_c/\sqrt{2}\,\{ j_{L\mu}\tilde{J}_L^{+\mu} + j_{R\mu}\tilde{J}_R^{+\mu} \}, \tag{6}$$

where

$$\tilde{J}_L^{+\mu} = J_L^{+\mu} + \kappa J_R^{+\mu}, \quad \tilde{J}_R^{+\mu} = \eta J_L^{+\mu} + \lambda J_R^{+\mu}. \tag{7}$$

In the non-relativistic impulse approximation, the hadronic currents are parametrized as

$$J_{\substack{L\\R}}^{+\mu}(\vec{x}) = \sum_n \tau_n^+ \left[\begin{pmatrix} g_V \\ g_V' \end{pmatrix} g^{\mu 0} + \begin{pmatrix} g_A \\ -g_A' \end{pmatrix} g^{\mu j} \sigma_n^j \right] \delta(\vec{x} - \vec{r}_n). \tag{8}$$

Then the effective currents are

$$\tilde{J}_L^{+\mu} = g_A \sum \tau_n^+ [G_V g^{\mu 0} + G_A g^{\mu j} \sigma_n^j] \, \delta(\vec{x} - \vec{r}_n)$$

$$\tilde{J}_R^{+\mu} = g_A \sum \tau_n^+ [\epsilon_V g^{\mu 0} - \epsilon_A g^{\mu j} \sigma_n^j] \, \delta(\vec{x} - \vec{r}_n)$$

where

$$G_V = (g_V + \kappa g_V')/g_A \simeq g_V/g_A, \quad \epsilon_V = (\lambda g_V' + \eta g_V)/g_A,$$

$$G_A = (g_A - \kappa g_A')/g_A \simeq 1, \quad \epsilon_A = (\lambda g_A' - \eta g_A)/g_A.$$

It should be noted that ϵ_V is assorciated with the Fermi-term and the ϵ_A with the Gamow-Teller term. Since the GT-matrix element is generally much larger than the F-term, ϵ_A is essentially the parameter to represent the right-current interaction.

By using the same approximation as in $(\beta\beta)_{2\nu}$ mode, the amplitude is obtained. I will not write the amplitude here because it is complicated, but I would mention that it is written explicitely in terms of the effective currents.

Now I will discuss the selection rule for the $0^+ \to J^+$ transition. At first, we notice that the emitted two electrons should be the S-wave states for the neutrino mass and q^0 parts because of the parity conservation. Thus the maximun angular momentum which two electrons can carry out is 1 so that the $0^+ \to 2^+$ transition is forbidden. On the other hand, for the $\vec{q}$ part which has the odd parity, one of the electron should be P-wave state and thus the $0^+ \to 2^+$ transition is allowed. As a result, the observation of the $0^+ \to 2^+$ transition directly implys the existence of the right-handed current as stressed in Refs.5,9,12. The fobiddenness of the $0^+ \to 0^+$ transition is derived in a similar manner as in $(\beta\beta)_{2\nu}$ mode as shown in Refs.5, 6,9. I shall summarize the result:

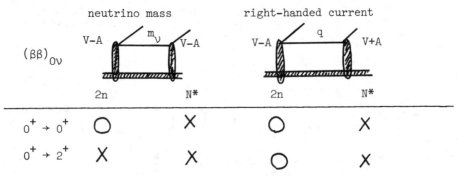

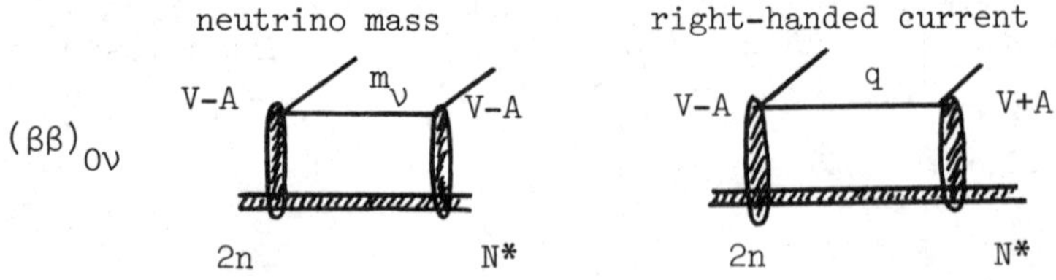

	neutrino mass 2n	neutrino mass N*	right-handed current 2n	right-handed current N*
$0^+ \to 0^+$	O	X	O	X
$0^+ \to 2^+$	X	X	O	X

RELATIVISTIC COULOMB DISTORTED ELECTRON WAVE FUNCTION

Traditionally the Coulomb correction of the emitted electrons from the nucleus is treated by multiplying the non-relativistic Fermi factor to the plane wave function (Primakoff-Rosen approximation):

$$\psi_s(E,\vec{r}) = \sqrt{F_{PR}(Z,E)}\sqrt{\frac{E+m}{2E}}\begin{pmatrix} \chi_s \\ \frac{\vec{\sigma}\cdot\vec{p}}{E+m}\chi_s \end{pmatrix} e^{i\vec{p}\cdot\vec{r}}, \tag{11}$$

where

$$F_{PR} = \frac{E}{P}\frac{2\pi\alpha Z}{1-e^{-2\pi\alpha Z}} \tag{12}$$

This simplified form has been used because the phase space integration can be done explicitely. However, as discussed by Haxton et al and Nishiura,[4] the relativistic correction is substancial for the medium-heavy nucleus. Nishiura used the form obtained by replacing F_{PR} with F_0 in Eq.(11), where

$$\sqrt{F_0(Z,E)} = 2\frac{|\Gamma(\gamma+iy)|}{\Gamma(1+2\gamma)}e^{\pi y/2}(2pR)^{\gamma-1}, \quad \gamma=\sqrt{1-(\alpha Z)^2}, \quad y=\alpha Z\frac{E}{p}. \tag{13}$$

In the relativistic treatement, the electron wavefunction is expressed in the spherical wave expansion:

$$\psi_s(E,\vec{r}) = \sum_{\kappa,\mu} 4\pi i^{l_\kappa} C(l_\kappa \tfrac{1}{2} j; \mu-s, s)\, Y^{\mu-s*}_{l_\kappa}(\hat{p}) \begin{pmatrix} \tilde{g}_\kappa(pr)\,\chi_{\kappa\mu}(\hat{r}) \\ i\tilde{f}_\kappa(pr)\,\chi_{-\kappa\mu}(\hat{r}) \end{pmatrix}. \tag{14}$$

The S- and P-wave radial function is explicitely expressed as

$$\psi_{S\text{-wave}} = \begin{pmatrix} \tilde{g}_{-1}\,\chi \\ \tilde{f}_1\,(\vec{\sigma}\cdot\hat{p})\,\chi \end{pmatrix}, \tag{15}$$

$$\psi_{P\text{-wave}} = \begin{pmatrix} [(\tilde{g}_1+2\tilde{g}_{-2})\hat{p}\cdot\hat{r} + i(\tilde{g}_1-\tilde{g}_{-2})\hat{r}\cdot(\hat{p}\times\vec{\sigma})]\,\chi \\ [3\tilde{f}_2\,(\hat{p}\cdot\hat{r})(\hat{p}\cdot\vec{\sigma}) - (\tilde{f}_{-1}+\tilde{f}_2)\hat{r}\cdot\vec{\sigma}]\,\chi \end{pmatrix}. \tag{16}$$

Note that* the total angular momentum is 1/2 for $\tilde{g}_{\pm1}$, $\tilde{f}_{\pm1}$, and 3/2 for $\tilde{g}_{-2}$, $\tilde{f}_2$. Therefore the combination of $\tilde{g}_{\pm1}$ and $\tilde{f}_{\pm1}$ appears in the $0^\mp \to 0^\mp$ transition, while $\tilde{g}_{-1}$, $\tilde{f}_1$, $\tilde{g}_{-2}$, and $\tilde{f}_2$ in the $0^+ \to 2^+$ transition.

For the radial wave function, we used

$$\begin{pmatrix} \tilde{g}_\kappa \\ \tilde{f}_\kappa \end{pmatrix}_{\text{present work}} = (pr)^{|\kappa|-1}\left[\begin{pmatrix} \tilde{g}_\kappa \\ \tilde{f}_\kappa \end{pmatrix} : \text{solution in an extended potential}\Big/(pr)^{|\kappa|-1}\right]_{r=0}, \tag{17}$$

where the uniform charge distribution in the nucleus is assumed.

In Fig.2, the comparison among the PR-approximation, F_0-approximation, and the present case is made.

* In the case of the PR-approximation, the radial parts are
$\tilde{g}_1 = -\tilde{f}_{-1} = \tilde{g}_{-2} = \tilde{f}_2 = \sqrt{F_{PR}}\, j_1(pr) \simeq \sqrt{F_{PR}}(1/3)pr.$

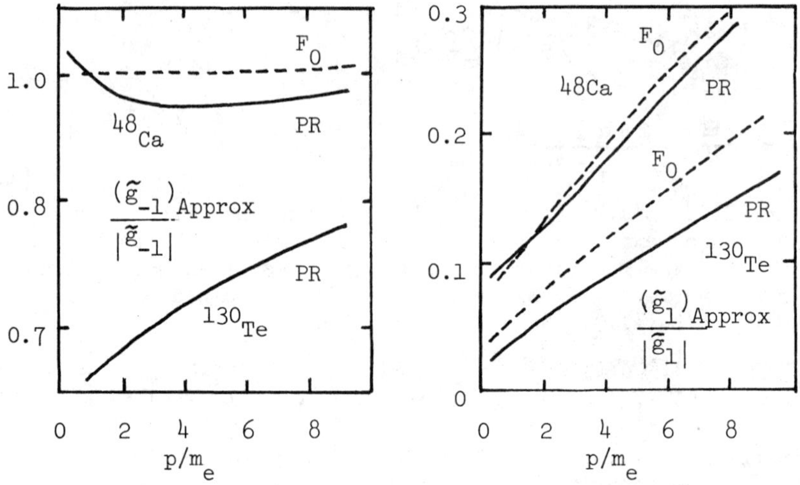

Fig.2: The comparison among three cases. The denominator is the present work (the relativistic one with the nuclear finite size effect). PR and F_0 indicate the approximation in Eqs.12 and 13.

For the S-wave function $\tilde{g}_{-1}$, the PR-approximation is valid only for the light nuclei, but the F_0 approximation is quite good. For the P-wave function $\tilde{g}_1$, both approximation fails. Since the right-handed part of the $0^+ \to 0^+$ transition and the $0^+ \to 2^+$ transition involve the P-wave function, the relativistic correction should be considered seriously.

DECAY FORMULAE

As I have stressed, the $0^+ \to 0^+$ transition in the 2n-mechanism dominates for the $(\beta\beta)_{2\nu}$ mode. The half-life is given by

$$[T_{2\nu}^{2n}(0^+ \to 0^+)]^{-1} = G_{GT}(T)\,|M_{GT}^{(2\nu)}/\mu_0|^2, \tag{18}$$

where

$$M_{GT}^{(2\nu)} = \langle \Sigma\, \tau_n^+ \tau_m^+ \vec{\sigma}_n \cdot \vec{\sigma}_m \rangle, \qquad \mu_0 = [\langle E_n\rangle - (M_i + M_f)/2\,]/m_e. \tag{19}$$

Here the Fermi type matrix element $M_F^{(2\nu)} = \langle \Sigma \tau_n^+ \tau_m^+ \rangle$ is neglected, because it is much smaller[6] than the Gamow-Teller type matrix element M_{GT}. The factor $G_{GT}(T)$ is essentially the phase space factor and is determined once the kinetic energy relese is known. The values for the typical nuclei are given in Appendix.

As for the $(\beta\beta)_{0\nu}$ mode, the situation is much more complicated. I refer the exact formula to our recent paper.[5] Here I adopt the following simplification: (a) Only the Fermi and Gamow-Teller terms are retained. (b) The higher order terms of ε_V and ε_A are neglected. (c) For the potentials due to the neutrino propagation H(r), H(r)$\sim$

$-rH(r)'$ is used. This is reasonable because $H(r)$ behaves as $1/r$ as $r \to 0$. Now I obtain

$$[T^{2n}_{0\nu}(0^+\to 0^+)]^{-1} = |M^{(0\nu)}_{GT}|^2 \{C_1|x|^2 + C_2 \mathrm{Re}\, x^* y + C_3 \mathrm{Re}\, x^* z + C_4|y|^2 + C_5|y\, z| + C_6|z|^2\}, \tag{20}$$

where

$$x = \left(\sum \frac{m_j}{m_e} U_{ej}^2\right)(1-\chi_F), \quad y = \epsilon_A \sum U_{ej} V_{ej}, \quad z = \epsilon_V \left(\frac{g_A}{g_V}\right) \sum U_{ej} V_{ej}\, \chi_F, \tag{21}$$

and

$$\chi_F = \left(\frac{g_V}{g_A}\right)^2 M_F^{(0\nu)} / M_{GT}^{(0\nu)}, \tag{22}$$

$$M_F^{(0\nu)} = \langle \sum \tau_n^+ \tau_m^+ R H(r_{nm}) \rangle, \quad M_{GT}^{(0\nu)} = \langle \sum \tau_n^+ \tau_m^+ \vec{\sigma}_n \cdot \vec{\sigma}_m R H(r_{nm}) \rangle. \tag{23}$$

The factor C_i's are essentially the phase space factors and are evaluated. The values of them for typical nuclei are given in Appendix. The formula can be rewritten in terms of η and λ using Eq.(10).

The formula for the $0^+ \to 2^+$ transition includes only the right-handed parameters ε_V and ε_A, but the detail expression is refered to Ref.5.

RATIO ARGUMENT

In Ref.9, we gave the following analysis of the data on $R_T = T_{1/2}(^{128}\mathrm{Te})/T_{1/2}(^{130}\mathrm{Te})$: First we observe that

$$(R_T)^{-1} = \frac{\Gamma^{2\nu}(^{128}\mathrm{Te})}{\Gamma^{2\nu}(^{130}\mathrm{Te})} \; \frac{1+(\Gamma^{0\nu}/\Gamma^{2\nu})(^{128}\mathrm{Te})}{1+(\Gamma^{0\nu}/\Gamma^{2\nu})(^{130}\mathrm{Te})} > \frac{\Gamma^{2\nu}(^{128}\mathrm{Te})}{\Gamma^{2\nu}(^{130}\mathrm{Te})} \tag{25}$$

$$\simeq \frac{\Gamma^{2\nu}(^{128}\mathrm{Te})}{\Gamma^{2\nu}(^{130}\mathrm{Te})} [1+(\Gamma^{0\nu}/\Gamma^{2\nu})(^{128}\mathrm{Te})]. \tag{26}$$

The inequality always holds if $\Gamma^{0\nu} \neq 0$ because $\Gamma^{0\nu} \propto T^5$ and $\Gamma^{2\nu} \propto T^{11}$ and $T = 1.7$ for $^{128}\mathrm{Te}$ and 5.0 for $^{130}\mathrm{Te}$. The argument follows in two steps: Once the data on R_T is given, the inequality should be tested. If this holds, (a) there exists the $(\beta\beta)_{0\nu}$ mode and thus the neutrino is the Majorana particle. Then, (b) the mass of the neutrino and the right-handed parameters are estimated. The important point is that the ratio of the decay rates of $(\beta\beta)_{2\nu}$ can be evaluated reliablly. There are two reasons for this: First, for the $(\beta\beta)_{2\nu}$ mode, we need to take care of the $0^+ \to 0^+$ transition in the 2n-mechanism. Secondly, the ratio of the nuclear matrix elements of $^{128}\mathrm{Te}$ and $^{130}\mathrm{Te}$ should be clouse to be unity.[13] Explicitely,

$$\frac{\Gamma^{2\nu}(^{128}\mathrm{Te})}{\Gamma^{2\nu}(^{130}\mathrm{Te})} = \frac{^{128}G_{GT}}{^{130}G_{GT}} \left(\frac{^{128}|M^{(2\nu)}_{GT}/\mu_0|}{^{130}|M^{(2\nu)}_{GT}/\mu_0|} \right)^2. \tag{27}$$

Therefore the ratio of G_{GT}'s essentially determines this. In the following, the values of G_{GT}'s are given.

	${}^{128}G_{GT}$	${}^{130}G_{GT}$	${}^{128}G_{GT}/{}^{130}G_{GT}$
PR-approx	$1.422\cdot10^{-22}$	$0.8976\cdot10^{-18}$	$6.31\cdot10^{3}$
Present work	$8.273\cdot10^{-22}$	$4.652\cdot10^{-18}$	$5.62\cdot10^{3}$

It should be noted that although the absolute magnitude enhances substancially with the correct treatment of electron wave function, the ratio of them does not change much. Thus our previous analysis in Ref.9 is still valid.

There are two data available now:

$R_T = (1.59 \pm 0.05)\cdot10^3$ Missouri[1]

$(R_T)^{-1} = (0.99 \pm 0.95)\cdot10^{-4}$ Heidelberg[7*]

Two different conclusions can be derived . If the Missouri data is considered seriously, there exists the $(\beta\beta)_{0\nu}$ mode and thus the neutrino is the Majorana particle. On the other hand, the Heidelberg data is consistent with no $(\beta\beta)_{0\nu}$ mode.

Now I turn the neutrino mass argument. Observe that

$$\Gamma^{0\nu}/\Gamma^{2\nu} = (C_1/G_{GT})\left|\Sigma \frac{m_j}{m_e} U_{ej}\right|^2 \left|M_{GT}^{(0\nu)}/(M_{GT}^{(2\nu)}/\mu_0)\right|^2 (1-\chi_F)^2. \tag{27}$$

The factor C_1/G_{GT} is determined unambiguously, but the problem is the evaluation of the ratio of the nuclear matrix elements of $(\beta\beta)_{0\nu}$ and $(\beta\beta)_{2\nu}$. In the previous analysis, we used $M_{GT}^{(0\nu)} \sim \langle RH(r)\rangle M_{GT}^{(2\nu)} \sim 0.55\cdot M_{GT}^{(2\nu)}$ and obtained $|\Sigma m_j U_{ej}^2| = 34$ eV from the Missouri data. After that, Haxton et al[6] have evaluated the nuclear matrix elements and obtained $M_{GT}^{(0\nu)} \sim 1.69\ M_{GT}^{(2\nu)}$ due to the existence of the 1/r factor in the $(\beta\beta)_{0\nu}$ mode, which reduced the neutrino mass value to 10 eV. Minkowski[15] assumed the short range spin-spin correlation and obtained much smaller mass value.

By analysing the Heidelberg data with $M_{GT}^{(0\nu)} \sim 1.69\ M_{GT}^{(2\nu)}$, we obtain the mass bound $|\Sigma m_j U_{ej}^2| < 4.7$ eV, requiring that the theory is within the two standard deviation of the data. The right-handed current parameter is also restricted as $|\varepsilon_A \Sigma U_{ej} V_{ej}| < 2.0\cdot10^{-5}$ by assuming $\varepsilon_V = 0$.

* In this conference, Prof. Kirsten has reported the final result $(R_T)^{-1} = (1.03 \pm 1.13)\cdot10^{-4}$.

BOUNDS ON MASS AND RIGHT-HANDED PARAMETERS

Once the data on the $0^+ \to 0^+$ transition of the $(\beta\beta)_{0\nu}$ mode is given in the direct measurement of electrons, the bounds on the neutrino mass and the right-handed parameters are in principle derived, by using the formula given in Eq.(20). The problem is the evaluation of the nuclear matrix elements. There is a vast discrepancies between the theory and the geochemical data as emphasized by Haxton et al.[6] In the following, the comparison is made:

	$(M_{GT}^{(2\nu)})_{\text{Theory}}$ Haxton et al	$\lvert M_{GT}^{(2\nu)} \rvert_{\text{Experiment}}$ Geochemical data
^{82}Se	-1.88	0.786
^{130}Te	-2.97	0.239

The experimental values are obtained by comparing the formula in Eq.(18) and the world averages of the geochemical data given by Kirsten: $T_{1/2}(^{82}Se) = (1.45 \pm 0.15)\cdot 10^{20}$ years and $T_{1/2}(^{130}Te) = (2.55 \pm 0.20)\cdot 10^{21}$ years. Although the difference for M_{GT} is only the factor 2 for ^{82}Se, the discrepancies for ^{130}Te is the order of the magnitude. Prof. Kirsten will discuss later that the life-times measured in the geochemical method by using the ores with the different locations and the different ages are consistent each other within the factor 2 and therefore it is hard to believe that the geochemical data are completely wrong. The maximal deviation would be at most by the factor of 2, i.e., in terms of the M_{GT}, the factor $\sqrt{2}$.

In Ref.5, we have taken the attitude that the theoretical evaluation of the absolute magnitude of the nuclear matrix elements is difficult, but the ratio of them would be relatively reliable. That is, we used the values of $M_{GT}^{(2\nu)}$ deduced from the geochemical data and the ratio of them from the theoretical estimates by Haxton et al.

The lower bounds of the half-lives measured by the Columbia group[16] for ^{82}Se and by Zdesenco[17] for ^{130}Te are used to determine the bounds. Fig.3 shows the constraint ellipses whose insides are the allowed domatin. For ^{76}Se, there is the data by Milano group[18] and the bound is obtained by assuming the value of M_{GT} deduced from the theoretical value by multiplying the same reduction factor as for Se.

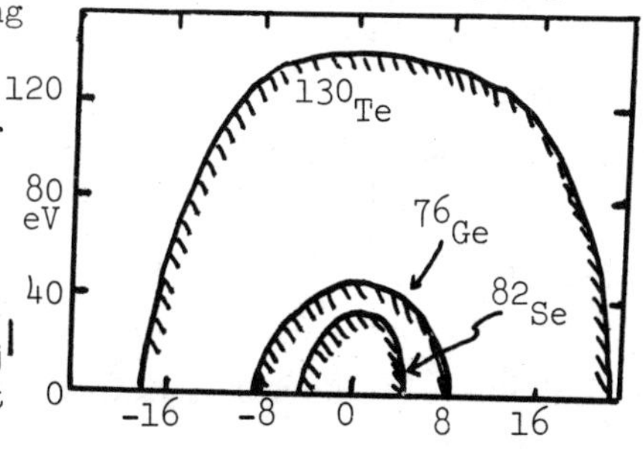

Fig.3; The constraint on the neutrino mass (vertical line in the unit of eV) and $\varepsilon_A |\Sigma U_{ej} V_{ej}|$ (the holizontal line in the unit of 10^{-5}).

CONCLUSION

The theoretical framework has become better shape now. Only problem left is the accuracy of the clousure approximation. Aside from that, the evaluation of the nuclear matrix element is problematic . If the geochemical data is correct, the theory should drastically be modified for the medium-heavy nuclei. Anyhow, the direct measurement of the $(\beta\beta)_{2\nu}$ mode gives the definite answer and fixes the nuclear matrix element. The background evaluation to the $(\beta\beta)_{2\nu}$ mode such as from the Majoron is important. The method to discriminate the background is under consideration.

Appendix

The numerical values for G_{GT} and C_i's are tabulated for the typical nuclei:

	^{76}Ge	^{82}Se	^{100}Mo	^{128}Te	^{130}Te	^{136}Xe	^{150}Nd	
	10^{-19}	10^{-18}	10^{-18}	10^{-22}	10^{-18}	10^{-18}	10^{-18}	$(yr)^{-1}$
G_{GT}	1.271	4.343	9.135	8.273	4.652	4.671	1.152	
	10^{-15}	10^{-14}	10^{-14}	10^{-15}	10^{-14}	10^{-14}	10^{-13}	
C_1	6.198	2.728	4.431	1.769	4.295	4.567	2.029	
C_2	-2.106	-1.176	-1.934	-0.2375	-1.687	-1.764	-0.9162	
C_3	-8.626	-4.303	-7.039	-1.663	-6.484	-6.845	-3.303	
C_4	9.108	8.059	13.62	0.6090	9.894	10.19	7.895	
C_5	29.43	25.35	42.79	2.065	31.37	32.32	24.60	
C_6	26.39	21.17	35.61	2.380	26.77	27.67	20.10	

References

1. D. Caldwell, the talk given at this Conference.
2. C. Liguori, the Milano group reported the data of ^{76}Ge in this Conference; $T_{1/2}^{0\nu}(0^+ \to 0^+) > 1.1\cdot10^{22}$, $T_{1/2}^{0\nu}(0^+ \to 2^+) > 0.7\cdot10^{22}$yr.
   F.T. Avignone,III, a talk given at the Conference on Science Underground held at Los Alamos, 1982; $T_{1/2}^{0\nu}(0^+ \to 0^+) > 1.3\cdot10^{22}$y.
3. R.N. Mohapatra and J.D. Vergados, Phys. Rev. Lett.47,1713(1982).
   J. Schechter and J.W.F. Valle, Phys. Rev. D25,2951(1982).
   L. Wolfenstein, Carnegie-Mellon Univ. Preprint, CMU-HEG82-5(1982)

4. The formal treatment was made by S.P. Rosen, but it is hard to use. S.P. Rosen,Can. Jour. Phys. 37,780(1959).

W.C. Haxton, G.J. Stepheson,Jr. and D. Strottman, Phys. Rev. Lett. 47,153(1981).

H. Nishiura, Ph.D Thesis; Kyoto Univ. Report No. RIFP-453(1981).

5. M. Doi, T. Kotani, H. Nishiura and E. Takasugi, Osaka Univ. Coll. Gen. Educ. preprint, OS-GE 82-43 (1982).

6. W.C. Haxton, G.J. Stephenson, Jr. and D. Strottman, Phys. Rev. D25, 2360 (1982); W.C. Haxton, Los Alamos National Laboratory preprint, LA-UR-82-1623 (1982).

7. T. Kirsten, private communication and a talk given at the Workshop on Low Energy Tests of High Energy Physics, Santa Barbara, (1982).

8. H. Primakoff and S.P. Rosen, Ann. Rev. Nucl. Part. Sci. 31,145(1981)

9. M. Doi, T. Kotani, H. Nishiura, K. Okuda and E. Takasugi, Phys. Lett. 103B, 219(1981):113B,513(1982)(E); Prog. Theor. Phys. 66, 1739 and 1765 (1981):68,347(E)(1982).

10. A. Zee, Phys. Lett. 93B, 389 (1980).

11. L. Wolfenstein, Phys. Lett. 107B, 77 (1981).

12. S.P. Rosen,Proceedings of Orbis Scientia 1981, Coral Gables 333 (1982).

13 B. Pontecorvo, Phys. Lett. 26B, 630 (1968).

14. E.W. Hennecke, O.K. Manuel and D.D. Sabu, Phys. Rev.C11,1378(1975)

15. P. Minkowski, Nucl. Phys. B201, 269 (1982).

16 B.T. Cleveland et al, Phys. Rev. Lett. 35, 757 (1975).

17. Yu.G. Zdesenko, JETP Lett. 32, 58 (1980).

18. E. Fiorini et al, Nuovo. Cimento 13A, 747 (1973).

GEOCHEMICAL DOUBLE BETA DECAY EXPERIMENTS

T. Kirsten
Max-Planck-Institut für Kernphysik, Heidelberg, Germany

ABSTRACT

Double Beta Decay provides a very sensitive test of lepton-quark symmetry, unfortunately, however, decay rates are so small that it has not yet been possible to unequivocally detect a Double Beta Decay event directly; so much the more important is the data obtained by the "geochemical" method where one detects the Double Beta Decay products which have accumulated in natural minerals during long geological time periods. In that order, I discuss

- the principles of the geochemical method, its strength and its limitations
- the available data
- new results on the ^{128}Te-^{130}Te system
- theoretical implications of the data concerning the electron neutrino restmass and lepton-number conservation.

A limit of $m_\nu \leq 5.6$ eV (95% confidence) has been obtained from our recent ^{128}Te-^{130}Te-measurements.

Series Editor's Note

The author has informed us that the same paper was presented at the conference on Science Underground at Los Alamos at the end of September 1982. The full paper has been published in AIP Conference Proceedings No. 96, "Science Underground (Los Alamos, 1982)", pages 396-410.

RECENT RESULTS OBTAINED IN EXPERIMENTS ON DOUBLE BETA DECAY OF ^{76}Ge AND OTHER NUCLIDES.

E. Bellotti, E. Fiorini, C. Liguori, A. Pullia, A. Sarracino and L. Zanotti

Dipartimento di Fisica dell'Università - Milano
I.N.F.N. - Sezione di Milano, Via Celoria 16 - 20133 Milano - Italy

ABSTRACT

Results are reported of an experimental search for lepton number violation in ^{76}Ge, ^{150}Nd, ^{148}Nd, ^{100}Mo, ^{58}Ni, and ^{92}Mo decays. The experimental set up has been running under M. Blanc ($\sim$ 5000 m w.e. depth). No evidence has been found so far for any type of neutrinoless process. Limits of 10^{22} years for ^{76}Ge transitions are reported.

INTRODUCTION

Other talks[1-5] in this Conference have outlined the topics of double beta decays. We will only remind that double beta processes (i.e. double electron emission, double positron emission, electron capture plus positron emission and double electron capture) may in principle occur in two channels, the first one in which two neutrinos are emitted and the second one with no neutrino emission. The first channel is a standard second order weak process which conserves lepton number, while the second channel violates lepton number conservation and can be due [6,7] either to a non zero neutrino mass or to a small admixture of right handed currents.

Let us remark that in neutrinoless double beta decays the two emitted electrons share the total available energy: the sum spectrum of their kinetic energies is therefore expected to be a line while, in the two neutrino mode, it should be distributed on a broad energy interval (see fig. 1).

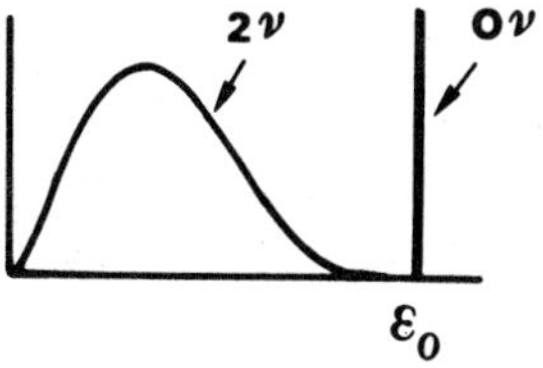

Fig. 1. Kinetic sum energy of the two emitted electrons.

EXPERIMENTAL DETAILS

In the present experiment a Ge(Li) crystal (P.G.T. DL 422) is used with an energy resolution better than 2 keV at 1 MeV. The active volume is 125 cm^3. For Ge double beta decay measurement, the crystal is both source and detector[8] of the transition, while, for the other nuclides, de-excitation gamma-rays were measured following double beta decay on excited level of the daughter nucleus, with the source samples external to the detector.

Special care has been devoted[9] in reducing background counting rate. The reduction has been obtained operating the detector in a laboratory placed close to the middle of the Mont Blanc road tunnel, connecting Italy to France, at a depth of 1600 m of rock, that is 5000 m of water equivalent shielding from cosmic radiation. Natural radioactivity is shielded surrounding the detector with consecutive layers of 4.5 cm of high purity Hg and 25 ÷ 30 cm of low activity Pb (see fig.2). Materials of the detector assembly have been carefully

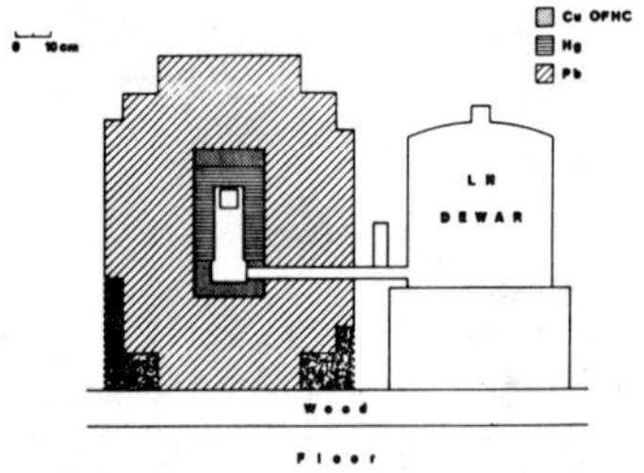

Fig. 2. Cross-section of the experimental set-up.

selected for their low contamination properties: the inner holder is made in OFHC Cu .5 mm thick and the outer end cap in Ti 1 mm thick. Remaining background is due to small residual contaminations. The total counting rate is shown in fig.3. The electronic chain is a conventional one with 8K channel ADC and MCA. A spectrum stabilizer keeps resolution at the nominal value over long lasting runs.

RESULTS

In this series of measurements the most interesting result has been obtained on ^{76}Ge (i.a. = 7.76%). The more likely transitions should occur from ground state (0^+) of the father nucleus to ground (0^+) or first excited state (2^+) of the daughter ^{76}Se. The energy release ε_0 is respectively of 2040.9 ± 2.5 keV and 1481.9 ± 2.5 keV[10]. The total run time up to now is of 3061 hours and the measure is still in progress.

No evidence has been obtained so far for double beta decay. Using a maximum likelihood method, we can set upper limits of .25 ± 5.60 counts and 0 ± 7.65 counts for the two transitions at 2040.9 and 1481.8 keV respectively (see fig.5). These limits account for energy resolution and nuclear mass uncertainty and correspond to

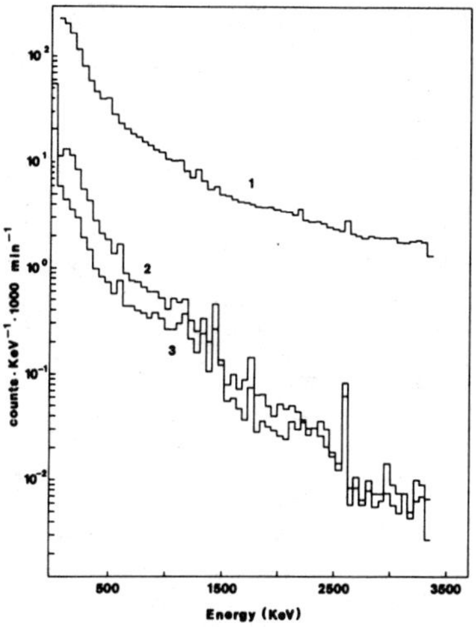

Fig. 3. Total counting rate. Curve 1 refers to a ground level measure; curves 2 and 3 refer to M. Blanc measures with Cu and Hg inner shieldings respectively.

$$T_{\frac{1}{2}}\left[(0\nu),\ 0^+\to 0^+\right] \geq \begin{array}{l} 1.8\cdot 10^{22} \text{ yrs at } 68\% \text{ c.l.} \\ 1.1\cdot 10^{22} \text{ yrs at } 90\% \text{ c.l.} \end{array} \tag{1}$$

$$T_{\frac{1}{2}}\left[(0\nu),\ 0^+\to 2^+\right] \geq \begin{array}{l} 7\cdot 10^{21} \text{ yrs at } 68\% \text{ c.l.} \\ 4.2\cdot 10^{21} \text{ yrs at } 90\% \text{ c.l.} \end{array} \tag{2}$$

where in (2) we have corrected for the probability of detecting the de-excitation gamma-ray emitted by ^{76}Se. Analyzing the total energy spectrum shape we can also quote a half-lifetime limit for the two neutrino mode:

$$T_{\frac{1}{2}}\left[(2\nu),\ 0^+\to 0^+\right] \geq \begin{array}{l} 4\cdot 10^{18} \text{ yrs at } 78\% \text{ c.l.} \\ 2.4\cdot 10^{18} \text{ yrs at } 90\% \text{ c.l.} \end{array} \tag{3}$$

Let us now compare the result given in (1) with theory. Following S.P. Rosen[5] we obtain limits on neutrino mass $\langle\langle m_\nu \rangle\rangle$ and on right handed current admixture parameter η:

$$\langle\langle m_\nu \rangle\rangle \lesssim 11 \text{ eV}; \qquad |\eta| \lesssim 2\cdot 10^{-5} \tag{4}$$

both at 90% c.l.. Following more recent calculation[1,11] which involves corrections for relativistic Coulomb wave function including the finite nuclear size effect, we obtain:

$$\langle\langle m_\nu \rangle\rangle \lesssim 29 \text{ eV}; \qquad |\eta| \lesssim 5.4\cdot 10^{-5} \tag{5}$$

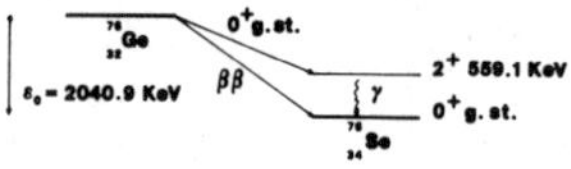

Fig. 4. Diagram of ββ transitions in ^{76}Ge.

A theoretical prediction for the $0^+ \to 2^+$ transition rate would be interesting since this transition can occur only in the presence of right handed currents. Unfortunately a precise one is not available at the moment and a rough estimate allows to set a limit of $|\eta| \lesssim 5 \cdot 10^{-5}$

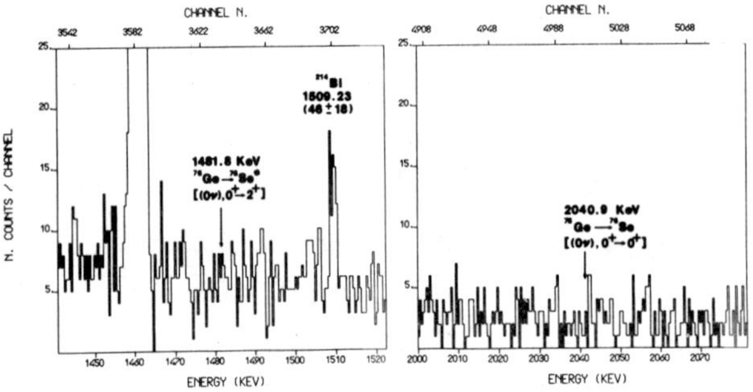

Fig. 5. Regions of interest for $0^+ \to 2^+$ and $0^+ \to 0^+$ transitions.

A summary of the preliminary measurements carried out on other nuclides is reported in table 1[12]. Note that the same sensitivity as for ^{76}Ge measure may be obtained only if one is able to regain the loss in detector efficiency using, for instance, isotopically enriched samples.

FUTURE PLANS

Milan group is now planning[13] a search for double beta decay of ^{136}Xe (i.a. = 8.9%). The measure should be performed using a Time Projection Chamber filled with Xenon at high pressure (5÷10 atm) (see fig. 6). The goal is to obtain a sensitivity of $5 \cdot 10^{22}$ years on neutrinoless $0^+ \to 0^+$ transition. A prototype will be assembled within one year. Preliminary measures on Xenon intrinsic activity and ionization properties are currently in progress.

Transition	Final state	Transition energy (keV)	Experimental limit (90% c.l.) (years)
$^{148}_{60}Nd - ^{148}_{62}Sm$	2^+ (550)	1379	$3 \cdot 10^{18}$
	2^+ (1455)	474	$2.7 \cdot 10^{18}$
$^{150}_{60}Nd - ^{150}_{62}Sm$	2^+ (334)	3033	$1 \cdot 10^{18}$
	0^+ (740)	2627	$1.5 \cdot 10^{18}$
	2^+ (1046)	2321	$1.7 \cdot 10^{18}$
	2^+ (1194)	2173	$2.7 \cdot 10^{18}$
	0^+ (1255)	2112	$2.1 \cdot 10^{18}$
$^{58}_{28}Ni - ^{58}_{26}Fe$	2^+ (811)	1116	$4 \cdot 10^{19}$
	2^+ (1675)	252	$4 \cdot 10^{19}$
$^{92}_{42}Mo - ^{92}_{40}Zr$	2^+ (935)	715	$3 \cdot 10^{18}$
	0^+ (1383)	266	$4 \cdot 10^{18}$
	4^+ (1496)	154	$6 \cdot 10^{18}$
$^{100}_{42}Mo - ^{100}_{44}Ru$	2^+ (540)	2492	$2 \cdot 10^{18}$
	0^+ (1130)	1902	$3 \cdot 10^{18}$
	2^+ (1362)	1670	$3 \cdot 10^{18}$

Table 1. Obtained limits on double beta processes.

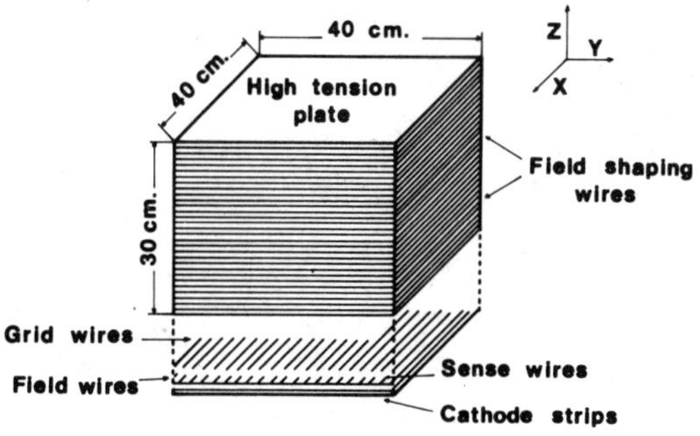

Fig. 6. Sketch of the designed Xenon TPC.

REFERENCES

1. E. Takasugi - Report to this Conference.
2. T. Kirsten - Report to this Conference.
3. D. Caldwell - Report to this Conference.
4. J.D. Vergados - Report to this Conference.
5. G.J. Stephenson - Report to this Conference.
6. S.P. Rosen, Lepton non conservation and double beta decay, invited paper to the 1981 Int. Conf. on Neutrino Phys. and Astrophys., Hawaii, July 1 - 8, 1981.
7. W.C. Haxton, D.J. Stephenson and D. Strottmann, Phys. Rev. Lett. 47 153 (1981).
8. E. Fiorini et al., Nuovo Cimento A13, 747 (1973).
9. C. Liguori et al., Nucl. Instr. and Methods 204, 585 (1983).
10. A.H. Wapstra, private communication.
11. W.C. Haxton, G.J. Stephenson and D. Strottmann, Phys. Rev. D25, 2360 (1982).
12. E. Bellotti et al., Lett. Nuovo Cimento 33, 273 (1982).
13. E. Bellotti et al., CERN report EP/EF/mm-0010 P - June 1982.

NEUTRINOLESS ββ-DECAY IN GAUGE THEORIES

by

J.D.Vergados

Department of Physics, University of Ioannina, Ioannina-Greece

A B S T R A C T

The lepton violating neutrinoless ββ-decay is investigated in the context of fashionable gauge theories. Various mechanisms are examined e.g. light or heavy neutrinos, with or without right-handed currents, intermediate doubly charged Higgs Particles, majoron emission etc. Numerical results have been obtained for the transitions $^{48}Ca \rightarrow {}^{8}Ti(\beta^-\beta^-)$ and $^{58}Ni \rightarrow {}^{58}Fe$ ($\beta^+\beta^+$, electron capture, double electron capture) employing realistic nuclear models.

1.- INTRODUCTION

Recent developments in gauge theories have rekindled interest in the question of whether lepton number (charge) is conserved[1-4], and revived attempts to observe neutrinoless double β decay (oν). Most of the theoretical and experimental efforts have hitherto focused on double electron emission

$$(A,Z) \longrightarrow (A,Z+2) + e^- + e^- \quad (o\nu) \tag{1}$$

which violates lepton number by two units. This process must be observed against the background originating from the two neutrino (2ν) decay

$$(A,Z) \rightarrow (A,Z+2) + e^- + e^- + \tilde{\nu}_e + \tilde{\nu}_e \tag{2}$$

0094-243X/83/990225-19 $3.00 Copyright 1983 American Institute of Physics which is allowed by lepton conservation. Experimentally, (0ν) can easily be distiguished from (2ν), given sufficiently fast rates, by measuring the energies of the two simultaneously emitted electrons[1,2]. In a world with maximal lepton violation ($\eta=1$), the $o\nu$ process is expected to be favoured over the 2ν one by large factors ($\sim 10^6$). In the actual world, however, with $\eta << 1$ the possibility of observing the $o\nu$ mode depends on the size of lepton violation. Neither process $o\nu$ nor 2ν has yet been observed in laboratory experiments[5], but the predicted rates for the 2ν process are within present experimental capabilities. Double $\beta\beta$ decay has, however, definitely been "seen" in geochemical measurements[6,7], which, unfortunately, cannot distinguish between $(o\nu)$ and (2ν) modes. Even from such total lifetime measurements one can, in principle, extract the lepton violating parameter η. In practice, however, the extraction of precise lepton violating parameters from the total lifetimes appears difficult. This can only come from laboratory measurements of $o\nu$ lifetimes. From such experiments we know that η must be small ($\eta \lesssim 10^{-5}$).

In recent years double $\beta\beta$ decay has become fashionable and experimentalists with high energy backgrounds are beginning to be involved. These people are expected to apply new techniques inlolving large target masses and focus their attention on reactions which emit positrons rather than electrons. Such processes are[8,9],

Double Positron Emission:

$$(A,Z) \longrightarrow (A,Z-2) + e^+ + e^+ \qquad (o\nu) \qquad (3)$$

Electron Positron Conversion

$$e_b^- + (A,Z) \longrightarrow (A,Z-2) + e^+ \qquad (4)$$

and

Double Electron Capture

$$e_b^- + e_b^- + (A,Z) \longrightarrow (A,Z-2)^* \underset{\hookrightarrow \gamma}{} \tag{5}$$

(The electron capture processes are always energetically allowed when double positron decay is.) These processes must be distinguished from

$$(A,Z) \longrightarrow (A,Z-2) + e^+ + e^+ + \nu_e + \nu_e \tag{6}$$

$$e_b^- + (A,Z) \longrightarrow (A,Z-2) + e^+ + \nu_e + \tilde{\nu}_e \tag{7}$$

$$e_b^- + e_b^- + (A,Z) \longrightarrow (A,Z-2)^* + \nu_e + \nu_e \tag{8}$$

which are allowed by lepton conservation and are always present. Again, reactions (3)-(5) have a different experimental signature than (6)-(9), since in the latter some energy is lost to the neutrinos. In reaction (3) one needs to measure the sum of the energies of the two positrons and in reaction (4) the energy of the positron as well as of the hard X-ray following the de-excitation of the atom. In reaction (5) one should measure γ-ray as well as the two hard X-rays following the de-excitation of the final system. In all cases, population of excited final nuclear states is preferred since then one has the extra signature of the emitted γ-ray. In the case of reaction (5) one should look for a γ-ray with energy characteristic of the lepton violating process which does not coincide with any of the γ-rays expected from the de-excitation of the final nucleus populated in the traditional fashion[9] (see ref.9).

The essential ingredients for computing the rates for the 0ν ββ-decays are: i) A lepton violating parameter, which contains the relevant parameters of the gauge model. ii) The nuclear matrix elements which are obtained with standard techniques of nuclear theory (shell model). iii) Proper calculation of kinematics and effects of distortions of the lepton wave functions.

In this paper our emphasis will be on the various aspects of different gauge theories as they manifest themselves in 0ν $\beta\beta$-decay. This gives rize to a variety of effective transition operators which require a large amount of nuclear physics. Thus, in order to keep the amount of nuclear computations to a minimum, we will specialize our results to the two simple systems $^{48}Ca \rightarrow {}^{48}Ti$ (negaton emission) and $^{58}Ni \rightarrow {}^{58}Fe$ (positron emission).

2.-LEPTON VIOLATING PARAMETERS

The lepton violating processes mentioned in the previous section are going to be discussed as second order weak interactions mediated by intermediate neutrinos or Higgs particles in the context of fashionable gauge theory models. The models we are going to employ can be grouped into two classes: Left-handed and right-left symmetric models.

2.1. Left-handed Models. Such models are suitable estensions of the standard model[10] so that the weak currents are strictly left-handed but lepton number is broken locally or globally. In such models 0ν $\beta\beta$-decay can occur via intermediate majorana neutrinos. We further distinguich two classes of such models.

2.1. The neutrino mass contribution. The relevant lepton violating parameter will in this case be proportional to the neutrino mass. We will not elaborate or how such a majorana mass can be generated[3]. We refer the interested reader to a recent review [11]. The important thing to emphasize is that one must distiguish between two kinds of neutrino eigenstates:

i) The weak eigenstates $\nu_0 = (\nu_e, \nu_\mu, \nu_e \ldots)$ put in left-handed isodoublets

$$\begin{pmatrix} \nu_e \\ e^- \end{pmatrix}_L \quad \begin{pmatrix} \nu_\mu \\ \mu^- \end{pmatrix}_L \quad \begin{pmatrix} \nu_\tau \\ \tau^- \end{pmatrix}_L \cdots$$

which may be accompanied by isosinglet neutrinos $N_0 = (N_e, N_\mu, N_\tau, \ldots)$ which do not appear in weak interactions.

ii) The physical mass eigenstates $\nu = (\nu_1, \nu_2, \ldots \nu_n)$, which are light, and $N = (N_1, N_2, \ldots N_n)$ which are heavy (n=Number of flavors). The two sets of states are related by a unitary tranformation

$$\begin{pmatrix} \nu_0 \\ N_0 \end{pmatrix} = \begin{pmatrix} X & Y \\ W & Z \end{pmatrix} \begin{pmatrix} \nu \\ N \end{pmatrix} \tag{10}$$

where (X),(Y),(W),(Z) are nxn matrices.

When the two mass scales are vastly different the matrices (Y) and (W) contain small elements while (X) and (Z) are approximately unitary. In our numerical colculations, in order to reduce the number of unknown parameters, we will further assume that $(X) \approx (Z)$ and thus we will be able to write $Y \approx \beta Z$ and $W \approx -\beta X$ where β is a small mixing parameter. Furthermore we will find it convenient to express the matrices (X) and (Z) as a product of a Kobayashi-Maskawa matrix[12], $U^{(1)}$ and $U^{(2)}$ respectively, and a matrix of phases $(e^{i\alpha})$, $(e^{i\varphi})$ respectively[13]. The latter appear in the majorana condition[13] e.g. $\nu^c = (e^{i\alpha})\nu$ etc.

Two relevant Feynmann Diagrams contributing to 0ν $\beta\beta$-decay are shown in fig.1.

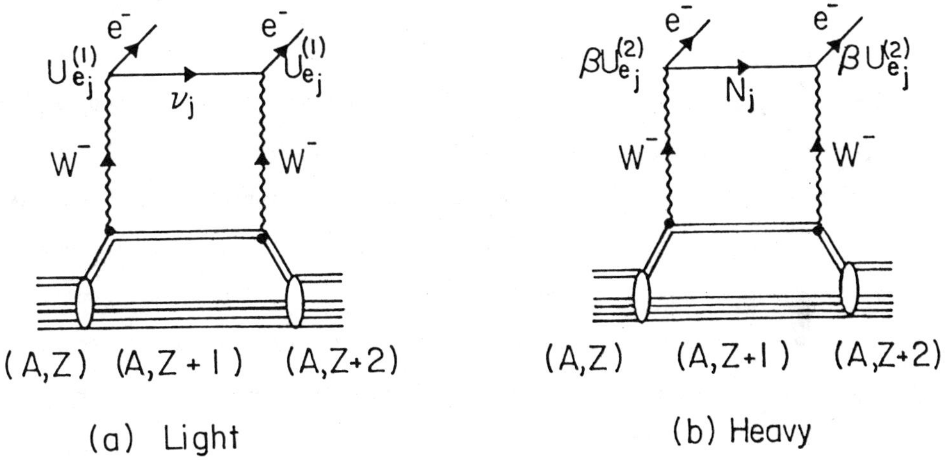

Figure 1 : The Majorana neutrino mediated neutrinoless double β decay when there exist left-handed currents only. (a) corresponds to light and (b) to heavy neutrinos.

The relevant lepton violating parameters are[14]

$$\eta_\nu = \frac{\langle m_\nu \rangle}{m_e}, \quad \langle m_\nu \rangle = \sum_{J=1}^{n} U_{ej}^{(1)} U_{ej}^{(1)} e^{i\alpha_j} m_j \quad \text{(light neutrino)} \tag{11a}$$

$$\eta_N = \beta^2 \sum_J U_{ej}^{(2)} U_{ej}^{(2)} e^{i\varphi_j} \frac{m_A}{m_j} \quad \text{(heavy neutrino)} \tag{11b}$$

where M_A = 0.85 Gev describes the dipole shape nucleon form factor . The inclusion of this form factor appears necessary in this case since otherwise the corresponding amplitude will be suppressed due to the short range nature of the interation potential[9,15]. We should also mention that for neutrino masses in the regime 10 MeV < m_ν < 10 GeV it is not possible to separate the gauge parameters from the nuclear matrix elements[16]

To further clarify the meaning of the above lepton violating parameters we will consider two generations. Then

$$\eta_\nu = \frac{m_1}{m_e}\left[\cos^2\theta + \frac{m_2}{m_1} e^{i\alpha} \sin^2\theta\right], \quad \eta_N = \frac{M_A}{m_1}\left[\cos^2\theta + \frac{m_1}{m_2} e^{i\varphi} \sin^2\theta\right] \tag{12}$$

In a CP invariant theory the quantity $e^{i\alpha}$ gives the ratio of the CP eigevalues[13,17] of the two mass eigenstates which are then ±1. We note that for Dirac neutrino, $m_1 = m_2 = m, \alpha = \pi, \theta = \pi/4$, we get $\langle m_\nu \rangle = 0$ ($\eta_\nu = 0$) as expected. Even in the case of majorana neutrinos, however, if one accepts the mass limit of the triton decay[18], 14 eV $\lesssim m_1 \lesssim$ 46 eV, one is forced to attribute the slowness of the 0ν $\beta\beta$-decay[19] ($|\langle m_\nu \rangle| < 5$ eV) to cancellations between the two neutrino states ($\alpha \approx \pi$). For a neutrino mass between 10 MeV and 10 GeV such a cancellation will require a conspiracy between nuclear and particle physics unlikely to materialize in all possible $\beta\beta$-decay systems[16].

To estimate η we will employ a horizontal gauge model[20] recently discussed by Zoupanos' and Papantonopoulos, which gives

$$U^{(1)} \approx \begin{pmatrix} 1 & \theta_1 & -\theta_3 \\ -\theta_1 & 1 & \theta_2 \\ \theta_3 & -\theta_2 & 1 \end{pmatrix}, \quad \theta_1 = \sqrt{\frac{m_e}{m_\mu}} \approx 0.07, \quad \theta_2 = 5.0\times10^{-5} \quad \theta_3 = 6.2\times10^{-4}$$

$\frac{m_2}{m_1} = \frac{m_\mu}{m_e}$, $\frac{m_3}{m_1} = \frac{m_\tau}{m_e}$. Taking $m_1 = 30$ eV we obtain

$$\eta_\nu = \begin{cases} 1.2\times10^{-4} & \text{(Same CP)} \\ 0 & \text{(opposite CP)} \end{cases} \tag{13}$$

The exact cancellation in the case of opposite CP eigenvalues is a special feature of this particular model. It may not persist in other models.

Proceeding in an analogous fashion and using[21] $U^{(2)} \approx U^{(1)}$, $\beta^2 \approx 10^{-3}$ $m_2 = 2m_1$, $m_1 = 10^3$ GeV we obtain

$$\eta_N \approx 8.5\times10^{-7} \quad \text{(Approximatly the same for both CP eigenvalues)} \tag{14}$$

In the case of heavy intermediate particles one may have a contribution arizing from the virtual double β-decay of pions inside the nucleus[22] (see fig.2). In this case the process is described by the same lepton violating parameter and kinematics but different nuclear physics.

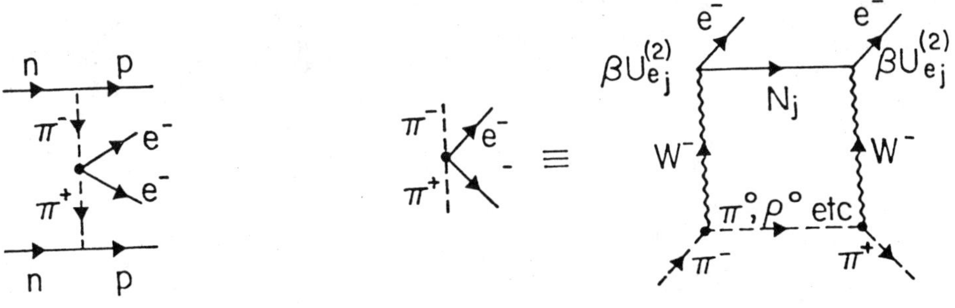

Figure 2 : A possible contribution to neutrinoless double β decay when the intermediate Majorana neutrinos are heavy. A pion in flight between two nucleons emits electrons. This process is negligible in the case of light Majorana neutrinos.

2.1.2. 0ν $\beta\beta$-decay accompanied by emission of Higgs particles.

One can construct gauge theory models which contain neutral Higgs particles that survive the Higgs mechanism as actual particles. One such model[4] has recently been proposed and extensively discussed. This model in addiiton to the standard isodoublet of Higgs scalars also contains an isotriplet. Two neutral scalars, one of which is massless and the other which is light, survive as physical particles and are designated as χ^0. (The model also contains doubly charged Higgs particles with small probability for decay into two leptons (see next section)). These particles can play the role of a mass term in the Lagrangian converting this way a neutrino to antineutrino via a coupling of the type

$$\mathcal{L}_{eff} = \frac{g_{ee}}{\sqrt{2}} \tilde{\nu}_e \nu_e \chi^0 \qquad (15)$$

which can break lepton number by two units (see fig.3)

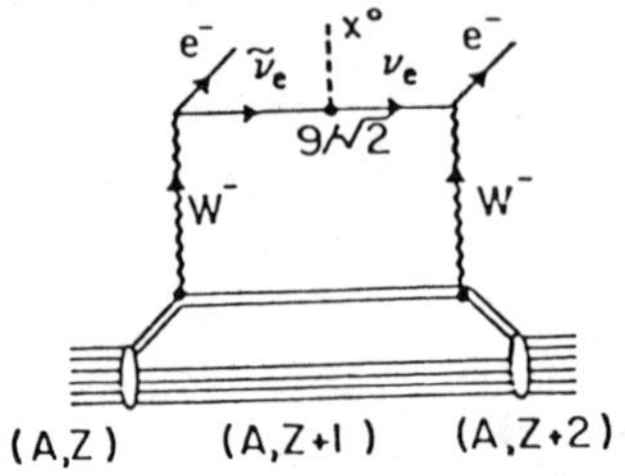

Fig. 3: A possible contribution to 0ν $\beta\beta$-decay which is accompanied by the emission of a massless (light) Higgs particle χ^0.

The effective lepton violating parameter is

$$\eta_g = \frac{g_{ee}}{\sqrt{2}} \simeq 1.0 \times 10^{-3} \qquad (16)$$

Before concluding this section we note that the cancellation between the various mass eigenstates is not the only mechanism for suppressing 0ν $\beta\beta$-decay while allowing other lepton violating processes. The zee model is another such example[23]

2.2 Right-Left Symmetric models.

It is obvious that such models[21] offer many more possibilities some of which we are going to explore.

2.2.1. Neutrino mediated processes. Now in addition to the left-handed currents discussed in the previous section one has right handed currents in weak interactions

$$\begin{pmatrix} N_e \\ e^- \end{pmatrix}_R \quad \begin{pmatrix} N_\mu \\ \mu^- \end{pmatrix}_R \quad \begin{pmatrix} N_\tau \\ \tau^- \end{pmatrix}_R \quad \cdots$$

mediated by vector bosons with much bigger mass. We will write $m^2_{W_R} = \frac{m^2_W}{\kappa}$ and take $\kappa \approx 0.02$. Once again one may have a contribution arizing from light neutrinos. In this case the dominant contrbution comes from the part of the amplitude which is independent of the neutrino mass, i.e. from the interference between the lepton R and L currents[9,24]. The hadronic current now can either be R-L, in which case the amplitude will be proportional to $\kappa = m^2_W / m^2_{WR}$, or R-R type in which case the amplitude is proportional to ϵ (ϵ is the mixing angle between the vector bosons mediating Right and Left handed currents). The situation is exhibited in fig.4

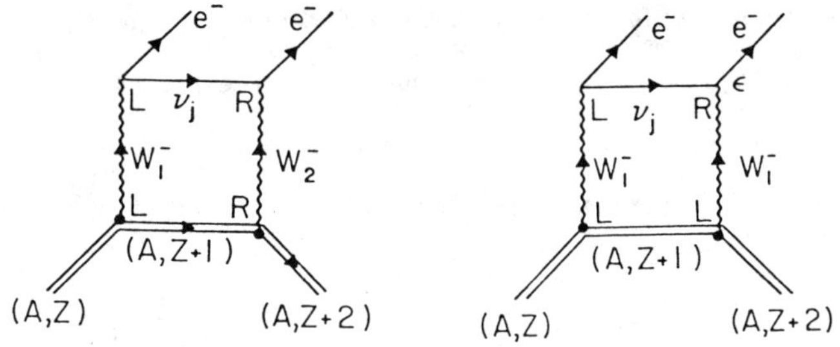

Figure 4: A Possible Majorana neutrino mediated contribution to bouble β decay in the presence of right-handed interactions. Lepton violation is in this case not explicity dependent on the neutrino masses. The left diagram is proportional to κ and the right one to ε

The pertinent lepton violating parameter now is

$$\eta_{RL} = -\beta \sum_{j=1}^{n} U^{(L)}_{ej} \, U^{(R)}_{ej} \, e^{i\alpha_j} \tag{17}$$

Proceeding exactly in 2.1.1 and using $\kappa = 0.02$, $\epsilon = 10^{-3}$ we obtain

$$\eta_{RL} \approx 2.0 \times 10^{-5} \qquad \text{(Approximately the same for both CP eigenvalues)} \qquad (18)$$

(We find it necessary to use a smaller value than previously estimated). One can also have a contribution arising from the heavy component of the right handed currents. One finds that the corresponding lepton violating parameter is

$$\eta_R = \kappa^2 \sum_j U_{ej}^{(1)} U_{ej}^{(2)} e^{i\varphi_j} \frac{m_A}{m_j} \qquad (19)$$

(one has the same situation as in fig. 1b with β replaced by κ) Proceeding in a fashion analogous to that discussed in section 2.1.1 for heavy neutrinos and taking $\kappa \approx 0.02$ we obtain

$$\eta_R \approx 3.4 \times 10^{-7} \qquad \text{(Approximately the same for both CP eigenvalues)} \qquad (20)$$

2.2.2. Intermediate doubly charged Higgs particles. It is well known that majorana neutrino mass can be generated via isotriplets of Higgs particles[11] whose doubly charged members can, in principle, decay into two leptons. Such particles, however, can couple to quarks only indiretly i.e. via their coupling to vector bosons or Higgs doublets. Such couplings are expected to be negligible in left-handed theories[25] but they may give sizable contributions in right-left symmetric theories. The most dominant contribution[9,26] is shown in fig. (5)

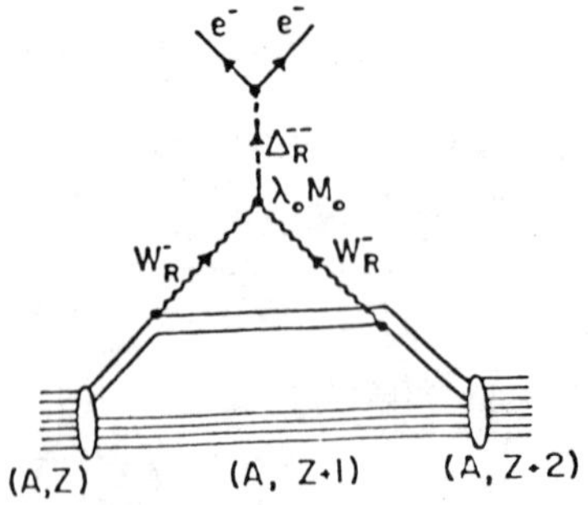

Fig.5: A possible contribution to 0ν ββ-decay arising from the decay into two electrons of a doubly charged Higgs particle Δ_R^{--}. Communication with the hadronic sector is achieved via the vector bosons associated with the right-handed interaction.

the relevant lepton violating parameter is given by

$$\eta_H = \frac{4}{g^2}\frac{\lambda_o m_o m_A}{m_\Delta^2}\kappa^2 h_{ee} \approx 6.5\times 10^{-7} \tag{21}$$

We have used $h_{ee} \approx 1$, $\lambda_o m_o/m_\Delta^2 = 2\times10^{-4}$ Gev^{-1}, $\kappa = 0.02$ consistent with 0ν $\beta\beta$-decay data[9], but quite a bit smaller than previously employed[26]. Another mechanism is exhibited if fig.6 . In this case the coupling of the doubly charged Higgs to quarks is achieved via ordinary Higgs[9,27]. The latter couple to quarks via a scalar and a psecdoscalar coupling[27] which was obtained from the work of Adler[28] et al i.e. α_s=0.25 and α_p=0.40

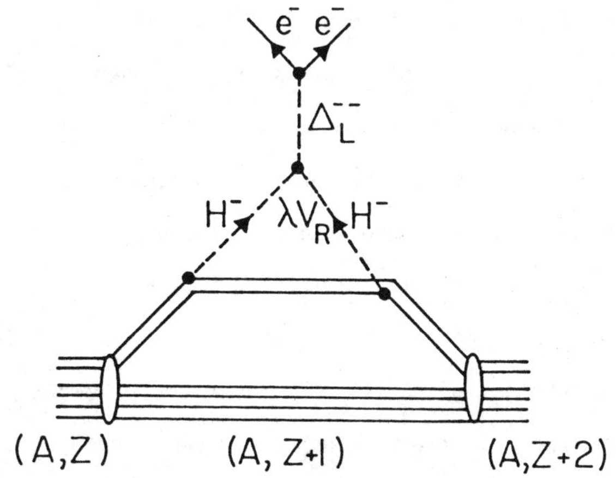

Figure 6 : A similar process with that of the previous figure except that now communication with the hadronic sector is achieved via the Higgs doublets.

The corresponding lepton violating parameter takes the form

$$\eta_\Delta = \frac{\lambda V_R\, m_A\, h_q^2\, h_{ee}}{m_H^4\, m_\Delta^2\, G_F^2} \approx 10^{-9} \tag{22}$$

This numerical value is was obtained with the following "reasonable" choice of the parameters[27]

$$\lambda V_R = 10^3 \text{ GeV}, \quad m_H \approx m_\Delta \approx 100 \text{ GeV}, \quad h_q^2 = 10^{-9}, \quad h_{ee} \approx 1$$

3. - RESULTS

For each of the mechanisms described in the previous section (see figs (1) - (6)) one can obtain the transition rates as discussed in some detail in ref. 9. The ov ββ-decay half-lives take the form

$$T_{1/2}(0\nu, \beta\beta) = \frac{K_{0\nu}(\beta\beta)}{|\eta \, ME|^2} \tag{23}$$

where $K_{0\nu}(\beta\beta)$ depends on the available energy[29] and the bulk nuclear properties (A and Z). The relevant expressions are given in ref. 9. Numerical values for the nuclear systems of interest to us here are given in table I. η is the lepton violating parameter discussed in the previous section which contains all the information of the gauge model. M E is the nuclear matrix element of the transition operator associated with each of the mechanisms described in the previous section. The form of these opereators was given in ref. 9. The matrix elements corresponding to the nuclear systems of interest to us are given in table II. They were obtained by standard shell model calculations as discussed in ref. 30.

TABLE I. Kinematical expressions $K_{0\nu}$, $K_{0\nu,\chi^0}$ and $K_{2\nu}$ for the A =48 and A = 58 systems. The available energy is also given.

SYSTEM	PROCESS	$K_{0\nu}$ (yrs)	$K_{0\nu,\chi^0}$ (yrs)	$K_{2\nu}$ (yrs)	Energy MeV
$^{48}Ca \rightarrow {}^{48}Ti$	$\beta^-\beta^-$	1.1×10^{13}	3.6×10^{14}	4.7×10^{18}	4.290
$^{58}Ni \rightarrow {}^{58}Fe$	$\beta^+\beta^+$	1.4×10^{20}	1.7×10^{25}	3.0×10^{36}	0.090
	(e^-,e^+)	9.7×10^{17}	6.2×10^{20}	3.5×10^{24}	1.100
	2e capture	4.6×10^{28}	4.4×10^{24}	2.8×10^{25}	1.315

Table II. The nuclear martix elements involved the in ββ-decay of the A=48 and A=58 systems. For notation see ref.9.(a) refers to the results of ref.30 and b to those of ref.9

SYSTEM	CALCULATION	μ_o	$-\vec{Y}\cdot\vec{Y}$	Ω_V	$\Omega_N+\Omega_\pi$	Ω_Δ	Ω_1	Ω_2	Ω_3
$^{48}Ca \rightarrow {}^{48}Ti$	a	9	0.250	1.05	142	284	3.67	1.33	-4.03
	b	12	0.130	0.63	122	260	3.90	0.56	-6.70
$^{58}Ni \rightarrow {}^{58}Fe$	a	6.5	0.300	1.17	160	146	1.77	0.79	-1.17
	b	11	1.36	2.22	112	213	0.46	1.08	0.50

In our numerical calculations we will use the set of nuclear matrix elements indicated by a in table II, which are improved over those employed previously[9]. Discrepancies such as those between a and b reflect, in an extreme sense, nuclear physics uncertainties. Thus the nuclear matrix elements are known with an accuracy better than factor of two.

The half-life of 0ν ββ-decay with majoron emission is given by equ.(23) except that now the Kinematical dependence, given by $K_{0\nu,\chi^0}$ (ββ), is different[9]. Numerical values of this quantity for the systems of interest to us are included in table I.

For comparison purposes we are also going to present results for the lepton allowed 2ν ββ-decay which serves as an undesirable background (Laboratory observation of this process will also be an experimental success). The relevant half-life now is given [9] by

$$T_{1/2}(2\nu,\beta\beta) = \frac{K_{2\nu}(\beta\beta)}{|\langle f|\vec{Y}\cdot\vec{Y}|i\rangle|^2} \qquad (24)$$

where $\vec{Y} = \sum_i \tau \pm (i)\, \vec{\sigma}(i)$ is the Gamow - Teller operator. The quantity $k_{2\nu}(\beta\beta)$ depends not only on the available energy, A and Z, as before, but on some suitable mean excitation energy μ_0 defined by

$$\sum_n \frac{\mu_0}{\mu_n} \langle f|\vec{Y}|n\rangle \cdot \langle n|\vec{Y}|i\rangle = -\langle f|\vec{Y}\cdot\vec{Y}|i\rangle \qquad (25)$$

where μ_n is essentially the excitation energy of the intermediate nuclear states (n). Numarical values for the relevant quantity $k_{2\nu}(\beta\beta)$ is also given in table I.

The thus obtained half-lives associated with each of processes (3)-(8) for the A = 48 and A = 58 systems are given in table III.We see 0ν that double

Table III. The $\beta\beta$-decay half-lives for the $^{48}Ca \rightarrow {}^{48}Ti$ and $^{58}Ni \rightarrow {}^{58}Fe$ transitions for the various gauge models discussed in this work and the nuclear matrix elements (a) of table II. The existing experimental limits[31] on the A=48 decay are $T_{1/2}(o\nu) > 2\times10^{21}$yr, $T_{1/2}(2\nu) > 3.6\times10^{19}$yr.

Mechanism	η	Double e-capture $T_{1/2}$ yrs	(e^-,e^+) $T_{1/2}$ yrs	$\beta^+\beta^+$ $T_{1/2}$ yrs	$\beta^-\beta^-$ $T_{1/2}$ yrs
ν +	1.2×10^{-4}	4.8×10^{34}	7.1×10^{25}	3.4×10^{27}	6.8×10^{28}
ν -	0	∞	∞	∞	∞
N	8.5×10^{-7}	3.9×10^{35}	5.6×10^{25}	7.4×10^{27}	2.1×10^{21}
g	1.0×10^{-3}	3.0×10^{25}	9.0×10^{24}	1.2×10^{31}	5.8×10^{21}
H	6.6×10^{-7}	6.4×10^{35}	1.0×10^{26}	1.2×10^{28}	1.8×10^{21}
RL	2.0×10^{-5}	—	2.1×10^{27}	3.3×10^{30}	2.5×10^{22}
R	3.4×10^{-7}	2.4×10^{36}	3.5×10^{26}	4.7×10^{28}	1.3×10^{22}
2ν (lepton allowed)		1.5×10^{25}	1.0×10^{26}	1.1×10^{37}	4.1×10^{19}

electron capture is unobservable. The positron decays are also much slower than the negaton decay of $^{48}Ca \rightarrow {}^{48}Ti$. The predicted half-lives for ^{48}Ca are close to the experimental lower limit set by the Columbia group[31]

$$T_{1/2}^{exp}(0\nu) > 2\times10^{21}\ yr$$

This should not come as a great surprise. We did not, of course, use the existing $\beta\beta$-decay limits to determine the lepton violating parameters as has hitherto been standard practice. The choise of some of the parameters, however, was influenced by such data (e.g. κ, m_1 etc). The combined effect of all considered mechanisms may violate the above limit. This, of course, is not disturbing in view of the existing uncertainties in the choice of the gauge theory parameters.

4. CONCLUSIONS

In the present paper we have discussed various double β decays with different experimental signatures like $\beta^-\beta^-$ and $\beta^+\beta^+$ decay, (e^-, e^+) conversion and double electron capture. Our discussion focussed on some attractive features of present-day gauge theories. We thus considered various lepton violating mechanisms including light and heavy majorana neutrino, Higgs mediated processes, possible contributions of right-handed currents and models with unconventional neutrino masses (majoron emission). Our results crucially depend on the choice of the gauge parameters.

Since the nuclear matrix elements are known with accuracy betten than a factor of 2 one hopes to reliably extract the lepton violating parameter η from the 0ν rates when they become available. Such accuracy may not be sufficient to extract η from the total geochemical rates[6,19]. The reason is that the measured rates are close to (and sometimes[32], unfortunately, less than) those calculated for the lepton allowed 2ν process.

It will be much harder from the 0ν data to discriminate against the various gauge models. The presence of the right-handed currents may be inferred from the R-L interference term which is in principle distiguishable experimentally.This, e.g.,can be done if one looks for $0^{+}\to J\neq 0$ transitions which occur only if there is a right-handed interaction, or by looking at systems with small available energies, for example ^{58}Ni where the R-L interference will be suppressed. The latter implicitly assumes that the disadvantages of the long lifetimes may be overcome by employing new experimetal techniques which allow the use of large and pure targets.

It appears that the traditional $\beta^{-}\beta^{-}$ decay experiments are close to their ultimate limits, i.e., $T_{1/2} \sim 10^{22}$ yr which constitutes the longest limit attained experimentally [33,34]. One should therefore look at the other processes discussed in the present work. Double electron capture appears to be hopelessly slow without providing the compensation of an additional desirable signature. Positron emission may prove to provide great advantages. The double positron decay of ^{58}Ni due to the tiny available energy proceeds with a lifetime of the order of 10^{27} yrs. If such a lifetime does not deter the experimental efforts, the experimentalists will get the bonus of a negligible 2v reaction background ($\sim 10^{36}$ yr). Furthermore we have seen that in this case one need not worry about the contribution of the right-handed interaction. Also at the same time one can look for the much faster single positrons of the electron capture in coincidance with the X-ray. The fact that the lifetimes associated with e-capture are about 10^{3} longer than typical $\beta^{-}\beta^{-}$ lifetimes may be compensated for by the signature of the positron. Transitions to excited states will provide the extra signature of the emitted de-excitation γ-ray to discriminate against background.

Last, but not least, we strongly suggest that such an experimental set up may be used to also look for high-energy positrons ($\sim$1 GeV) which are expected from double proton decay $(A,Z) \rightarrow (A-2,Z-2) + e^+ + e^+$. This unlike ββ decay, is a B-L conserving process, favored in some models, expected to proceed with a lifetime[35] of 10^{34} yrs.

Finally, we stress again that it will be very difficult to distinguish among themselves the various mechanisms discussed in this work (light neutrino, heavy neutrino, right-handed currents, Higgs contribution, χ^0 production). But such questions may be premature at this point since the observation of lepton violation, regardless of the mechanism, will make big news.

$\beta^-\beta^-$ decay has by now a history of 50 years (see the Introduction of Zdesenko's article in ref. 2). It is not easy to assess the theoretical and experimental progress made on this complicated subject in the above period. Given the expectations from double β decay experiments in this post-gauge-theory era, we hope that the experimental and theoretical efforts will intensify. Decays involving positrons may contribute in this direction.

REFERENCES

1) See, for example:
H.Primakoff and S.P.Rosen, Phys.Rev. 184 (1969) 1925 and references therein.

2) For recent reviews, especially on the experimental side, see:
Yu G.Zdesenko, Sov.J.Part.Nucl. 11(6) (1980) 542;
D.Bryman and C.Piccioto, Rev.Mod.Phys. 50 (1978) 11.

3) For mechanisms for neutrino mass see, for example:
R.Barbieri, D.V.Nanopoulos, G.Morchio and F.Strocchi, Phys.Lett.90B (1980) 91;
E.Witten, Phys. Rev.Lett. 91B (1980) 81.

4) H.Georgi, S.L.Glashow and S.Nussinov, Nucl.Phys.B193 (1981) 297;
G.B.Gelmini and M.Roncadelli, Phys. Lett. 99B (1981) 411.

5) Observation of 2ν ββ decay in the A = 82 system has been reported by M.K.Moe and D.Lowental, Phys.Rev. C22 (1980) 2186.
The reported rate, however, is an order of magnitude faster than the accepted geochemical rate.

6) T.Kirsten, W.Gentner and O.A.Schaeffer, Zeit.für Physik 202 (1967) 203; T.Kirsten, O.A.Schaeffer,E.Morton and R.W.Stoenner,Phys.Rev.Lett.20(1968)1300.

7) E.W.Hennecke, O.K.Manuel and D.D. Sabu, Phys.Rev. C11 (1975) 1378.

8) See, for example:
S.P.Rosen and H.Primakoff, Alpha, Beta and Gamma Ray Spectroscopy, ed.by K.Siegbahn, North Holland Publishing Co. (1965)

9) J.D.Vergados,Lepton Violating $\beta^-\beta^-$,$\beta^+\beta^+$ Decays,(e^-e^+) Conversion and Double e-Capture in Gauge Theories, TH 3396-CERN, August (1982).

10) S.Weinberg, Phys. Rev.Lett.19 (1967) 264. A.Salam, in Elementary Particle Theory: Relativistic Groups and Analyticity (Nobel Symbosioum No.8), Edited by N.Svarthholm (Almqvist and Wiksell, Stockholm, 1968).S.L.Glashow, Nucl. Phys. 22 (1961) 579.

11) See e.g. P.Langacker, Physics Reports 72 (1981) 185.

12) M.Kobayashi and K.Maskawa, Prog.Theor.Phys. 49 (1973) 652.

13) J.Bernabeu and P.Pascual, CP properties of the leptonic sector for majorana Neutrinos, TH 3393-CERN, August 1982.

14) In ref. 9 we use a different notation. The $U^{(1)}$ and $U^{(2)}$ used there coincide with our present (X),(Z).

15) J.D.Vergados, Phys.Rev. 24C (1981) 640

16) J.D.Vergados,to be published.

17) L.Wolfenstein, Phys.Lett. 107B (1981) 72. Also contribution to this Conference.

18) V.A.Lubimov, E.G.Novikov, V.Z.Nozik, E.F.Tretyakov, V.S.Kosik Phys.Lett. B94 (1980) 266.

19) T.Kirsten, H.Richter and E.Jessberger, M.P.I Preprint, Heidelberg, West Germany (1982). Aslo T.Kirsten, Contribution to this conference.

20) Papantonopoulos and G.Zoupanos to be published in Z.Für Physik.

21) R.N.Mohapatra and G.Senjanovic, Phys. Rev. 23D (1981) 165.Riazuddin, R.E.Marshak and R.N.Mohapatra Phys.Rev. 24D (1981) 1310.

22) J.D.Vergados, Phys.Rev.D25 (1982) 914

23) A.Zee, Phys.Lett. 93B (1980) 389. J.D.Vergados, Phys.Rev. 23D(1981) 703.

24) M.Doi et al, Prog.Theor.Phys. 66(1981) 1739; 66 (1981) 1765.

25) J.Schechter and J.W.F.Velle, Phys.Rev. 25D (1982) 2951.
L.Wolfenstein, Carnegie Mellon University Preprint (MU HEG 82-5 (1982)

26) C.E.Piccioto and M.S.Zahir, Neutrinoless β-decay in left-right symmetric models, University of Victoria Preprint (1982)

27) R.N.Mohapatra and J.D.Vergados, Phys.Rev. Lett. 47 (1981) 1713.

28) S.L.Adler et al, Phys.Rev. D11 (1975) 3309

29) In the case of the interference between left and right handed currents (See subsection 2.2.1) there is additional kinematical dependence which is included in our calculations but not discussed here (see ref.9 for details).

30) L. D. Skouras and J.D.Vergados, Nuclear matrix elements, for the A=48 and 58 systems,University of Ioannina Preprint (1983)

31) B. J.Cleveland, W.R.Leo, C.S.Wu, P.J.Callon and J.D.Ullman, Phys.Rev.Lett. 35 (1975) 757.
R.K.Burdin, P.J.Callon, J.D.Ullman and W.S.Wu,Nucl.Phys.A158 (1970)337.

32) W.C.Haxton, G.J.Stephenson and D.Strottman, Phys.Rev. D25 (1982) 2360.

33) E.Bellioti, E.Fiorini, C.Liguori, A.Pullia, A.Sarracino and L.Zanotti, Lett. Nuovo Cimento 33 (1982) 273.

34) E.Bellotti, E.Fiorini,C.Liguori, A.Pullia, A.Sarrachino and L.Zannoti, XXI International Conference on High Energy Physics, Paris (July 1982). Also C.Liguori, This Conference.

35. J.D.Vergados, Phys.Lett. 118B, (1982) 107.

FUTURE DOUBLE BETA DECAY EXPERIMENTS

David O. Caldwell
University of California
Santa Barbara, California 93106

ABSTRACT

Double beta decay laboratory experiments in progress or planned are surveyed. Particularly searches for neutrinoless double beta decay are emphasized and present limits on lifetimes and corresponding neutrino masses given. Especially discussed is a UCSB-LBL ^{76}Ge experiment which may reach limits $\sim 10^{24}$ years or ~ 1 eV for m_ν.

INTRODUCTION

Double beta decay in general and geochemical measurements of it in particular are the subjects of other talks, so here we shall emphasize laboratory experiments, especially the search for neutrinoless double beta decay. We will be mainly concerned with experiments in progress or being planned, but will mention a few published results for comparison. This opportunity will be taken to discuss particularly a UCSB-LBL experiment which is on a larger scale than any others presently being worked on.

PAST LABORATORY EXPERIMENTS

The laboratory experiments looking fot the decay $[A,Z] \rightarrow [A,Z+2] + 2e^-$ (which will be referred to as the 0ν decay) are very difficult. One is looking for a process with a lifetime probably exceeding 10^{23} years in a background of many competing processes. This requires of the order of 10^2 grams of an isotope, usually not having a very large abundance, to get a few counts per year. Thus the source must be large, and yet the β energies are low (~ 1 MeV), so ranges are short. All materials have some natural radioactivity which give confusing backgrounds, and in addition, cosmic ray interactions can cause problems.

To overcome these difficulties, two techniques are used: (1) tracking devices look for two electrons, and (2) energy measuring devices look for a peak due to the summed electron energies. The more copious decay $[A,Z] \rightarrow [A,Z+2] + 2e^- + 2\bar{\nu}_e$ (which will be referred to as the 2ν decay) would, of course, give a broad energy spectrum below that peak and a different energy and angular distribution of electrons.

Since the 0ν decay requires lepton number violation and at least one of the following: (1) that the electron neutrino have a non-zero mass or (2) that right-handed currents (RHC) exist, we can compare the accuracies of different experiments in terms of limits on m_ν and RHC. Both m_ν and RHC limits should be given because the processes have a different dependence on the energy available in the decay.

For $m_\nu \neq 0$, $\Gamma(0\nu) \sim (m_\nu/m_e)^2$, but if $m_\nu = 0$ and the leptonic current is

$J = J_L + \eta J_R$, then $\Gamma(0\nu) \sim \eta^2 = \left[\frac{M_{W_L}}{M_{W_R}}\right]^4$, where W_L is the usual

intermediate vector boson and W_R would be that giving a right-handed current. Note that in principle the m_ν and RHC effects could be separated, since either can contribute to a $0^+ \rightarrow 0^+$ ground state transition, while only RHC could produce the $0^+ \rightarrow 2^+$ decay to the first excited state.

The best published measurements are given below, along with the Los Alamos-Purdue calculated[1] limits on m_ν (for $\eta = 0$) and for η (with $m_\nu = 0$). Note that the matrix elements used by the Japanese group[2] give m_ν and η values about three times larger, because their matrix elements are based on the geochemical lifetimes.

Source	$T_{1/2}(0_\nu)>$	$m_\nu>$	$\eta<$	Exp. Ref.
^{48}Ca	$10^{21.3}y$	30 eV		3
^{76}Ge	$10^{21.7}y$	15 eV	3×10^{-5}	4
^{82}Se	$10^{21.5}y$	12 eV	1×10^{-5}	5

For the Ca and Se measurements, a thin source in a spark chamber was placed between two counters. The β's hitting the counters were to trigger the spark chamber which could measure the curvature of the tracks in a magnetic field. Another Se measurement[6] is the first reported laboratory observation of the 2ν decay, and for this a thin source was used in a cloud chamber. Incidentally, the Ca decay is especially favorable energetically, but the nuclear matrix element is apparently unusually small because of a K selection rule, so that the expected rate is about the same for Ca and Se or Ge.

The Ge measurement utilizes an energy determination, since Ge makes an ideal β counter. The Milan[4] group had 0.2% resolution in energy in a 69 cm^3 counter which had 7.67% ^{76}Ge, the apparatus being located in the Mt. Blanc tunnel.

PRESENT AND FUTURE EXPERIMENTS

With the recent revived interest in the questions that ββ decay can give answers to, it is natural that a number of new experiments are underway or being planned. A brief description will be given here of work going on, starting with the USSR. At INR (Moscow) the $0^+ \rightarrow 0^+$ 0ν and 2ν decays have been searched for using thin, planar sources of ^{58}Ni, ^{124}Sn, ^{130}Te, and ^{150}Nd sandwiched between hodoscoped scintillation counters which give coincidence information, as well as electron energies and spatial positions. To look for $0^+ \rightarrow 2^+$ decays a NaI well counter is used to detect the de-excitation γ-ray in separate experiments. The limits set so far are in the 10^{19} to a few times 10^{20} year range.

In Japan work is going on at Tokyo Metropolitan University on Ca^{48} and at Osaka on ^{100}Mo and ^{150}Nd.

Turning now to the Western Hemisphere, we discuss the experiments according to the decay, starting with ^{82}Se. The University of California at Irvine group (A. Hahn, M. Moe, and F. Reines) is replacing the cloud chamber with a Time Projection Chamber (TPC) and expects to get serious data within a year. They hope to reach a 0ν lifetime of 3×10^{22} years, limited by practical source size. A quite different kind of experiment is being planned by the Brookhaven-SUNY-Oak Ridge group (W. Chen, B. Cleveland, S. Hurst, and J. Ullman) which will do a radiochemical experiment using 10 kg of Se, by means of single-atom counting of ^{82}Kr. This should settle the question of the lifetime of the 2ν decay which has arisen because of a discrepancy of 20-30 between the laboratory[6] measurement and those obtained geochemically. However, it will require a rather uncertain calculation to obtain any 0ν contribution to the measured lifetime.

The second decay is $^{136}Xe \rightarrow {}^{136}Ba$, which has not previously been utilized. The Xe has the interesting advantage that it scintillates and hence can provide some energy measurement suitable for triggering, and yet it can be used as the working substance in a TPC. Thus, except for expense, the source volume is not a problem. The U.C. Irvine group of H.H. Chen, P.J. Doe, and H.J. Mahler, plans to use one liter of liquid xenon, and they have already made a liquid argon TPC function. Their feasibility studies should be completed by about the end of 1982. A Lawrence Berkeley Laboratory group (R. Muller, R. Tripp, R. Kenney, etc.) investigated using a high-pressure gas TPC. Their measurements of backgrounds on the TPC at PEP this summer were discouraging. Now they are considering using separated ^{100}Mo as a very thin source between counters capable of giving energy, position, and timing information. Fortunately two moles of ^{100}Mo are available from Oak Ridge, giving a factor of ten enhancement.

The most popular $\beta\beta$ source is ^{76}Ge, with the Milan group (E. Bellotti, E. Fiorini, C. Liguori, A. Pullia, Q. Sarracino, and L. Zanotti) again putting a shielded Ge crystal in the Mt. Blanc tunnel. Although Dr. Liguori discussed this experiment in detail, the highlights will be mentioned here in order to make comparisons with other experiments easier. As he has stated, their Ge crystal is twice as big (135 cm^3) as in their previous experiment, and they have also made some improvement in the materials so as to reduce radioactivity in the cryostat (which keeps the Ge at liquid nitrogen temperature) and also in the surrounding shielding. In September 1982, they had been counting for 3000 hours, getting a lifetime limit of 1.8×10^{22} years, or a neutrino mass limit in the absence of right-handed currents of about 10 eV. They also set a limit of 1.5×10^{22} years on the lifetime of the electron. Their future plans probably involve building a gas Xe TPC, but if tests are not encouraging, they will go in the direction of increased quantities of Ge.

The CEN group (Bordeaux-Gradignan) use a 110 cm^3 Ge counter with 6 NaI crystals in coincidence to look for the $0^+ \rightarrow 2^+$ transition. In September, after 900 hours of counting they had a limit $T_{1/2} = 1.0 \times 10^{21}$ years.

An experiment which probably will be counting in Fall 1982 is that of F. Boehm's group at Caltech. They have a 100 cm^3 Ge detector surrounded by an active plastic scintillator shield and plan to reach a limit of 5×10^{22} years.

The collaboration of South Carolina (F.T. Avignone) and Batelle Pacific Northwest (R. Brodzinski and W. Wogman) has been using an existing Ge low-level counting facility to reach a limit which in September 1982 was about 10^{22} years. They plan to obtain a large Ge crystal (perhaps 250 cm^3), place it in an active NaI shield, and locate it in a deep mine.

The group at Guelph (J.L. Campbell, P. Jagam, and J.J. Simpson) is collaborating with a maker of Ge detectors (H. Malm of Aptec, Toronto) and has obtained a 200 cm^3 crystal with which they plan to achieve a lifetime limit of 3×10^{22} years (at the 95% confidence level). They have hopes of using up to 500 cm^3 of Ge, possibly with active shielding.

UCSB-LBL EXPERIMENT

Finally, there is the collaboration of the University of California at Santa Barbara group (D.O. Caldwell, R.M. Eisberg, and M.S. Witherell) with the Ge experts of the Instrumentation Science Division of Lawrence Berkeley Laboratory (F. Goulding, N. Madden, R. Pehl and A. Smith). Already ordered are about 1400 cm^3 of Ge (eight 170 cm^3 crystals in a close-packed array) and sufficient NaI to provide a 6"-thick shield surrounding it, so it is a bigger scale experiment than the others. Considerable development work has also been done on finding materials with a low level of radioactivity and in improving the information obtained from the Ge detector. The arrangement of the apparatus is shown in Fig. 1.

It is worth mentioning the advantages this experiment will have over the 1973 Milan work, and some of these apply to other new experiments as well. First, there is 20 times as much Ge, providing a proportional increase in lifetime limit. Second, there is an active NaI shield to veto the largest source of background, the 2.6 MeV γ in ^{208}Pb at the end of the ^{228}Th decay chain. This γ (and a few others of less importance) can deposit 2.041 MeV in the Ge by Compton scattering, but if the rest of the γ energy goes into the NaI (or another Ge detector), that background can be vetoed. This

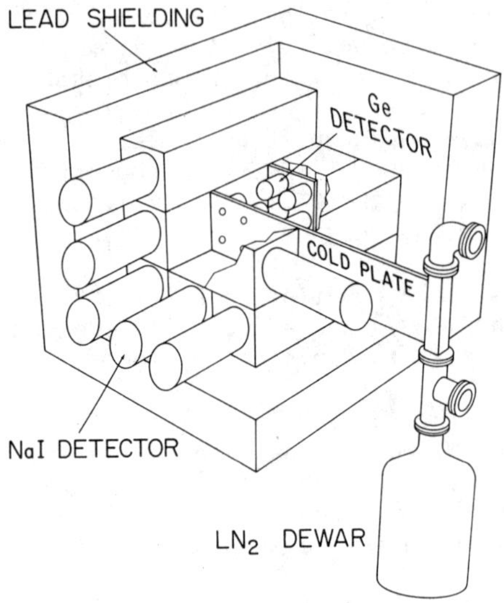

Fig. 1 - Schematic lay-out of the apparatus

is one reason inert material between the Ge and NaI must be kept to a minimum. Note that the NaI shield (or again some other Ge detector) is very useful in providing a coincidence signal when trying to detect the $0^+ \rightarrow 2^+$ transition, since one looks for 1.486 MeV in one Ge detector from the 2 β's and 0.555 MeV from the de-excitation γ in the NaI (or sometimes in another Ge detector). The third advantage is that materials can be used, particularly inside the NaI well containing the Ge (as shown in Fig. 2), which have lower natural radioactivity. We have done extensive counting of materials and find zone-refined Si, quartz, Mg, and BeO particularly useful, especially as we want low-Z materials to avoid absorbing background γ's which would otherwise be vetoed. Note that the zone-refined Ge in the detector has impurities at the 10^{-13} level. A fourth advantage is a factor of two improvement in resolution (2.5 KeV full width at 2.041 MeV), which reduces the background by that factor. A fifth advantage is that we have learned how to discriminate against β's that come from outside the Ge and against multiple γ interactions in the Ge. These advances, which have surprisingly not been known before, will be of use in many low-level counting applications.

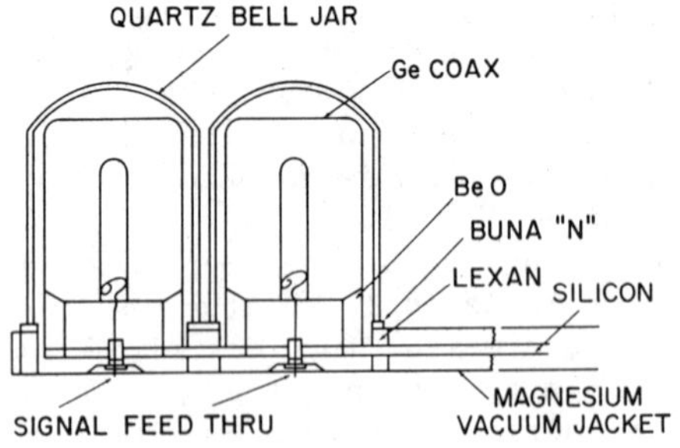

Fig. 2 - Arrangement of two of the eight Ge detectors.

With these advantages, we should reach a lifetime limit of about 10^{24} years in one year of counting, but the old Milan result can be reached in a couple of days. If the lifetime were 3×10^{23} years, we would get about a count a month. If the results are negative, this will provide a limit of $\sim$ 1 eV on the neutrino mass for no right-handed currents, or $\sim$ 60 TeV for the mass of the right-handed W boson, based on Ref. 1. For reaching these limits, it is important that for Ge the $0^+ \rightarrow 0^+$ and $0^+ \rightarrow 2^+$ transitions have about equal probabilities, as some calculations indicate, although this is still being investigated. It is also interesting to note that the present limit on M_{W_R} is about 210 GeV, but that applies to Dirac or Majorana neutrinos, while the $\beta\beta$ limit exists only if neutrinos are of the Majorana type.

These are rather impressive limits on m_ν and M_{W_R}, as well as on lepton-number conservation, and it will be difficult and expensive to get appreciably better limits. To get to $\sim$ 0.3 eV for m_ν would require at least a million dollars worth of Ge, for instance, so one hopes that these important questions will be answered by the present round of experiments.

REFERENCES

1. W.C. Haxton, G.J. Stephenson, Jr., and D. Strottman, Phys. Rev. Lett. 47, 153 (1981).
2. M. Doi, T. Kotani, H. Nishiura, and E. Takasugi, Osaka-Kyoto preprint OS-GE 82-43, 1982, unpublished.
3. R. Bardin, P. Gollon, J. Ullman, and C.S. Wu, Nucl. Phys. A158, 337 (1970).
4. E. Fiorini, A. Pullia, G. Bertolini, F. Capellani, and G. Rostelli, Nuovo Cimento A13, 747 (1973).
5. B.T. Cleveland, W.R. Leo, C.S. Wu, L.R. Kasday, A.M. Rushton, P.J. Gollon, and J.O. Ullman, Phys. Rev. Lett. 35, 737 (1975).
6. M.K. Moe and D. Lowenthal, Phys. Rev. C22, 2186 (1980).

CONCLUDING REMARKS ON DOUBLE BETA DECAY

S.P. Rosen
Physics Division, National Science Foundation,
Washington, D.C. 20550
and
Physics Department, Purdue University
West Lafayette, Indiana 47907

Let me try to summarize what has been a full and frank discussion of double beta decay in the following set of remarks.

1) The two-neutrino double beta decay is expected to occur as a second-order effect of the same Hamiltonian as is responsible, in first order, for ordinary single β-decay. Theoretical estimates of the lifetime for the 2ν-process are in the range $10^{21\pm2}$ years, with recent calculations by Haxton, Stephenson, and Strottman tending towards the faster times. Geochemical experiments on ^{130}Te and ^{82}Se fall well within the range, while the laboratory experiment of Moe and Lowenthal on ^{82}Se falls close to the faster edge.

2) The big question is whether or not the no-neutrino process is also present. We have no direct evidence that it is, and the only indirect evidence for it, based upon the Tellurium Ratio argument, has gotten weaker in the past year. The recent finding by Professor Kirsten and his colleagues that the ratio of ^{128}Te to ^{130}Te lifetimes is much larger than was previously thought to be the case suggests that the no-neutrino decay mode does not occur at all. However this argument depends upon some "reasonable" assumptions, and so one might ask: Is the actual situation an "unreasonable" one?

It has always been assumed that the nuclear matrix elements for the two isotopes are roughly equal to one another, and model calculations appear to support this assumption at the 10% level. However, Haxton, Stephenson, and Strottman have raised the possibility that large cancellations in the matrix element might negate the presumed equality. They observe that when it comes to small differences between large numbers, the fact that $A \approx A^1$ and $B \approx B^1$ does not necessarily imply that $(A-B) \equiv 1(A^1-B^1)$; indeed, when $(A-B)$ is very much smaller than A or B, one could easily find that $(A-B) \simeq 2$ or $3\ (A^1-B^1)$!

Should the nuclear part of the argument prove correct, then something strange could be happening on the particle physics side. Wolfenstein has observed that, should the no-neutrino decay be engendered by the exchange of the two Majorana neutrinos with opposite C eigenvalues [In the language of reference (1), $C^{(1)} = +D^{(1)}$ for neutrino (1) and $C^{(2)} = -D^{(2)}$ for neutrino (2)], then the two contributions to the total amplitude would tend to cancel one another. This observation enables Wolfenstein to reconcile the apparent conflict between the Kirsten et al limit, $m_\nu < 5.6$ ev, and the ITEP tritium β-decay result, $14 < m_\nu < 46$ ev, without saying that νe is a pure Dirac particle.

In principle, there are at least two possibilities for the masses of the two neutrinos. They could both be light (less than 50 ev in mass) in which case both neutrinos are emitted in β-decay; or one could be light and the other heavy (greater than, say, 100 Mev in mass) in which case only the light one is emitted in β-decay and the constraints from neutrino oscillation experiments would not apply. There could, however, be some subtle effects upon universality.

Different in spirit, but, I suspect, not terribly different in practice, is the suggestion by Valle (and also Petcov) for an "almost Dirac" neutrino. The idea is that the neutrino is represented by a dominantly pure particle field ψ_ν but carries a small admixture ξ of the charge conjugate field $\psi_{\nu c}$ [In the language of reference (1) $D_\lambda = \xi C_\lambda$]. In this case the Kirsten result places a bound on the product $m_\nu \xi$ rather than on m_ν itself. Once ξ is sufficiently small, the conflict with tritium β-decay vanishes.

3) There is a great need for us to "see" actual ββ events in the laboratory, that is to obtain events the interpretation of which leaves no doubt that they are double beta decay. Seeing such events would be a great experimental achievement just as the experiment of Kirsten, Schaeffer, Norton, and Stoenner was a great achievement. Once such events have been found, we can begin to answer the kinds of questions raised above, for example: how good are the calculations of nuclear matrix elements? And, does the no-neutrino decay mode occur?

4) If the no-neutrino process is actually discovered, then we will have to undertake a major program of experiments to unravel the mechanism for it. This will involve careful studies of the lifetimes as functions of the energy release, measurements of the energy sum spectrum, the angular correlation between electrons, and the search for transitions to low-lying 2^+ excited states.

REFERENCES

(1) H. Primakoff and S. P. Rosen, Ann. Rev. Nucl. Sci. 31, 145 (1981).

(2) L. Wolfenstein, this session.

NEW LIMITS ON OSCILLATION PARAMETERS FOR ELECTRON ANTINEUTRINOS

CALTECH - SIN - TUM Collaboration
Presented by J.-L. Vuilleumier
Caltech and SIN

ABSTRACT

The positron spectrum from the reaction $\bar{\nu}_e + p \rightarrow e^+ + n$ has been measured at L = 37.9 m from the core of the Goesgen power reactor to search for neutrino oscillations of the type $\bar{\nu}_e \leftrightarrow x$ in the low mass parameter range. The results are consistent with the no oscillation hypothesis. Upper limits of $\Delta^2 < 0.016$ eV^2 (90 % C.L.) for full mixing, and $\sin^2 2\theta < 0.17$ (90 % C.L.) in the limit of large mass parameters were determined. For the ratio of the integrated experimental yield to that predicted for the case of no oscillations a value of 1.05 ± 0.02 (statistical) ± 0.05 (systematics; 68 % C.L.) was obtained.

1. INTRODUCTION

A search for neutrino oscillations of the type $\bar{\nu}_e \rightarrow x$ has been performed at a distance of L = 37.9 m from the core of the 2806 MW (thermal) power reactor at Goesgen, Switzerland[1)]. This study succeeds a previous experiment carried out at L = 8.76 m from the core of the ILL reactor[2)].

Oscillations may occur if physical neutrinos (eigenstates of the weak interaction) are superpositions of the mass eigenstates. In such a case, neutrinos of one flavor, e.g. electron antineutrinos emitted by a fission reactor, could undergo intensity oscillations in space. In the most simple two neutrino model, these oscillations are characterized by the parameters $\Delta^2 = |m_1^2 - m_2^2|$ and $\sin^2 2\theta$, where θ is the mixing angle. The intensity of the emitted neutrinos oscillates with distance L according to

$$P(E_\nu,L,\Delta^2,\theta) = 1 - 0.5 \sin^2 2\theta \, [1 - \cos(2.5\,\Delta^2 L/E_\nu)] \qquad (1)$$

$$(E_\nu \text{ in MeV, L in m, } \Delta^2 \text{ in eV}^2).$$

If the oscillation length is of the same order or shorter than the distance L, this leads to an observable modulation of the neutrino energy spectrum. If the oscillation length is too short, however, the modulation will completely be washed out by the finite spatial and energy resolution of the system. In this case one only observes a global energy and distance independent rate reduction. At ILL (L = 8.76 m, core radius r = 20 cm) one would have seen modulations for $0.1 < \Delta^2 < 10$ eV^2 . At Goesgen (L = 37.9 m, r = 3.2 m) this range becomes $0.01 < \Delta^2 < 2$ eV^2 , and the experiment is sensitive to smaller values of Δ^2. In addition, because of the better statistics and of the better known overall spectrum normalization improved limits in the region of larger mass parameters were determined.

2. NEUTRINO DETECTOR

The neutrino detector uses the reaction $\bar{\nu}_e + p \rightarrow e^+ + n$, which has an energy threshold of 1.8 MeV. The basic detector system consists of 5 planes of 6 target cells of liquid scintillator (377 l total volume of NE235C), alternated with four ^{3}He wire chambers. The scintillation cells (86.7 cm long, 18.7 cm high, 7.7 cm thick) serve as proton target, positron detector, and neutron moderator. The wire chambers provide gamma-insensitive neutron detection. Low natural radioactivity was maintained through the exclusive use of stainless steel and teflon in the chamber walls and wire support frames. A 250 μs time coincidence between positron and neutron detection is required as the signature for a neutrino event. Pulse shape discrimination is used in the target cells to suppress fast neutron backgrounds. The detector system has been modified for Goesgen to enable position-sensitive detection of positron and neutron events, allowing an effective supression of accidental coincidence events. The position of an event along the length of a target cell is obtained from the time difference of photo-multiplier pulses at either end of the cell. Each ^{3}He multiwire proportional chamber now has 16 coplanar groups of four resistive wires arranged with 1.7 cm spacing running paralell to the target cells.[3] The position along a wire is determined by charge division. The requirement of a position correlation of 24 cm between target cell and wire chamber events allowed a factor of seven reduction in the accidental background rate, while retaining a 92 % acceptance efficiency for neutrino events.

The detector assembly is completely surrounded by veto counters followed successively by 5 mm of boron carbide, 20 cm of deionized water, 15 cm of steel, and 2 m of concrete with an additional 2 m of concrete overhead. The entire assembly was positioned outside the reactor containment building providing an extra 8 m of concrete shielding from the core.

3. NEUTRINO SPECTRA

Neutrino oscillations according to eq. (1) depend both on energy E_ν and distance L. An oscillation parameter analysis of neutrino rates measured as a function of energy at one fixed distance L requires a knowledge of the reactor antineutrino spectrum (e.g. L=0). Uncertainties in this spectrum will propagate into the oscillation parameters.

The Goesgen reactor (2806 MW thermal power) produces energy essentially from the fission of ^{235}U, ^{239}Pu, ^{238}U, and ^{241}Pu. The information on the reactor power and the relative fission contributions of the different isotopes were obtained from the staff of the power station[5]. Averaged over the duration of the experiment the relative fission contributions of the above four isotopes were 61.1 %, 27.8 %, 6.7 %, and 4.3 % , respectively. The fission rates were determined from the values of the energy released per fission.[6] The composite time-averaged neutrino spectrum from the beta decay of the fission products was obtained by combining these fission rates with the neutrino yields per fission.

Calculated yields were taken for the relatively small contribution of ^{238}U and ^{241}Pu.[7] For the two dominant isotopes ^{235}U an ^{239}Pu reliable spectra were obtained at ILL.[4] They were derived from the measured β-spectra. Targets of these isotopes (~1 mg), sandwiched between two Ni cover foils to contain the fission fragments, were placed near the core of the reactor, in a flux of $3.3 \cdot 10^{14}$ neutron $cm^{-2}s^{-1}$. The electrons were analyzed on-line with the magnetic spectrometer BILL. The electron spectra were converted into the neutrino spectra taking into account the charge distribution of the fission fragments. Conversion uncertainties (90 % C.L.) of 3 % (3 MeV) to 4 % (7 MeV) are quoted. They have to be added to the normalization uncertainty (90 % C.L.) of 4.8 % for ^{235}U and 3.3 % (2.5 MeV) to 4.2 % (6.5 MeV) for ^{239}Pu.

4. MEASUREMENTS

A total of 10,930 ± 220 neutrino-induced events were recorded during a six months reactor-on period (3441 h of data acquisition). Background was measured during a reactor-off period (551 h). The deadtime, largely due to the cosmic ray rate in the veto counters, was 15.8 %. The measured spectra are plotted in Fig. 1a. The errors shown are statistical. The accidental background was monitored continuously by delayed coincidence measurement, and turned out to be stable throughout the experiment. Fig. 1b shows the experimental positron spectrum.

5. DATA ANALYSIS

In the two-neutrino model, the neutrino-induced positron yield is given by

$$Y(E_{e^+},L,\Delta^2,\theta) = Y_{no\ osc}(E_{e^+},L) \cdot P(E_{\bar{\nu}},L,\Delta^2,\theta) \quad (2)$$

with $E_{e^+} = E_{\bar{\nu}} - 1.804$ MeV the positron kinetic energy, and L the core-to-detector distance. P is the oscillation function of eq. (1). $Y_{no\ osc}(E,L)$ is the neutrino yield in the absence of neutrino oscillations, and is given by

$$Y_{no\ osc}(E_{e^+},L) = n_p \cdot \sigma(E_{\bar{\nu}}) \cdot \varepsilon \cdot S(E_{\bar{\nu}}) \cdot 1/4\pi L^2 \quad (3)$$

Here, $n_p = 2.41 \times 10^{28}$ is the number of free protons in the liquid scintillator, and $\sigma(E_{\bar{\nu}})$ is the reaction cross-section. The net detection efficiency $\varepsilon = 0.168 + 0.005$ (S.D.) contains the neutron detection efficiency (measured with a calibrated Sb(Be) source) and the acceptance efficiencies of the time coincidence and position correlation windows imposed for target cell-neutron counter events. $S(E_{\bar{\nu}})$ is the neutrino yield at L = 0 .

In actuality, eq. (1) must be folded with the energy response of the detector and averaged over reactor core and detector volumes:

$$\tilde{Y}(E_{e^+},L,\Delta^2,\theta) = \int Y(E,L',\Delta^2,\theta) \cdot a(E_{e^+}) \cdot r(E_{e^+},E) \cdot h(L,L')\,dE\,dL'. \quad (4)$$

The variable E_{e^+} is the total observed energy deposited by the positron in the detector. The factor $a(E_{e^+})$ represents the acceptance of the pulse shape discrimination window and is weakly energy dependent (94 % to 97 %). The function h is the weighting factor for the finite core and detector sizes. The positron response function of the detector, $r(E_{e^+},E)$, was determined by Monte Carlo calculation. Fig. 2 illustrates the effective detector response to monoenergetic positrons of 1, 3, and 5 MeV. This function reflects the energy resolution of the detector, the effect of positron annihilation in flight and at rest (including multiple scattering of the annihilation gamma rays), the energy deposition of the escape positrons, and the added contribution of the positrons from neutrino interactions in the lucite walls of the target cells. A small correction to the cross-section and positron kinetic energy due to the neutron recoil is included in the analysis.

We now consider the ratio R of the positron yield $\tilde{Y}$ to the yield $\tilde{Y}_{no\ osc}$ predicted for the case of no oscillations. Fig. 3 shows the ratio R^i_{exp} of the experimental data points Y^i_{exp} to $Y^i_{no\ osc}$ plotted as a function of L/E_ν. For the integral yields ($E_{\bar{\nu}} > 2.5$ MeV) we obtain the ratio

$$\int Y_{exp}(E_{e^+}, 37.9\ \mathrm{m})dE_{e^+} / \int \tilde{Y}_{no\ osc}(E_{e^+}, 37.9\ \mathrm{m})dE_{e^+} =$$

$$= 1.05 \pm 0.02\ (\mathrm{S.D,\ statistical}) \pm\ 0.05\ (\mathrm{S.D,\ systematic}).$$

The systematic error arises primarily from uncertainties in the neutrino yield per fission and the detector efficiency. A chi-square function is defined as follows:

$$\chi^2 = \sum_{i=1}^{16} \frac{1}{\sigma_i^2} [R^i_{exp} - N \cdot R(g, E_{e^+}, L, \Delta^2, \theta)]^2 + \left(\frac{N-1}{\sigma_N}\right)^2 + \left(\frac{g-1}{\sigma_g}\right)^2 \quad (6)$$

with $R(g, E_{e^+}, L, \Delta^2, \theta) = \tilde{Y}(g \cdot E_{e^+}, L, \Delta^2, \theta) / \tilde{Y}_{no\ osc}(E_{e^+}, L)$. The parameters N and g are introduced to allow for variations of the absolute normalization and of the energy scale, which have experimental uncertainties ($\sigma_N = 5.3$ % and $\sigma_g = 1.2$ %). X^2 was minimized for a large array of fixed Δ^2 and θ values, allowing N and g to vary. For $\theta = 0$ (no oscillations) the resulting value is $X^2_0 = 13.1$ for 16 degrees of freedom. The lowest value ($\chi^2_{min} = 11.8$) was realized for $\Delta^2 = 0.18$ eV^2 and $\sin^2 2\theta = 0.04$. To obtain probabilities for the parameters Δ^2 and θ a maximum likelihood ratio test was performed. The function

$$\lambda(\Delta m^2, \theta) = 1/2[\chi^2(\Delta m^2, \theta) - \chi^2_{min})] \quad (7)$$

was computed to determine the significance of a particular set of parameters. χ^2 represents the minimized (with respect to N and g) function of eq. (6). From the expected distribution for λ (calculated by Monte Carlo simulation) confidence contours in the Δ^2 vs. $\sin^2 2\theta$ plane were determined. The 68 % and 90 % confidence limits are indicated by the curves labeled GO in Fig. 4. The regions to the right of the contours are excluded.

This procedure is also applied to the ratio of the ILL data to the present data $[R^i_{exp}(L=8.76\ m)/R^i_{exp}(L=37.9\ m)]$. These ratios are rather insensitive to the uncertainties in the neutrino spectrum following the fission of ^{235}U, and are essentially independent of the energy gain uncertainty (g is held fixed at g = 1). However, the experimental conditions at ILL and at Goesgen are not equivalent because reactors with different contributions of fissioning isotopes were used and because modifications introduced to the detector between experiments resulted in somewhat different efficiencies. In addition, the ILL data is of only moderate statistical accuracy. The result of the maximum likelihood test, labelled ILL/GO in Fig. 4, is therefore not very selective.

We conclude from the present experiment that on a 90 % confidence level neutrino oscillations with parameters larger than those given by the contour plots labeled GO in Fig. 4 are excluded. In the limit of large mass parameters we find $\sin^2 2\theta < 0.17$ (90 % C.L.). For the case of maximum mixing our limit is $\Delta^2 < 0.016\ eV^2$ (90 % C.L.). These new limits are substantially more restrictive than the bounds established in the previous experiment of ref. 1. The disagreement with the results of ref. 8 is further enhanced. The experiment is presently continued at a distance of 46 m from the core of the Goesgen reactor so as to have two measurements at different distances under otherwise equivalent conditions.

The increased distance will extend the range studied to include smaller mass parameters.

REFERENCES

1) Goesgen: CALTECH-SIN-TUM Collaboration, J.L. Vuilleumier, F. Boehm, J. Egger, F.v.Feilitzsch, K. Gabathuler, J.L. Gimlett, A.A. Hahn, H. Kwon, R.L. Mossbauer, G. Zacek and V. Zacek, Phys Lett., 114B (1982) 298.
2) ILL: CALTECH-ISN-TUM Collaboration, H. Kwon, F. Boehm, A.A. Hahn, H.E. Henrikson, J.L. Vuilleumier, J.F. Cavaignac, D.H. Koang, B. Vignon, F.v. Feilitzsch and R.L. Mossbauer, Phys. Rev. D24 (1981) 1097.
3) G. Zacek, et al., to be published.
4) K. Schreckenbach, H.R. Faust, F.v. Feilitzsch, A.A. Hahn, K. Hawerkamp and J.L. Vuilleumier, Phys. Lett. 99B (1981) 251; F.v. Feilitzsch, A.A. Hahn and K. Schreckenbach, Prepr. SP. 82-110, ILL, Grenoble, France (to be published in Phys. Lett.).
5) G. Meier and W. Sauser, KKW Goesgen, private communication.
6) M.F. James, J. Nucl. Energy 23 (1969) 517; J.M. Paratte, Technische Mitteilung Nr. TM-45.81-19, Swiss Institute for Reactor Research, Wurenlingen, Switzerland.
7) P. Vogel, G.K. Schenter, F.M. Mann and R.E. Schenter, Phys. Rev. C24 (1981) 1543.
8) F. Reines, H.W. Sobel and E. Pasierb, Phys. Rev. Letters 45, (1980) 1307.

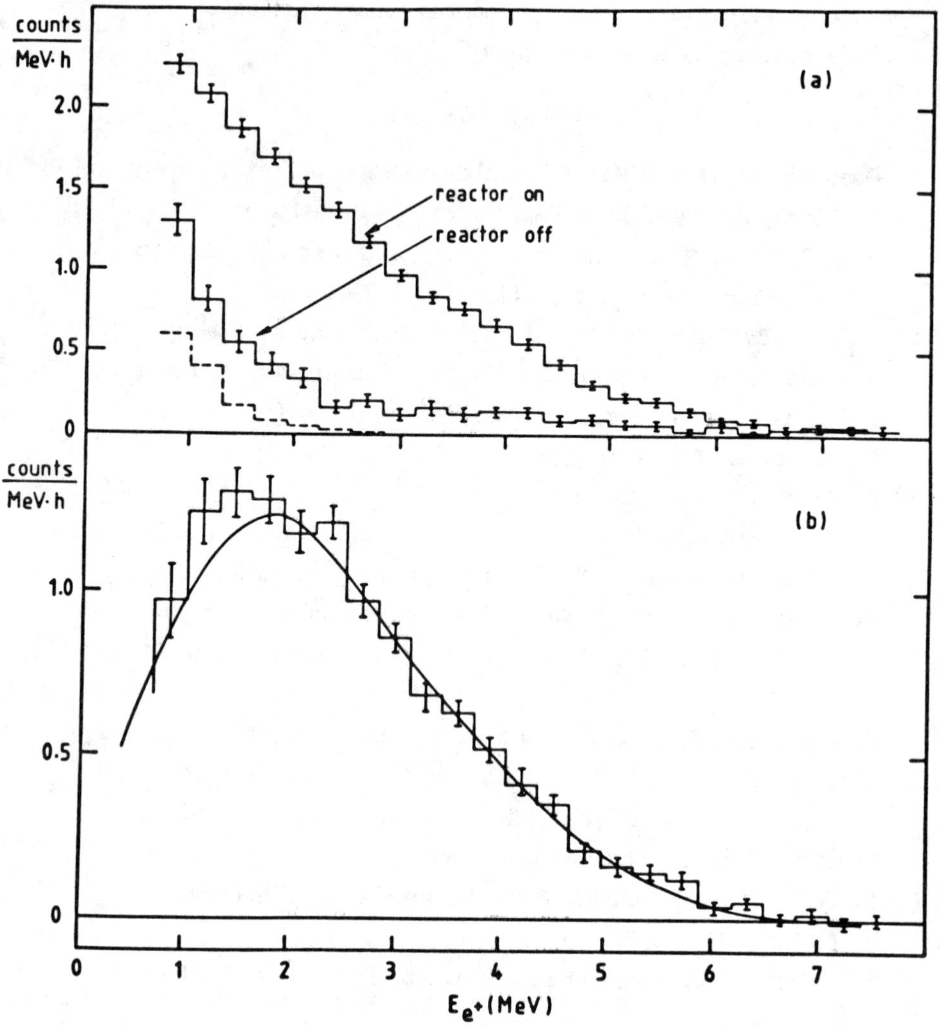

Fig. 1. Experimental results: coincidence rate (e^+,n) as a function of observed positron energy E_{e^+}. Bin width = 0.305 MeV. The errors shown are statistical. a) Reactor-on and reactor-off spectra. The dashed curve indicates the contribution of the accidentals. b) Experimental positron yield Y_{exp} (reactor-on minus reactor-off spectrum). The solid curve represents the predicted positron spectrum assuming no neutrino oscillations $\tilde{Y}_{no\ osc}$.

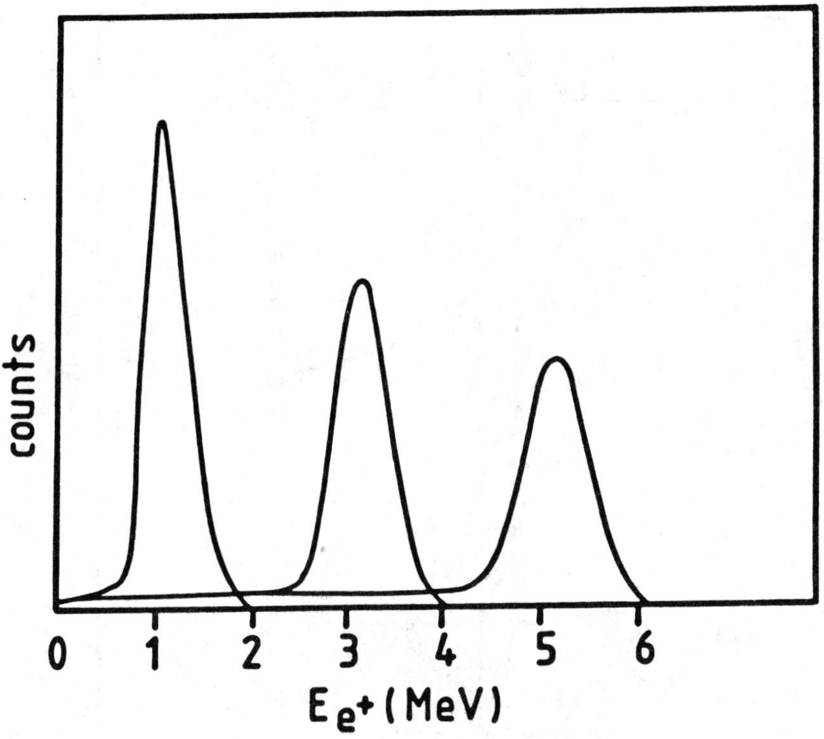

Fig. 2. Examples of the response function $r(E_{e^+},E)$ for monoenergetic positrons of 1, 3, and 5 MeV. E_{e^+} is the positron energy observed in the detector.

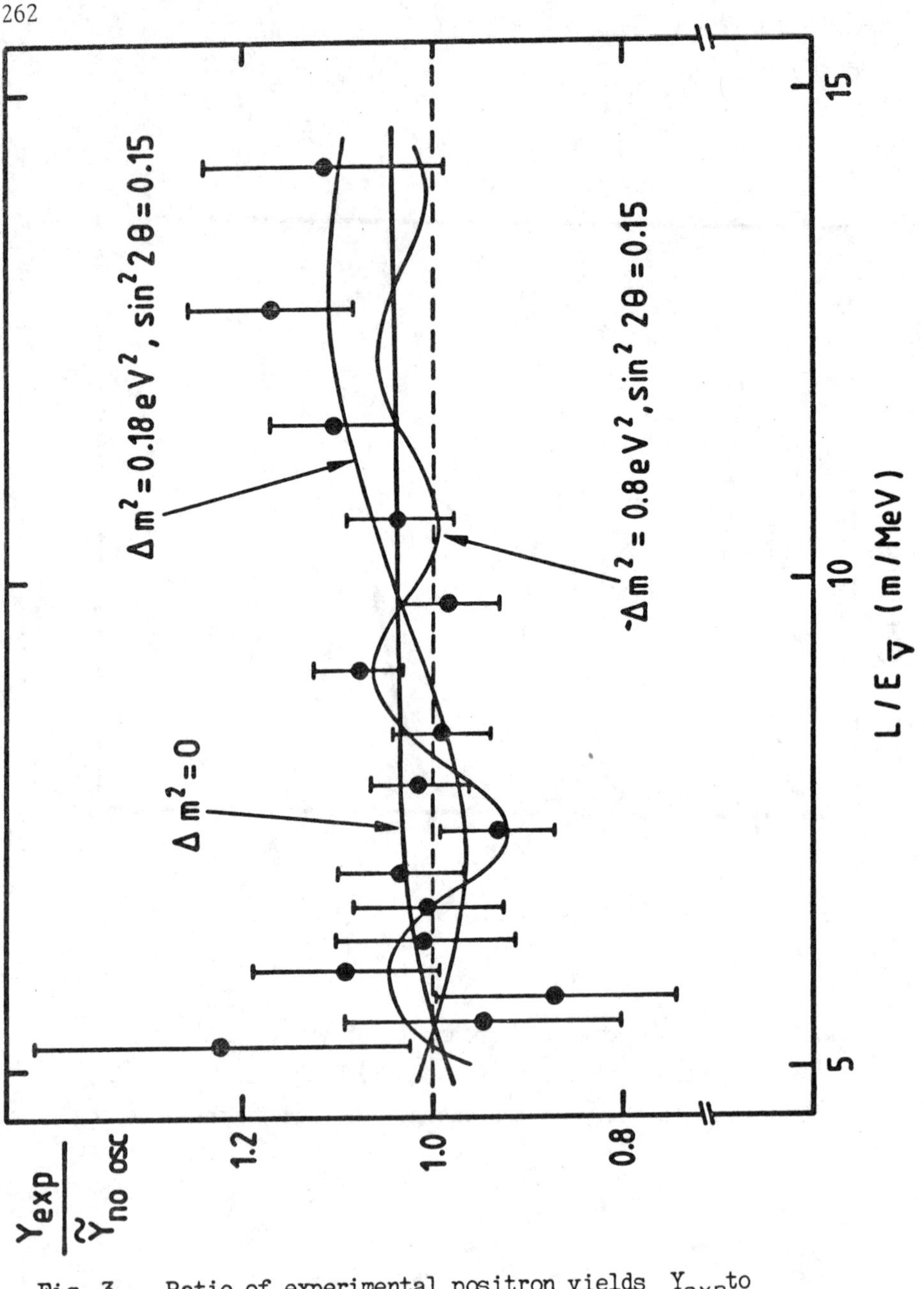

Fig. 3. Ratio of experimental positron yields Y_{exp} to predicted no oscillation yields $Y_{no\ osc}$, for the 16 energy bins used in the experiment. The errors shown are statistical. Oscillation functions obtained from parameter fits (see text) are illustrated for several sets of parameters.

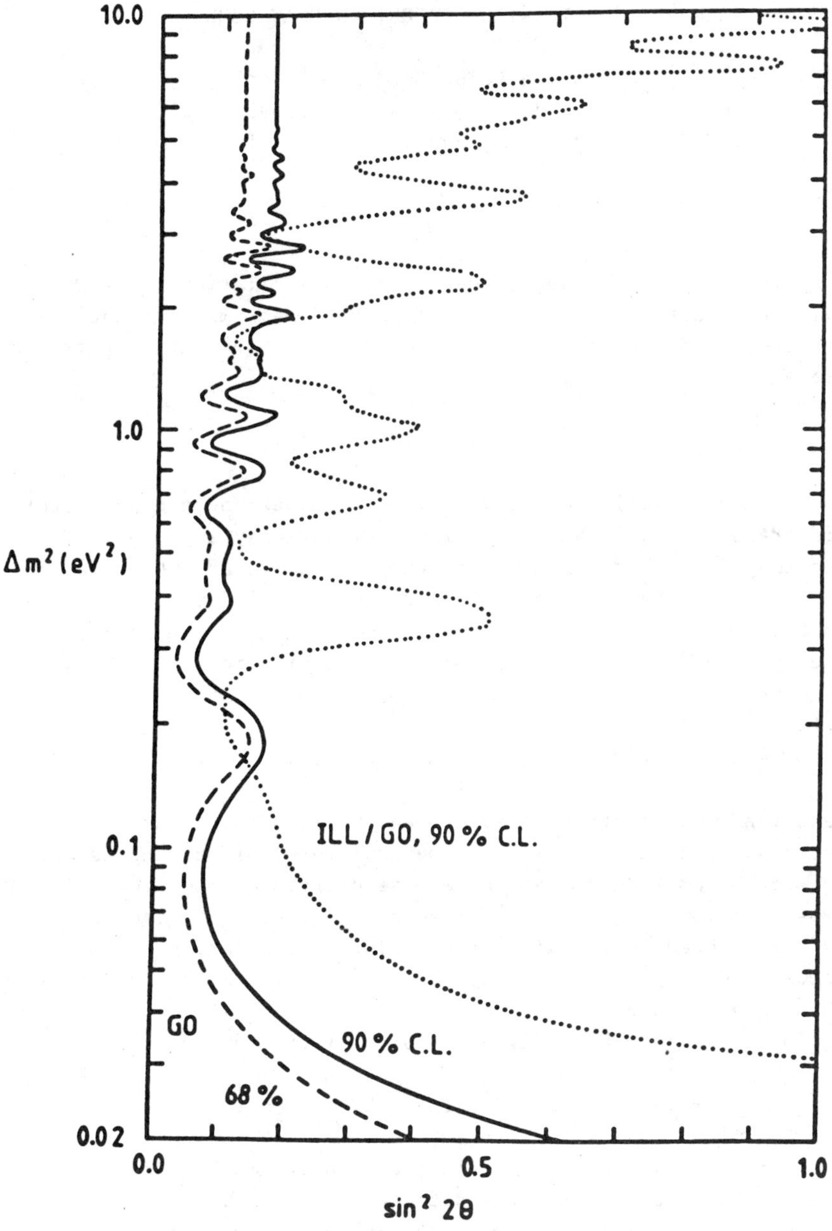

Fig. 4. Limits on the neutrino oscillation parameters Δ^2 and $\sin^2 2\theta$. The regions to the right of the curves are excluded at the confidence level indicated. Limits obtained from the ratio of the ILL experiment to the Goesgen experiment are indicated by the dotted curve labelled ILL/GO.

NEUTRINO OSCILLATION EXPERIMENTS AT CERN

Milla Baldo-Ceolin
Istituto di Fisica dell'Università, Padova, Italy
I.N.F.N., Sezione di Padova, Padova, Italy

ABSTRACT

A review is given of the three neutrino oscillation experiments which will be started next Spring at CERN-PS. Their main characteristics are presented. Possible analyses with three oscillation parameters are also discussed.

INTRODUCTION

The non conservation of flavour lepton number and the non zero mass difference allow flavour neutrino oscillations, a process which has been long discussed at length in the literature and which continues to be an important object of experimental activity.

In the presence of ν oscillations, ν_e, ν_μ and ν_τ are no more eigenstates of the mass matrix, but a linear combination of the new mass eigenstates ν_i, such that

$$|\nu_\alpha\rangle = \Sigma_i\, U_{\alpha i}|\nu_i\rangle \qquad (L = e,\mu,\tau\ \ldots)$$

Assuming three neutrino flavours and CP invariance, neutrino oscillations are described by 6 parameters: three mixing angles for the oscillation amplitudes and three mass differences squared related to the oscillation frequencies. For semplicity, however, the data are usually represented in two dimensional plots, considering the oscillation between two types of neutrinos at a time, so that one deals with one mixing angle Θ and one mass difference squared Δm^2. For ν_μ, ν_e mixing, for instance, the oscillation probability can be expressed as

$$P(\nu_\mu \to \nu_e) = P(\nu_e \to \nu_\mu) = \sin^2 2\Theta\, \sin^2\left(1.27 \cdot \frac{\Delta m^2 \cdot L}{E}\right)$$

where $\Delta m^2 = |m_1^2 - m_2^2|$ in eV^2, E is the ν energy in MeV and L the distance of the ν source from the detector in meters.

There are essentially two methods for detecting neutrino oscillations:

a) the "appearance method" where one searches for a different flavour component in an initially flavour defined ν-beam;
b) the "disappearance method" where the neutrino beam intensity is measured at different distances from the source: a reduction of the neutrino flux demonstrates the presence of neutrino oscillations.

0094-243X/83/990264-08 $3.00

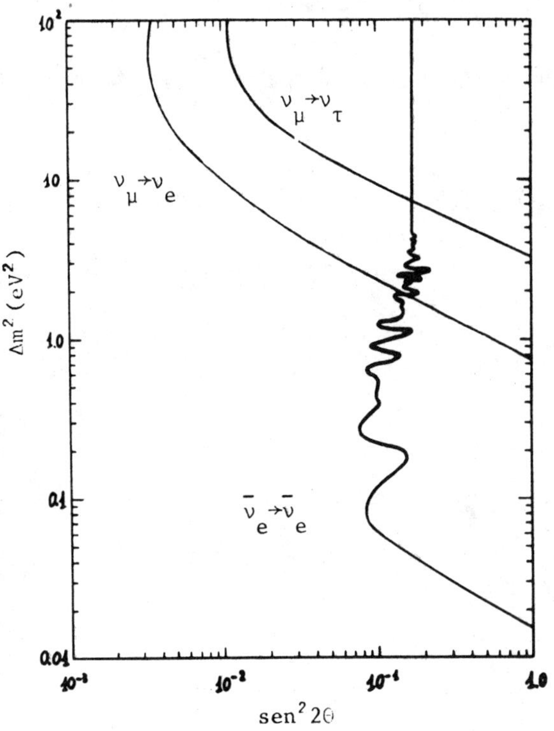

Fig. 1 - Present limits on neutrino oscillation parameters Δm^2 and $\mathrm{sen}^2 2\Theta$.

The sensitivity of the experiments is usually defined by quoting the minimum values for Δm^2 and for $\sin^2 2\Theta$ which the experiments could detect: it is limited by the statistical and systematic errors as well as by the incertitude on the ν-energy and on the dimensions of the ν-source and of the detector. The present experimental situation is illustrated in Fig. 1, where data from reactors as well as from accelerators are represented(1).

THE CERN ν-OSCILLATION PROGRAM

Three experiments will be started next Spring at CERN, searching for neutrino oscillations. They will use the new neutrino facility, a low energy neutrino beam from CERN-PS, whose primary protons will have 12-19 GeV/c momentum range.

Protons will hit a 80 cm long Be target followed by a 45 m long decay tunnel and a 4 m thick iron absorber. The beam, monitored by several μ counters placed inside the iron shield, will cross the three existing large neutrino detectors, BEBC, CDHS and CHARM, at their present position: a magnetic horn can focus the parent pions, thus increasing the neutrino beam intensity at the detectors by more than a factor ten (Fig. 2).

Before discussing the three experiments in some detail, it is worth-while to mention that two more experiments have been proposed at CERN. One of the two projects[2] will study in BEBC a neutrino beam originated from K_L° decays at SPS, the other[3] will detect and measure ν_μ and ν_e interactions in high resolution calorimeters placed in two positions along the SPS neutrino beam, the second detector being quite far away on the Jura-Mountains: it will be able to detect ν_μ disappearance as well as ν_e and ν_τ appearance effects.

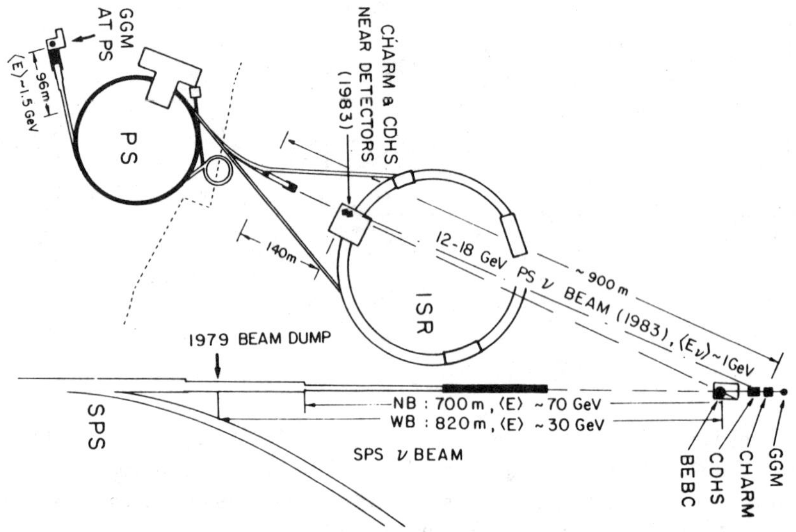

Fig. 2 - Layout of the new neutrino beam.

BEBC experiment

The Padova-Pisa-Athens-Wisconsin collaboration[4] will search for neutrino oscillations detecting and measuring in BEBC electron-neutrino interactions originated by the initially almost pure muon-neutrino beam.

The bubble chamber, filled with mixture of 75% neon in hydrogen, will be irradiated by the horn focused neutrino beam. Electron neutrino interactions will be identified with an average efficiency of 90% and their energy measured with ∿15% resolution; the detection efficiency for muons is practically 1, with an energy resolution of a few per cent.

The lower limit on the sensitivity will be set by the number of events due to ν_e initially present in the beam or simulating electron neutrino interactions; these events have been evaluated to be less than 0.5 per cent of ν_μ events. Furthermore, if a ν_e signal should be observed, it would be possible to distinguish the effects of the oscillation amplitude from that of the oscillation frequency by means of the ν_e interactions energy measurement as shown in Fig. 3.

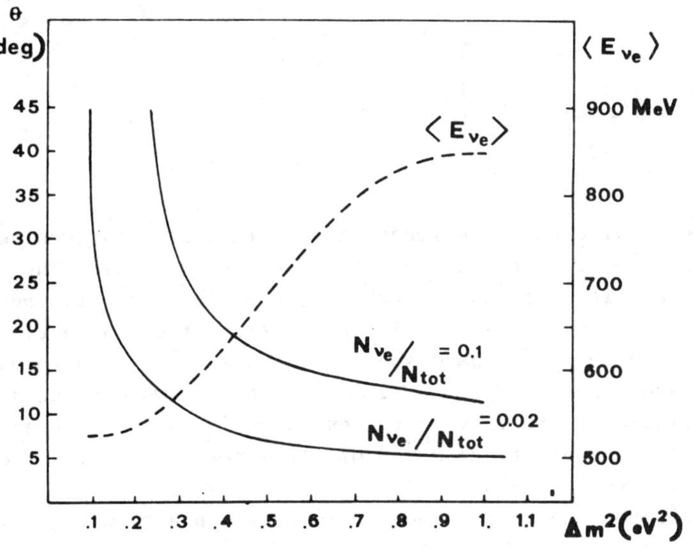

Fig. 3
Correlations between mixing angle Θ, mass difference Δm^2 and the average energy of the electron-neutrino events.

It will also be possible to obtain the value of Δm^2 involved in the $\nu_\mu \to \nu_\tau$ process from the ratio of the neutral to the charged current events: the limit on the sensitivity in this case is mainly due to the uncertainties on the expected ratio NC/CC (less than 20 per cent).

The counter experiments

The CDHS(5) and CHARM(6) collaborations aim at measuring the rate of disappearance of μ-neutrinos between two distances from their production point.

The experiments will be carried out with existing apparatus of the CDHS and CHARM detectors respectively with the addition and the displacement of a few modules immediately behind the pion decay region. Muon tracks will constitute the event signature. The overall detection efficiency for muon events is shown for each detector in Fig. 4.

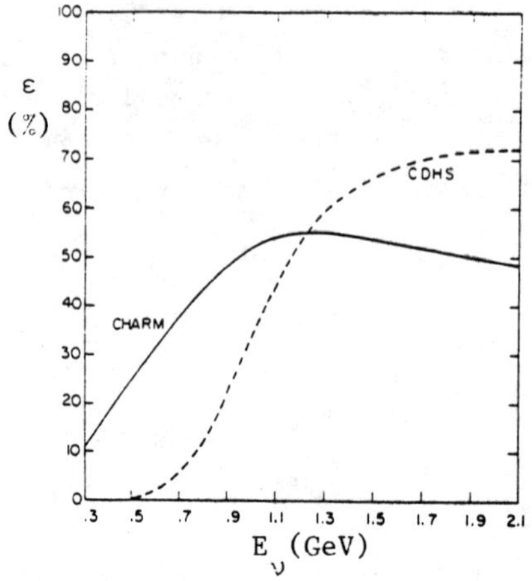

Fig. 4 - Detection efficiency for CC events in the counter experiments.

The sensitivity of these experiments to neutrino oscillations depends mostly on the uncertainty in the far to near event ratio. The corresponding systematic error has a number of contributions from beam geometry, pion production model, radial event distribution, detection efficiency and background. The beam should be run without horn focusing in order to have a simple spectrum relation at the two distances. The overall systematic uncertainty in the event ratio is for each experiment ≃0.05. With a statistical error of a few per cent, a 10% oscillation effect will then be observable.

Fig. 5 illustrates the limits the experiments can attain in a two parameter analyses.

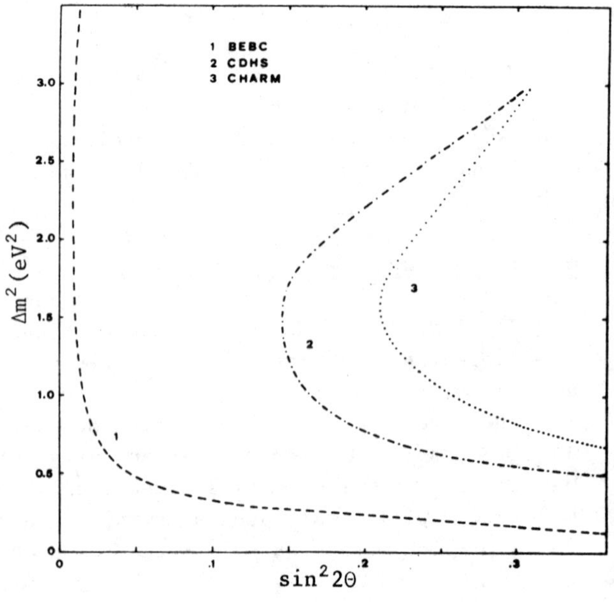

Fig. 5 - Limits on Δm^2 and $sen^2 2\Theta$ obtainable in the CERN projects.

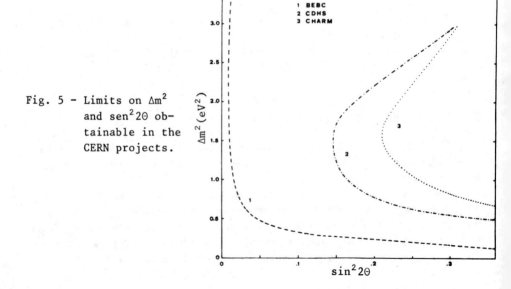

The analysis

The final analysis of the results, taking advantage from the fact that BEBC measuring an appearance effect can reach for the oscillation probability $\nu_\mu \to \nu_e$ a sensitivity about one order of magnitude better than counter experiments, will hopefully allow to reach a deeper insight on the oscillation phenomena: under the assumption $P(\nu_\mu \to \nu_\mu) = 1 - [P(\nu_\mu \to \nu_e) + P(\nu_\mu \to \nu_\mu)]$ counter experiments will mainly measure $P(\nu_\mu \to \nu_\tau)$.

The experimental results can be then analyzed involving more parameters than the two, $\mathrm{sen}^2 2\Theta$ and Δm^2, usually used.

a) In particular if it is assumed that the neutrino mixing matrix is the same as the quark mixing matrix which can be assumed to be

$$U \simeq \begin{pmatrix} 1-\frac{\varepsilon^2}{2} & \varepsilon & \varepsilon^2 \\ -\varepsilon & 1-\varepsilon^2 & -\varepsilon \\ -\varepsilon^2 & \varepsilon & 1-\frac{\varepsilon^2}{2} \end{pmatrix}$$

where $\varepsilon^2 \simeq (\Theta_{\text{Cabibbo}})^2 \simeq 0.05$ and CP violating effects are neglected, the oscillations probability involve one mixing angle ε, and two oscillatory terms depending on Δm^2_{21} and Δm^2_{32}.

In this hypothesis the experiments are sensitive to $\Delta m^2_{21} \gtrsim 0.2$ and $0.7 \gtrsim \Delta m^2_{32} \gtrsim 2\ \mathrm{eV}^2$.

b) If, on the other hand, it is assumed that two of the m_i masses almost coincide, the oscillation probabilities involve one oscillatory term and two mixing angles α and β.

Fig. 6 shows the limits the experiments can reach for α and β mixing angles.

Finally in case that counter experiments will detect oscillation effects, the hypothesis that the effects seen by the counters are due to $\nu_\mu \to \nu_\tau$ oscillation, can be probably checked measuring in BEBC the NC/CC ratio.

Table I summarizes the most relevant characteristics of the previously discussed experiments.

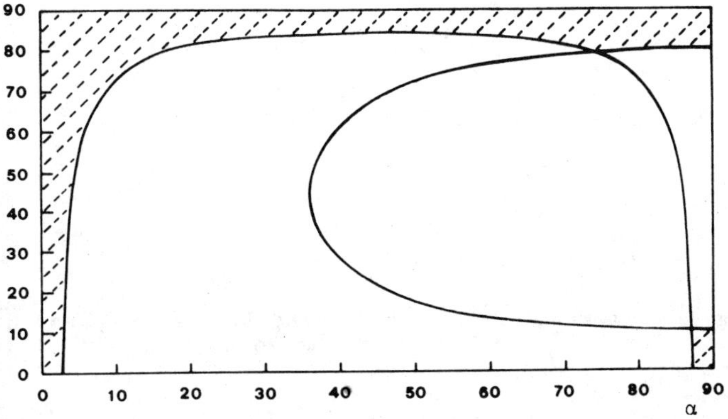

Fig. 6 - Limits on mixing angles α and β. The allowed region is the dashed one.

TABLE I

Experiment	BEBC	CDHS	CHARM
ν_μ beam	horn focused	bare target	bare target
Detectors	bubble chamber 75% Ne in H_2	scintillation counters and Fe slabs	scintillation and proportional tube counters, marble plates
ν_μ interaction detection efficiency	∿1	see Fig. 4	see Fig. 4
far detector: distance (m)	840	870	900
mass (tons)	12	1200	135
N° ν_μ CC events	900	2000	200
near detector distance (m)	-	140	130
mass (tons)	-	350	40
N° ν_μ CC events	-	25000	2000
Processes studied	$\nu_\mu \to \nu_e$	$\nu_\mu \to \nu_\mu$	$\nu_\mu \to \nu_\mu$
	$\nu_\mu \to \nu_\tau$	-	-
Sensitivities from 2 parameters analysis, $\Delta m^2, sen^2 2\Theta$			
$\nu_\mu \to \nu_e$	0.1,0.01	-	-
$\nu_\mu \to \nu_\mu$	-	0.25,0.12	0.25,0.20
$\nu_\mu \to \nu_\tau$	0.4,0.20	-	-

Sensitivities from 3 parameters analysis

a) for $\Theta \simeq 0.23$, $\Delta m^2_{21} \lesssim 0.2$, $0.7 \lesssim \Delta^2_{32} \lesssim 2$ eV^2;

b) with a leading mass difference: $\Delta m^2 \lesssim 0.1$, the limits for α and β mixing angles are given in Fig. 6.

REFERENCES

1. R.L. Mössbauer, Proceedings of Neutrino '82 Conference, Vol. 1, page 1.
   N.J. Baker et als., Phys. Rev. Lett. 47, (1981), 1576.
   N. Ushida et als., Phys. Rev. Lett. 47, (1981), 1694.
2. B. Jongejans et als., CERN/SPSC 80-77, (P150).
3. T.C. Bacon et als., CERN/SPSC 82-20, (P178).
4. M. Baldo-Ceolin et als., (Padova-Pisa-Athens-Wisconsin Collaboration), CERN/SPSC/P77 - PSCC/P33.
5. H. Abramowicz et als., (CDHS Collaboration), CERN/PSCC/P30.
6. F. Bergsma et als., (CHARM Collaboration), CERN/PSCC/P30.

RESONANT NEUTRINO ACTIVATION and NEUTRINO OSCILLATIONS

William P. Kells
Fermi National Accelerator Laboratory, Batavia, Illinois 60510

ABSTACT

Low Q value weak nuclear decays are considered which have two body final states(electron captures and bound state beta decays, BSD). This permits an analogy with the Mössbauer effect, where the emitted (anti)neutrinos will resonantly activate daughter nuclei in a suitable absorber. Candidates for such a process are examined and the relevant solid state host problems are discussed. We point out that resonant line widths as large as the narrowest observed in Mössbauer spectroscopy suffice to greatly extend the sensitivity of ν [disappearance] oscillation experiments.

INTRODUCTION

The fact that electron captures, in certain circumstances, might provide resonant neutrinos in analogy to the Mossbauer effect was pointed out long before current astrophysical and elementary particle interest in the neutrino mass developed.[1] The outgrowth of a study of Tritium decay in light of recent topical interest in m_ν measurement lead to a proposal with respect to beta decay.[2]

We consider the general features of a Mössbauer-like resonant absorbtion of nuclear decay ν ($\bar{\nu}$) by daughter nuclei. Considering this activation as a "signal" it will be shown that potentially superior neutrino oscillation experiments can be constructed. It is with this tantilizing possibility in mind that we discuss the discouraging solid state[host matrix] requirements posed.

We specify this discussion by considering the "experimental" arrangement of Figure 1. The spherical source,S, consists of some solid matrix including radioactive atoms (Z,N). Weak decay, $(Z,N) \to (Z\pm 1, N) + e^{\mp} + \nu$, may proceed via a channel which leaves the daughter in a total electronic ground state (GS), and which is a zero phonon transition(i.e. "recoilless"). We refer to these as the resonant decay fraction, η, of the total parent decay rate since the accompanying ν will have a resonantly enhanced cross section to activate daughter nuclei ($Z\pm 1$) in the absorber, A, provided that the compound A contains the daughters at sites equivalent(no chemical shift) to those occupied by the parent atoms in S.

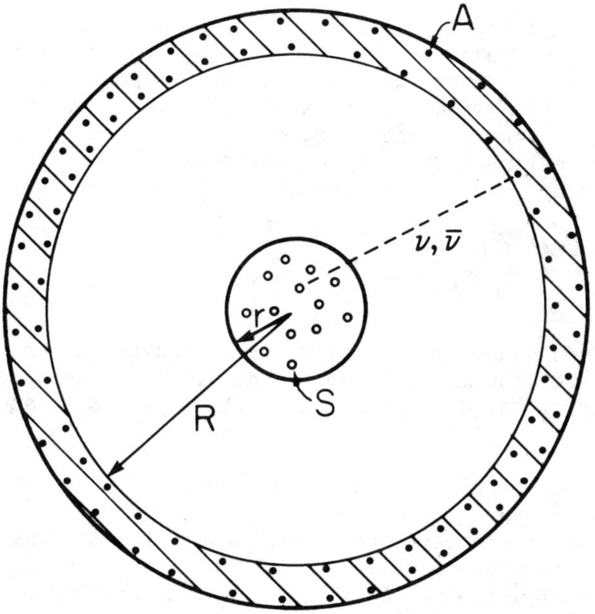

Fig. 1 Conceptual arrangement for resonant absorbtion. A spherical source (S) and absorber shell (A) contain parent and daughter nuclei respectively within some host matrix.

RESONANT ABSORBTION

An experiment would consist of counting the atoms ($Z\pm 1$, N) activated within A during some time interval after the introduction of S. Thus A is to be initially uncontaminated by (Z,N). With regard to questions of such contamination; backgrounds; and methods of counting the extremely low induced ($Z\pm 1$, N) signal(e.g. whether by chemical/mass separation or by radiochemistry) this situation is similar to those successfully dealt with in the solar neutrino searches.[3] We return later to some of these questions. The rate of activation ($Z\pm 1$, N) $\rightarrow$ (Z,N) in A may be written:

$$\lambda_A = \left(\frac{\lambda_S}{4\pi R^2}\right) N_A\ \eta_A \eta_S\ \sigma_u\ \frac{\Gamma_n}{\Gamma_e} \tag{1}$$

where λ_s is the net source activity; N_A is the number of absorbing nuclei; Γ_n is the natural width of the decay; Γ_e is an effective experimental line width; and σ_u is the unitarity cross section, $c^2h^2/2Q^2$.

In the Mossbauer effect it is quite possible for $\eta_{S,A} \simeq 1$ and $\Gamma_n/\Gamma_e \simeq 1/2$ with proper source/absorber preparation.[4] The same inhomogeneous broadenings observed in Mossbauer studies(e.g. HFS and QSF field smearing due to an imperfect lattice or impurity distribution) contribute here to Γ_e. For the weak decays of interest here, $\Gamma_n < 8\times10^{-11}$ eV (narrowest Mossbauer $\Gamma_e \simeq 5\times10^{-11}$ eV) whereas the gravitational shift limit[5] for an absorber of characteristic size R < 10cm is $\Gamma_e < 4\times10^{-13}$ eV.

We now discuss η as the product of two factors f_R(recoilless fraction) and f_G(electronic GS branching ratio). f_R is the familiar Debye-Waller factor , necessarily modified to reflect the fact that the initial and final states of the lattice(S or A) are different even if there is no recoil[6]:

$$f_{R,j} = (\zeta_j/\mu_j)^3 f_j^{\zeta} \qquad j = S,A$$

where (2)

$$\zeta_j \equiv [2\mu_j^2/(1+\mu_j^2)]$$

$$\mu_j \equiv \langle r_f^2\rangle_j / \langle r_i^2\rangle_j$$

The quantities $\langle r_{i,f}^2\rangle$ are the mean square lattice positions of the initial/final nucleus, while f_j is the usual Mossbauer fraction for the initial nucleus($= \exp[-k^2\langle r^2\rangle_j]$, with $k = Q/\hbar c$). The suprising conclusion from (2) is that if f_S is large, for example, a much more weakly bound daughter ($\mu_S >> 1$) only causes a μ_S^{-3} dependence in $f_{R,S}$. Since $\langle r_j^2 \rangle \propto M$ we expect Q^2/M, for candidates with usefully large f_R, to be limited (0.1keV-c^2 for all known Mossbauer decays). Also, host matrix structure can considerably influence $\langle r_j^2 \rangle$. Very light isotopes, specifically Tritium(which violates this rule), would not be candidates unless bound in a much heavier host lattice, where $\langle r^2 \rangle \propto \langle r_{host} \rangle M_T/M_{host}$ (such embedding tricks are commonly employed in Mossbauer work).

Low Q is further favored, below recoil energies allowing $f_R \simeq 1$, by the weak decay/capture kinematics dependence of f_G. For beta decay the BSD branch grows with decreasing Q such that

$$f_G \propto Q^{-1.5} \tag{3}$$

for allowed transitions[7] (at least for Q where the β^- branch still dominates). For electron capture f_G ordinarily has no Q dependence, giving a λ_A also with no Q dependence. However, all extremely low Q captures may be anomalous in that K,L,M,... shell capture will be successively excluded as Q decreases. Since f_G is essentially the fraction of valence electron captures, its value dramatically increases in such anomalous cases.

f_G is also sensitive to the parent/daughter electronic GS configeration (which determines the nuclear overlap $\propto |\psi(0)|^2$ with the relevent valence electrons). The nuclear overlap Z dependence (even including the influence of host matrix binding) is, for electron captures, no better than

$$f_G \propto |\psi_{s,valence}(0)|^2 \Big/ \sum_n |\psi_{ns}(0)|^2 \sim Z_{nucl} \Big/ Z_{core}^3 \tag{4}$$

since higher ℓ orbitals, or various corrections to the Fermi-Segre expression(e.g. relativity) will tend to decrease f_G. In the anomalously low Q captures Z_{core} is effectively reduced from Z_{nucl}. Similarly, BSD will at best be given by the s orbital Fermi-Segre overlap: $f_G \propto Z_{nucl}$ (and in addition a substantial relativistic enhancement occurs, amounting to a factor 3-9 for the period 6 elements[8]).

Unfortunately, these systematic trends serve little purpose in selecting candidate decays. Isotopes selected according to a criteria of possibly large f_R (plus the requirement of a stable daughter) amount to only the few unique cases described in table 1 (plus a few for comparison). A figure of merit ,U(assuming $f_R \equiv 1$), is tabulated, which is λ_A specified per source and absorber

Table 1. Candidates for resonant absorbtion (^{35}S, ^{55}Fe, ^{181}W, and ^{145}Pm for comparison only). All daughters are stable isotopes, with naturally occuring abundance shown. The relative (to ^{3}H) figure of merit, U, for detection assumes that all daughters are counted.

Decay (%)	SPIN	Q keV	Q^2/M eV	τ yr	GS VALENCE CONFIGURATION	f_G	U
^{3}H-He(*)	$\frac{1}{2}^- \rightarrow \frac{1}{2}^-$	18.6	11.5	12.3	$1s \rightarrow 1s^2$	6 E-3	≡1.0
^{35}S-Cl(76)	$\frac{3}{2}^+ \rightarrow \frac{3}{2}^+$	167	79.6	0.26	$3s^2p^4 \rightarrow 3s^23p^5$	4.8E-3	<E-9
^{63}Ni→Cu(69)	$\frac{1}{2}^- \rightarrow \frac{3}{2}^-$	68	6.8	92	$3d^94s^2 \rightarrow 3d^{10}4s^2$	3.9E-3	4.3E-4
^{107}Pd→Ag(52)	$\frac{5}{2}^+ \rightarrow \frac{1}{2}^-$	33	10.2	6.5E+6	$4d^{10} \rightarrow 4d^{10}5s$	0.12	4.1E-3
^{151}Sm→Eu(48)	$\frac{5}{2}^- \rightarrow \frac{5}{2}^-$	76	3.8	90	$4f^66s^2 \rightarrow 4f^76s^2$	8 E-2	0.31
^{171}Tm→Yb(14)	$\frac{1}{2}^+ \rightarrow \frac{1}{2}^-$	97	5.5	1.9	$4f^{13}6s^2 \rightarrow 4f^{14}6s^2$	8.0E-2	42.0
^{193}Pt→Au(100)	$\frac{1}{2}^- \rightarrow \frac{3}{2}^+$	61	1.92	50	$5d^96s \rightarrow 5d^{10}6s$	2.2E-2	0.31
^{187}Re→Os(1.6)	$\frac{5}{2}^+ \rightarrow \frac{3}{2}^-$	2.6	.0036	4 E+10	$5d^56s^2 \rightarrow 5^66s^2$	0.5	4.3E-4
^{55}Fe→Mn(100)	$\frac{3}{2}^- \rightarrow \frac{5}{2}^-$	231	97	2.7	$3d^64s^2 \rightarrow 3d^54s^2$	2.E-4	
^{145}Pm→Nd(8)	$\frac{5}{2}^+ \rightarrow \frac{7}{2}^-$	160	17.6	17.7	$4f^56s^2 \rightarrow 4f^46s^2$	1.1E-4	3.8E-6
^{157}Tb→Gd(16)	$\frac{3}{2}^+ \rightarrow \frac{3}{2}^-$	57.6	2.06	150	$4f^85d6s^2 \rightarrow 4f^75d6s^2$	2.6E-3	1.8E-3
^{163}Ho→Dy(25)	$\frac{7}{2}^- \rightarrow \frac{5}{2}^-$	2.6	.002	7000	$4f^{11}6s^2 \rightarrow 4f^{10}6s^2$	0.	330
^{179}Ta→Hf(14)	$\frac{7}{2}^+ \rightarrow \frac{9}{2}^+$	115	7.38	1.7	$5d^36s_2 \rightarrow 5d^26s_2$	1.1E-4	6.3E-5
^{181}W→Ta(100)	$\frac{9}{2}^+ \rightarrow \frac{7}{2}^+$	188	19.5	0.33	$5d^46s \rightarrow 5d^36s$	1.6E-3	1.8E-2
^{205}Pb→Tl(71)	$\frac{5}{2}^- \rightarrow \frac{1}{2}^+$	60	1.74	1.4E+7	$6s^26p^4 \rightarrow 6s^26p^3$	8.2E-4	1.6E-9

*For ^{3}He the extremely low natural abundance is irrelevant due to the large quantities which can be artificially produced.

to calculate U are either taken from published $|\psi(0)|$ data (mostly for the electron captures) or deduced from a refined Fermi-Segre model.[8] However a drastic oversimplification is employed by assuming that the electron orbital created/destroyed in the GS is, in all cases, the ns level nearest(in energy) to the true GS. This prescription is accurate only for ^{3}H and ^{107}Pd. For the others it might be supposed that U is actually much less due to the poor nuclear overlap(d or f wave) of the created/destroyed electron. If configuration interaction is taken into account the situation may be considerably better, especially for the transition elements(filled d states).[9] For example it is well known that the true G.S. of Ni has a large $3d^{10}4s$ admixture (similarly, the metal has 20% overlap of 4s with 3d bands). Since the exact ns wave admixture in the valence state will depend on the particular solid state matrix wave functions we do not include this factor. Obviously the true U values for the rare earth elements would be considerably lower.

Taking Tritium as the absolute example (where f_G is accurately known[7]) we find λ_A =1.6x10^{-8}/s/Curie/Mole for the geometry of Fig.1 (R =10cm, and assuming Γ_e =10^{-10}eV). It is interesting to compare this to the non-resonant ^{37}Ar activation rate expected in the Brookhaven C_2Cl_4 tank. They propose to conduct calibration and oscillation experiments with a MegaCurie ^{65}Zn source which will yield λ_A =1.5x10^{-17}/s/Curie/Mole.

Realistic experiments would need >10^{3}Ci of Tritium(80mg) which can indeed fit very compact source matricies (even if very dilute). This has important resolution consequences for oscillation measurements, where $\lambda_A(R)$ is compared for various R. Using the above λ_A(10cm) and comparing with λ_A(100cm) could give, for 10^{4}Ci ^{3}H and 10^{4}Moles ^{3}He, a statistical limit on δm_ν (maximal mixing) of 1.4x10^{-4}eV^{2}. Such a Tritium experiment depends on two formidable experimental problems. First, since ^{3}He is obtained from T decay (unlike all others in table 1), there is a severe natural contamination of absorber material. Second is the ability to measure small induced T "contaminations" (<10^{-20} concentration)[10], a field in which much encouraging recent work has been done. Radiochemical techniques would suffice (for the shorter half-lives in table 1) to initially detect the resonant signal.

We turn now to the problems presented by the solid state binding. In the Mossbauer effect, linewidths(Γ_e) and "chemical shifts" can be small because the decay final state(the nucleus) has a small volume overlap with its environment. For the weak decay f_G branch of interest here the final state(in S or A) is the nucleus plus a "valence" electron. A much larger environment is sampled. Figure 2a represents the most straightforward arrangement: pure parent S; pure daughter A. Aside from the fact that this may not give large f_R for some species(e.g. ^{3}H) this

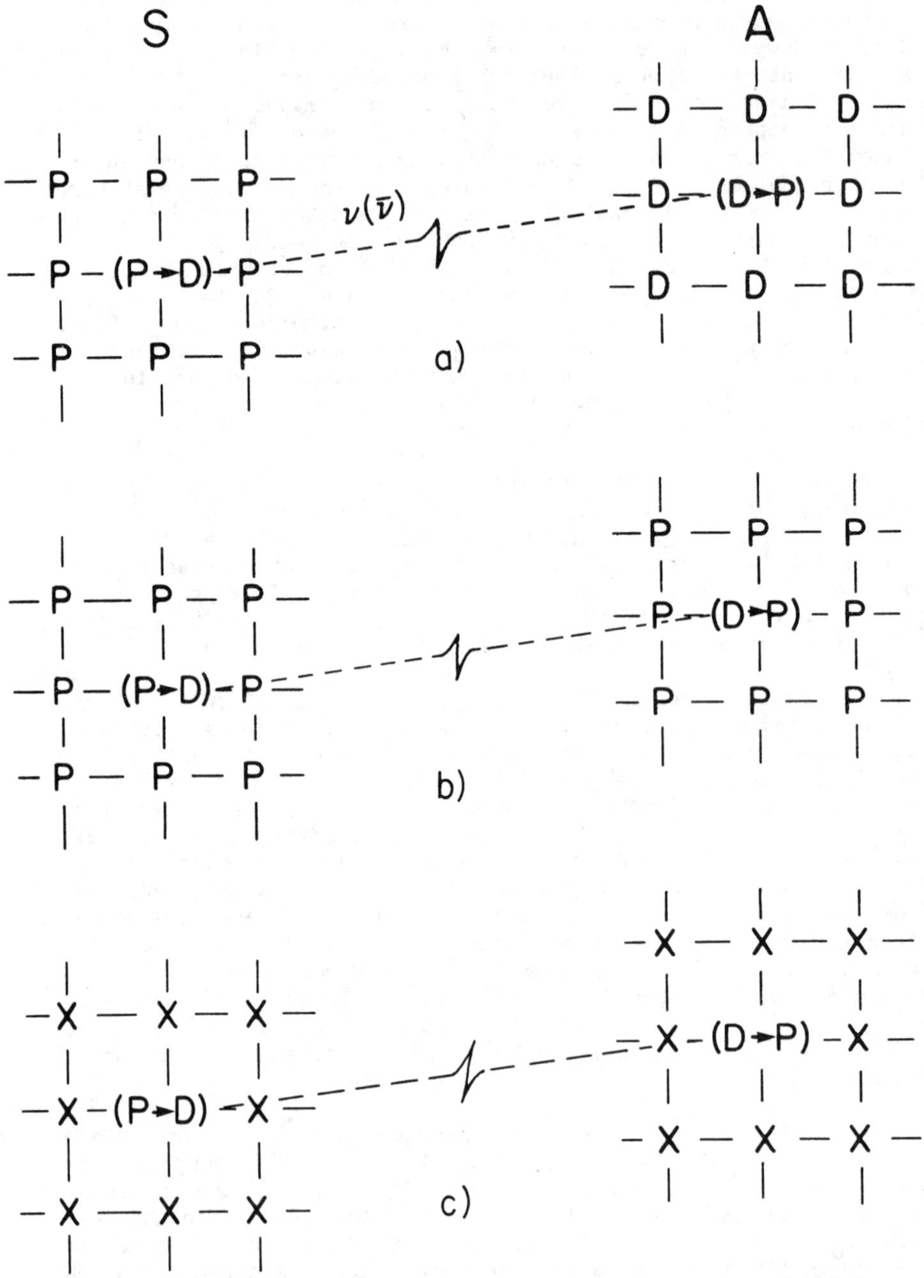

Fig. 2 Substitutional impurity host matrix possibilities. a) Pure parent S and pure daughter A. b) Pure parent S and dilute daughter (in host X=P) A. c) Both S and A are silute solutions in host X lattice.

that this may not give large f_R for some species(e.g. ^3H) this arrangement will not work on two counts, though it would provide the most compact apparatus. First, since all the table 1 candidates are metals(excepting Tritium), the fact that a capture(decay) electron must leave(enter) a Block state band would typically give Γ_e >0.1eV. Second, even if the bands were sufficiently narrow, the environment containing the transmuting atom is different in the absorber than in the source (e.g. different chemical potentials). The resultant chemical shift between emission and absorbtion line centers would certainly be $>$ 10^{-3}eV. Such shifts in the Mossbauer effect are $<10^{-8}_{eV}$; easily compensated for by doppler drive.

These difficulties may be overcome at the cost of having to prepare the parent(P), daughter(D) or both as impurities in a host lattice. Initially consider a simple picture of substitutional impurities. Figure 2c illustrates the type of S/A symmetry necessary to eliminate[ideally] chemical shifts. A degenerate case is possible(Fig.2b) where the host ,X, is one of P,D.

This impurity problem is closely related to one much studied in the theory of solids: the "X-ray problem"[concerning the state of a core level excited by X-rays].[11] It is possible for the β^- (from P$\rightarrow$D, Fig.2b) to be trapped in a discreet localized state at the positive impurity daughter site. In the case of metals (Fig. 2b, necessarily) we hope that screening causes exponential localization, allowing a high impurity electron density (with low Imp-Imp interaction probability). It seems, however, that this localized transition is effectively smeared in energy via the Anderson "I.R. catastrophy"[12], by an amount proportional to the occupation fraction of the conduction band (just what gives good plasma shielding!). Furthermore the static screening will not, in general, be exponential (as in a simple single particle band-impurity description) but will have long range "Friedel oscillations".[13]

On the other hand insulating hosts imply the structure Fig. 2c and S/A's much less concentrated than in the case of Fig. 2a. For many insulating hosts (typically the commercial semiconductors) the localized impurity levels are well characterized. For sufficiently low impurity concentrations (e.g.$<$ 10^{14}/cc in Si) these lines can be extremely narrow ($<$ 10^{-6}eV).[14] The unbroadened concentrations are expected to be low because the Bohr radii of these shallow impuriity levels are large($>$ 100Å). There are other, less well understood, impurity classes in insulators which are not shallow in this sense. A famous example is the iso-electronic Nitrogen(substituting for Phosphoros) impurity in GaP.[15] Further study is necessary to determine the limiting concentrations possible with such deep impurity levels, for this application.

Highly localized impurity states are evidently the key to realizing the ideas presented here. One approach is embedding in a highly inert host. Mossbauer studies of P, D species in noble gas ices show that the P, D atoms behave very much like free atoms.[16] Whether this is so to the extent needed for weak decay resonance is an open question (an indication that such atoms are not exactly "free" is the fact that, in these amorphous ices, Γ_e is typically $\gg\Gamma_n$). Many other disordered alloy and interstitial impurity matricies could usefully be studied from our point of view. A natural example[for Tritium] are the heavy metal hydrides. Here the problem is with equvalently embedding ^3He in an absorber.

Acknowledgements

Portions of this work were done in collaboration with J. Schiffer with help from several of his colleages at Argonne National Labs. I would like to thank H. Lipkin for informative discussions, especially his elucidation of the appropriate transition density expression for f_R (Eq.2). Thanks also go to J. Bahcall, R. Lewis, and P. McIntyre for their helpful comments.

REFERENCES

[1] William M. Visscher, Phys. Rev. 116 , 1581 (1959).

[2] William P. Kells, FermiLab internal report FN-340 (July, 1981).

[3] Bruce T. Cleveland, et.al., in Proceedings of the VPI grand unification workshop, edited by G.B. Collins, et.al(AIP, N.Y.1981), AIP proceedings 72, p154.

[4] G. Kaindl and D. Salomon Phys. Lett. 42A , 333 (1972).

[5] H. Lipkin, Ann. Phys.(N.Y.) 9 , 51 (1 9 60).

[6] H. Lipkin, unpublished note, 1981.

[7] J. Bahcall, Phys. Rev. 124 , 495 (1961).

[8] M.A. Bouchiat and C.C. Bouchiat, J. Phys.(Paris) 35 , 899 (1974). The f_G for ^{187}Re is special. It may be derived from a comparison of the geological and radio-counting methods of measuring the lifetime. See R.L. Brodzinski and D.C. Conway, Phys. Rev. B138 ,1368(1965).

[9] J.C. Slater, Quantum Theory of Molecuels and Solids, (McGraw Hill, N.Y., 1963).

[10] P.F. Smith,et.al., Nucl. Phys. B206, 333 (1982).

[11] M. Combescot and P. Nozieres, J. Phys.(Paris) 32 , 913 (1971).

[12] P.W. Anderson, Phys. Rev. Lett. 18 , 1049 (1967).

[13] W.H. Kohn and S.H. Vosko, Phys. Rev. 119 , 912 (1960).

[14] C. Jagannath, et.al., Phys. Rev. B23 , 2082 (1981).

[15] D.G. Thomas, in Localized Excitations in Solids , edited by R.F. Wallis (Plenum,N.Y., 1968), p239.

[16] P.H. Barrett,et. al., Phys. Rev. B12 , 1676 (1975).

PARTICIPANTS

BAER, Howard	University of Wisconsin
BALDO-CEOLIN, Milla	University of Padova, Italy
BALL, Robert C.	University of Michigan
BARGER, Vernon	University of Wisconsin
BARNETT, Bruce	Johns Hopkins University
BLECHER, Marvin	Virginia Polytechnic Institute
BOBISUT, F.	University of Padova, Italy
BOWLES, Tom	Los Alamos National Laboratory
CALDWELL, David	University of California-Santa Barbara
CLARK, Gregory J.	IBM Research
CLINE, David	University of Wisconsin
CUDELL, Jean	University of Wisconsin
CUTTS, David	Brown University
DELFINO, Manuel	University of Wisconsin
DESHPANDE, N.G.	University of Oregon
DIETERLE, Byron	University of New Mexico
DURAND, Bernice	University of Wisconsin
EBEL, Marvin	University of Wisconsin
FACKLER, Orrin	Rockefeller University
FREEDMAN, Stuart	Argonne National Laboratory
FRISCH, Margaret	IBM Research
FRY, William	University of Wisconsin
GRAHAM, Robert L.	Atomic Energy of Canada Ltd., Canada
HAMPEL, Wolfgang	Max Planck Institut, Germany
HUERTA, Rodrigo	Fermi National Accelerator Laboratory
JOVANOVIC, Drasko	Fermi National Accelerator Laboratory
KABIR, P.K.	University of Virginia
KAYSER, Boris	National Science Foundation
KELLS, William	Fermi National Accelerator Laboratory
KIRSTEN, Till A.	Max Planck Institut, Germany
KOLB, Edward	Los Alamos National Laboratory
LANGACKER, Paul	University of Pennsylvania
LANOU, Robert E.	Brown University
LEUNG, Chung Ngoc	University of Minnesota
LEVEILLE, Jacques P.	University of Michigan
LIGUORI, Cesare	INFN, Milano, Italy
LOVELESS, Richard	University of Wisconsin
MA, Ernest	University of Hawaii
MANN, Al	University of Pennsylvania
MCKELLAR, Bruce	University of Melbourne, Australia
MILTON, K.A.	Oklahoma State University
MOHAPATRA, R.N.	City College of New York
MORSE, Robert	University of Wisconsin
MUGGE, Marshall	Fermi National Accelerator Laboratory
NAPOLITANO, Jim	Argonne National Laboratory
OLSSON, Martin	University of Wisconsin
PETERSON, Vincent	University of Hawaii
PEVSNER, Aihud	Johns Hopkins University
PRIMAKOFF, Henry	University of Pennsylvania

PROCARIO, Michael	University of Wisconsin
RAGHAVAN, R.S.	Bell Laboratories
RAVN, H.L.	CERN, Switzerland
REEDER, Don	University of Wisconsin
ROBERTSON, Hamish	Los Alamos National Laboratory
ROBINETT, Richard	University of Wisconsin
ROBINSON, Barry	University of Pennsylvania
ROSEN, S.P.	National Science Foundation
ROSNER, Jonathan	Enrico Fermi Institute
SAMUEL, Mark	Oklahoma State University
SEILER, Paul	ETH, Zurich, Switzerland
SENJANOVIC, Goran	Brookhaven National Laboratory
STEPHENSON, Gerry	Los Alamos National Laboratory
TAKASUGI, Eiichi	Osaka University, Japan
VERGADOS, John	University of Ioannina, Greece
VUILLEUMIER, Jean-Luc	California Institute of Technology
WACHSMUTH, Horst	CERN, Switzerland
WEILER, Thomas J.	University of California-San Diego
WHISNANT, Kerry	Iowa State University
WOLFENSTEIN, Lincoln	Carnegie-Mellon University
ZEPEDA, Arnulfo	IPN, Mexico

AIP Conference Proceedings

		L.C. Number	ISBN
No.1	Feedback and Dynamic Control of Plasmas	70-141596	0-88318-100-2
No.2	Particles and Fields - 1971 (Rochester)	71-184662	0-88318-101-0
No.3	Thermal Expansion - 1971 (Corning)	72-76970	0-88318-102-9
No.4	Superconductivity in d-and f-Band Metals (Rochester, 1971)	74-18879	0-88318-103-7
No.5	Magnetism and Magnetic Materials - 1971 (2 parts) (Chicago)	59-2468	0-88318-104-5
No.6	Particle Physics (Irvine, 1971)	72-81239	0-88318-105-3
No.7	Exploring the History of Nuclear Physics	72-81883	0-88318-106-1
No.8	Experimental Meson Spectroscopy - 1972	72-88226	0-88318-107-X
No.9	Cyclotrons - 1972 (Vancouver)	72-92798	0-88318-108-8
No.10	Magnetism and Magnetic Materials - 1972	72-623469	0-88318-109-6
No.11	Transport Phenomena - 1973 (Brown University Conference)	73-80682	0-88318-110-X
No.12	Experiments on High Energy Particle Collisions - 1973 (Vanderbilt Conference)	73-81705	0-88318-111-8
No.13	π-π Scattering - 1973 (Tallahassee Conference)	73-81704	0-88318-112-6
No.14	Particles and Fields - 1973 (APS/DPF Berkeley)	73-91923	0-88318-113-4
No.15	High Energy Collisions - 1973 (Stony Brook)	73-92324	0-88318-114-2
No.16	Causality and Physical Theories (Wayne State University, 1973)	73-93420	0-88318-115-0
No.17	Thermal Expansion - 1973 (lake of the Ozarks)	73-94415	0-88318-116-9
No.18	Magnetism and Magnetic Materials - 1973 (2 parts) (Boston)	59-2468	0-88318-117-7
No.19	Physics and the Energy Problem - 1974 (APS Chicago)	73-94416	0-88318-118-5
No.20	Tetrahedrally Bonded Amorphous Semiconductors (Yorktown Heights, 1974)	74-80145	0-88318-119-3
No.21	Experimental Meson Spectroscopy - 1974 (Boston)	74-82628	0-88318-120-7
No.22	Neutrinos - 1974 (Philadelphia)	74-82413	0-88318-121-5
No.23	Particles and Fields - 1974 (APS/DPF Williamsburg)	74-27575	0-88318-122-3
No.24	Magnetism and Magnetic Materials - 1974 (20th Annual Conference, San Francisco)	75-2647	0-88318-123-1
No.25	Efficient Use of Energy (The APS Studies on the Technical Aspects of the More Efficient Use of Energy)	75-18227	0-88318-124-X

No.26	High-Energy Physics and Nuclear Structure - 1975 (Santa Fe and Los Alamos)	75-26411	0-88318-125-8
No.27	Topics in Statistical Mechanics and Biophysics: A Memorial to Julius L. Jackson (Wayne State University, 1975)	75-36309	0-88318-126-6
No.28	Physics and Our World: A Symposium in Honor of Victor F. Weisskopf (M.I.T., 1974)	76-7207	0-88318-127-4
No.29	Magnetism and Magnetic Materials - 1975 (21st Annual Conference, Philadelphia)	76-10931	0-88318-128-2
No.30	Particle Searches and Discoveries - 1976 (Vanderbilt Conference)	76-19949	0-88318-129-0
No.31	Structure and Excitations of Amorphous Solids (Williamsburg, VA., 1976)	76-22279	0-88318-130-4
No.32	Materials Technology - 1976 (APS New York Meeting)	76-27967	0-88318-131-2
No.33	Meson-Nuclear Physics - 1976 (Carnegie-Mellon Conference)	76-26811	0-88318-132-0
No.34	Magnetism and Magnetic Materials - 1976 (Joint MMM-Intermag Conference, Pittsburgh)	76-47106	0-88318-133-9
No.35	High Energy Physics with Polarized Beams and Targets (Argonne, 1976)	76-50181	0-88318-134-7
No.36	Momentum Wave Functions - 1976 (Indiana University)	77-82145	0-88318-135-5
No.37	Weak Interaction Physics - 1977 (Indiana University)	77-83344	0-88318-136-3
No.38	Workshop on New Directions in Mossbauer Spectroscopy (Argonne, 1977)	77-90635	0-88318-137-1
No.39	Physics Careers, Employment and Education (Penn State, 1977)	77-94053	0-88318-138-X
No.40	Electrical Transport and Optical Properties of Inhomogeneous Media (Ohio State University, 1977)	78-54319	0-88318-139-8
No.41	Nucleon-Nucleon Interactions - 1977 (Vancouver)	78-54249	0-88318-140-1
No.42	Higher Energy Polarized Proton Beams (Ann Arbor, 1977)	78-55682	0-88318-141-X
No.43	Particles and Fields - 1977 (APS/DPF, Argonne)	78-55683	0-88318-142-8
No.44	Future Trends in Superconductive Electronics (Charlottesville, 1978)	77-9240	0-88318-143-6
No.45	New Results in High Energy Physics - 1978 (Vanderbilt Conference)	78-67196	0-88318-144-4
No.46	Topics in Nonlinear Dynamics (La Jolla Institute)	78-057870	0-88318-145-2
No.47	Clustering Aspects of Nuclear Structure and Nuclear Reactions (Winnepeg, 1978)	78-64942	0-88318-146-0
No.48	Current Trends in the Theory of Fields (Tallahassee, 1978)	78-72948	0-88318-147-9
No.49	Cosmic Rays and Particle Physics - 1978 (Bartol Conference)	79-50489	0-88318-148-7

No.50 Laser-Solid Interactions and Laser Processing - 1978 (Boston)	79-51564	0-88318-149-5
No.51 High Energy Physics with Polarized Beams and Polarized Targets (Argonne, 1978)	79-64565	0-88318-150-9
No.52 Long-Distance Neutrino Detection - 1978 (C.L. Cowan Memorial Symposium)	79-52078	0-88318-151-7
No.53 Modulated Structures - 1979 (Kailua Kona, Hawaii)	79-53846	0-88318-152-5
No.54 Meson-Nuclear Physics - 1979 (Houston)	79-53978	0-88318-153-3
No.55 Quantum Chromodynamics (La Jolla, 1978)	79-54969	0-88318-154-1
No.56 Particle Acceleration Mechanisms in Astrophysics (La Jolla, 1979)	79-55844	0-88318-155-X
No. 57 Nonlinear Dynamics and the Beam-Beam Interaction (Brookhaven, 1979)	79-57341	0-88318-156-8
No. 58 Inhomogeneous Superconductors - 1979 (Berkeley Springs, W.V.)	79-57620	0-88318-157-6
No. 59 Particles and Fields - 1979 (APS/DPF Montreal)	80-66631	0-88318-158-4
No. 60 History of the ZGS (Argonne, 1979)	80-67694	0-88318-159-2
No. 61 Aspects of the Kinetics and Dynamics of Surface Reactions (La Jolla Institute, 1979)	80-68004	0-88318-160-6
No. 62 High Energy e^+e^- Interactions (Vanderbilt , 1980)	80-53377	0-88318-161-4
No. 63 Supernovae Spectra (La Jolla, 1980)	80-70019	0-88318-162-2
No. 64 Laboratory EXAFS Facilities - 1980 (Univ. of Washington)	80-70579	0-88318-163-0
No. 65 Optics in Four Dimensions - 1980 (ICO, Ensenada)	80-70771	0-88318-164-9
No. 66 Physics in the Automotive Industry - 1980 (APS/AAPT Topical Conference)	80-70987	0-88318-165-7
No. 67 Experimental Meson Spectroscopy - 1980 (Sixth International Conference , Brookhaven)	80-71123	0-88318-166-
No. 68 High Energy Physics - 1980 (XX International Conference, Madison)	81-65032	0-88318-167-
No. 69 Polarization Phenomena in Nuclear Physics - 1980 (Fifth International Symposium, Santa Fe)	81-65107	0-88318-168-
No. 70 Chemistry and Physics of Coal Utilization - 1980 (APS, Morgantown)	81-65106	0-88318-169-
No. 71 Group Theory and its Applications in Physics - 1980 (Latin American School of Physics, Mexico City)	81-66132	0-88318-170
No. 72 Weak Interactions as a Probe of Unification (Virginia Polytechnic Institute - 1980)	81-67184	0-88318-171
No. 73 Tetrahedrally Bonded Amorphous Semiconductors (Carefree, Arizona, 1981)	81-67419	0-88318-172
No.74 Perturbative Quantum Chromodynamics (Tallahassee, 1981)	81-70372	0-88318-17

No. 75	Low Energy X-ray Diagnostics-1981 (Monterey)	81-69841	0-88318-174-6
No. 76	Nonlinear Properties of Internal Waves (La Jolla Institute, 1981)	81-71062	0-88318-175-4
No. 77	Gamma Ray Transients and Related Astrophysical Phenomena (La Jolla Institute, 1981)	81-71543	0-88318-176-2
No. 78	Shock Waves in Condensed Matter - 1981 (Menlo Park)	82-70014	0-88318-177-0
No. 79	Pion Production and Absorption in Nuclei - 1981 (Indiana University Cyclotron Facility)	82-70678	0-88318-178-9
No. 80	Polarized Proton Ion Sources (Ann Arbor, 1981)	82-71025	0-88318-179-7
No. 81	Particles and Fields - 1981: Testing the Standard Model (APS/DPF, Santa Cruz)	82-71156	0-88318-180-0
No. 82	Interpretation of Climate and Photochemical Models, Ozone and Temperature Measurements (La Jolla Institute, 1981)	82-071345	0-88318-181-9
No. 83	The Galactic Center (Cal. Inst. of Tech., 1982)	82-071635	0-88318-182-7
No. 84	Physics in the Steel Industry (APS.AISI, Lehigh University, 1981)	82-072033	0-88318-183-5
No. 85	Proton-Antiproton Collider Physics - 1981 (Madison, Wisconsin)	82-072141	0-88318-184-3
No. 86	Momentum Wave Functions - 1982 (Adelaide, Australia)	82-072375	0-88318-185-1
No. 87	Physics of High Energy Particle Accelerators (Fermilab Summer School, 1981)	82-072421	0-88318-186-X
No. 88	Mathematical Methods in Hydrodynamics and Integrability in Dynamical Systems (La Jolla Institute, 1981)	82-072462	0-88318-187-8
No. 89	Neutron Scattering - 1981 (Argonne National Laboratory)	82-073094	0-88318-188-6
No. 90	Laser Techniques for Extreme Ultraviolt Spectroscopy (Boulder, 1982)	82-073205	0-88318-189-4
No. 91	Laser Acceleration of Particles (Los Alamos, 1982)	82-073361	0-88318-190-8
No. 92	The State of Particle Accelerators and High Energy Physics(Fermilab, 1981)	82-073861	0-88318-191-6
No. 93	Novel Results in Particle Physics (Vanderbilt, 1982)	82-73954	0-88318-192-4
No. 94	X-Ray and Atomic Inner-Shell Physics-1982 (International Conference, U. of Oregon)	82-74075	0-88318-193-2
No. 95	High Energy Spin Physics - 1982 (Brookhaven National Laboratory)	83-70154	0-88318-194-0
No. 96	Science Underground (Los Alamos, 1982)	83-70377	0-88318-195-9

No. 97	The Interaction Between Medium Energy Nucleons in Nuclei-1982 (Indiana University)	83-70649	0-88318-196-7
No. 98	Particles and Fields - 1982 (APS/DPF University of Maryland)	83-70807	0-88318-197-5
No. 99	Neutrino Mass and Gauge Structure of Weak Interactions (Telemark, 1982)	83-71072	0-88318-198-3